Unconjugated Pterins and
Related Biogenic Amines

Unconjugated Pterins and Related Biogenic Amines

Proceedings of the
First International Workshop
Flims, Switzerland
February 28 – March 7, 1987

Editors
H.-Ch. Curtius · N. Blau · R. A. Levine

Walter de Gruyter · Berlin · New York 1987

Editors
Prof. Dr. Hans-Christoph Curtius
Dr. Nenad Blau

Division of Clinical Chemistry
Dept. of Pediatrics
University of Zurich
Steinwiesstr. 75

CH-8032 Zurich
Switzerland

Dr. Robert A. Levine
Laboratory of Molecular Neurobiology
Lafayette Clinic
951 East Lafayette Street

Detroit, MI 48207
U.S.A.

Library of Congress Cataloging in Publication Data

Workshop on Unconjugated Pterins and Related Biogenic Amines
(1st : 1987 : Flims, Switzerland)
Unconjugated Pterins and related biogenic amines.

Bibliography: p
Includes index.
1. Pteridines--Metabolism--Congresses. 2. Biogenic amines--Metabolism--Congresses. 3. Metabolic conjugation--Congresses. I. Curtius, H. Ch. Hans-Christoph) II. Blau, N. (Nenad), 1946- .
QP801.P69W67 1987 612'.01575 87-22341
ISBN 0-89925-368-7 (U.S.)

CIP-Kurztitelaufnahme der Deutschen Bibliothek

Unconjugated pterins and related biogenic amines : proceedings of the internat. workshop, Flims, Switzerland, February 28 - March 7, 1987 / ed. H.-Ch. Curtius ; N. Blau. - Berlin ; New York : de Gruyter, 1987.
ISBN 3-11-011341-4

NE: Curtius, Hans-Christoph [Hrsg.]

 Printing: Gerike GmbH, Berlin. Binding: Dieter Mikolai, Berlin. - Printed in Germany.

Dedicated to the Memory
of Alois Niederwieser

PREFACE

This volume contains the proceedings of the First Workshop on Unconjugated Pterins and Related Biogenic Amines, held in Flims, Switzerland, from February 28 to March 7, 1987. Contributors from all over the world reported on the latest developments in the field of pteridines and biogenic amines.

The workshop brought together pteridine researchers in the fields of chemistry, biochemistry, immunology, pediatrics, and internal medicine from 10 countries; there were 61 participants and 38 papers were presented.

Biochemical, analytical, immunological, and pediatric problems were the main themes discussed, but we also heard interesting contributions from the fields of biogenic amines, chemistry, and medical applications of pteridines. Each topic was introduced by a plenary lecture.

On February 5, 1987 our colleague and friend, Alois Niederwieser, passed away after a long illness. We dedicate the proceedings to him for his important and lasting contributions in the field of pteridines.

Finally we wish to express our gratitude to the speakers, chairmen, and other participants who helped to make the workshop a success; to the authors and all those who helped directly or indirectly with the production of this volume; to M. Killen and M. Stucki for excellent secretarial assistance; to the sponsors, listed separately, for their financial support, which made publication of the proceedings possible; and to Walter de Gruyter & Co., for their professional assistance with publication.

H.-Ch. Curtius
N. Blau
R.A. Levine

ACKNOWLEDGEMENTS

We gratefully acknowledge contributions from the following sponsors:

Almedica AG, Galmiz, CH
Brechbühler AG, Schlieren, CH
Burkard Instrumente AG, Zurich, CH
Ciba-Geigy AG, Basel, CH
Digitana AG, Horgen, CH
Henning Berlin GmbH, Berlin, FRG
Hewlett Packard (Schweiz) AG, Widen, CH
LKB Instrument AG, Littau, CH
Nestlé Produkte AG, Zurich, CH
Spectronex AG, Basel, CH
Technicon Schweiz, Zurich, CH
VG Instruments GmbH, Wiesbaden, FRG
Milupa AG, Friedrichsdorf, FRG

CONTENTS

In Memoriam: Alois Niederwieser (1935 - 1987)
H.-Ch. Curtius 1

SECTION A: CHEMISTRY AND ANALYSIS OF PTERINS

Homogeneous Catalytic Hydrogenation of L-Biopterin: Influence of the Chirality of the Side Chain and the Catalyst on its Diastereoselective Induction
S. Antoulas, M. Viscontini 7

Synthesis of D-Neopterin-3'-Phosphates
B. Zagalak, F. Neuheiser, R. Bosshard, U. Redweik 13

5-Alkyl-5,6,7,8-Tetrahydrobiopterins
A. Kaiser 23

Mass Spectrometric Investigations of Pteridines
Th. Kuster, A. Niederwieser 31

The Structural Characterization of 6-Pyruvoyl Tetrahydropterin by Fast Atom Bombardment Mass Spectrometry with One- and Two-Dimensional Mass Analysis (FAB/MS and FAB/MSMS)
W.J. Richter, F. Raschdorf, R. Dahinden, Th. Kuster,
A. Niederwieser, W. Leimbacher, H.-Ch. Curtius 39

SECTION B: BIOCHEMISTRY AND METABOLISM OF PTERINS

Tetrahydrobiopterin: Other Physiological Roles in Addition to Aromatic Amino Acid Hydroxylation?

S. Milstien 49

^{1}H-NMR Studies with Tetrahydrobiopterins, Evidence for the Structure of 6-Pyruvoyl Tetrahydropterin, an Intermediate in the Biosynthesis of Tetrahydrobiopterin

S. Ghisla, P. Steinerstauch, Th. Hasler, N. Blau, H.-Ch. Curtius .. 67

Purification of 6-Pyruvoyl Tetrahydropterin Synthase from Salmon Liver

Th. Hasler, A. Niederwieser, H.-Ch. Curtius 81

Purification of 6-Pyruvoyl Tetrahydropterin 2'-Keto Reductase from Human Liver

H.-Ch. Curtius, P. Steinerstauch, W. Leimbacher, U. Redweik, Sh.-I. Takikawa, S. Ghisla 89

Human GTP Cyclohydrolase I: Purification and Preparation of Monoclonal Antibodies

G. Schoedon, A. Niederwieser, H.-Ch. Curtius 99

Dihydromonapterin Triphosphate: Occurrence, Analysis, and Effect on Tetrahydrobiopterin Biosynthesis *in vivo* and *in vitro*

N. Blau, P. Steinerstauch, U. Redweik, K. Pfister, G. Schoedon, L. Kierat, A. Niederwieser 105

Biosynthesis of Pteridines and Deaza Flavins in Methanogenic Bacteria

B. Schwarzkopf, Q. Le Van, W. Eisenreich, P.J. Keller, H.G. Floss, A. Bacher 117

Tetrahydrobiopterin Biosynthesis and Pteridine Metabolism in Cells of the Blood System, Lymphoid Tissues and in Biological Fluids

B. Andondonskaja-Renz, H.-J. Zeitler 125

Modified Pteridine Synthesis in Transformed Cell Lines and during Heat Shock

J. Seidl, M. Przybylski, Ch. Meinschad, I. Ziegler 137

SECTION C: PTERINS IN IMMUNOLOGY

Disease Associated Alterations of Pterin Biosynthesis: Enhancement in Inflammatory Disease, Hemopoietic Regeneration and Endotoxinemia

Ch. Huber, M. Herold, J. Troppmair, H. Rokos 149

Metabolism of Pterins in the Cellular Immune System of Man and Mouse

G. Schoedon, A. Niederwieser, H.-Ch. Curtius, A. Fontana, J. Troppmair, Ch. Huber 161

Tetrahydrobiopterin Synthesis is Triggered by Interleukin 2 Binding and Modulates Receptor Affinity

I. Ziegler 169

Kinetics of Production and Release of Neopterin. Comparison of Human PBMC with the Permanent Monocytic Cell Line U937 and its Subclones

K. Rokos, R.O.F. Kunze, M.A. Koch, H. Rokos, K. Nilsson 177

SECTION D: PTERINS IN DIFFERENT DISEASES

Variations of Neopterin, Dihydroneopterin and Immunological Parameters during Physical Exercise (Jogging)

H. Rokos, K. Rokos, R.O.F. Kunze 187

Biopterin and Neopterin in Plasma and Lympho-Mononuclear Biopterin in Graft versus Host Disease and Infection after Bone Marrow Transplantation

M. Fink, V. Nüssler, K. Demtröder, E. Höniges, H.J. Kolb, I. Ziegler 197

Effects of Pirenzepine, Aciclovir and Trimethoprim on Pterins in Blood Cells and Plasma

M. Fink, V. Nüssler, K. Demtröder, E. Höniges, M. Goldberg 207

SECTION E: TETRAHYDROBIOPTERIN DEFICIENCIES

Heterogeneity of Tetrahydrobiopterin Deficiency

J.-L. Dhondt 219

Prenatal Diagnosis of Tetrahydrobiopterin Deficiency

N. Blau, A. Niederwieser, H.-Ch. Curtius, W. Leimbacher, L. Kierat, A. Matasovic, W. Staudenmann, I. Ozalp 237

Sepiapterin Reductase in Human Amniocytes, Skin Fibroblasts, Chorionic Villi and Stimulated Mononuclear Blood Cells

J. Ferré, E.W. Naylor 247

Impaired Biopterin Synthesis in a Patient with Mild Hyperphenylalaninemia. A New Variant?

J.L. Dhondt, G. Forzy, J.M. Hayte, P. Guibaud, M.O. Rolland, C. Dorche, S. Andre 257

Dihydropteridine Reductase Activity in Fetal Tissues

G. Bracco, A. Iavarone, S. Pagliardini, F. Levis, O. Guardamagna, S. Ferraris, A. Ponzone 265

Molecular Analysis of Human Dihydropteridine Reductase and Dihydropteridine Reductase Deficiency

R.G.H. Cotton, W.M. Hutchison, S. Wake, I.G. Jennings, W.J. McAdam, D.M. Danks, H.-H.M. Dahl 275

Fate of Peripherally Administered Tetrahydrobiopterin in Congenital Tetrahydrobiopterin Deficiency

S. Ferraris, O. Guardamagna, G. Bonetti, G. Bracco, L. Leone, A. Ponzone, A. Niederwieser 283

6-Pyruvoyl-Tetrahydropterin Synthase Deficiency: Therapeutic Trial with Two Different Synthetic Pterin Analogues in Three Patients

D. Leupold, H. Lehmann, H.-Ch. Curtius, A. Niederwieser 293

Placental Barrier in Mother-to-Fetus Transfer of Tetrahydrobiopterin in Humans

A. Ponzone, S. Ferraris, S. Biasetti, I. Dianzani, O. Guardamagna, L. Kierat, A. Niederwieser 303

SECTION F: MONOAMINE NEUROTRANSMITTERS AND ENZYMES

Therapeutic Potential of Tetrahydrobiopterin Therapy in Neurological and Psychiatric Illness

R.A. Levine 315

Organization of the Catalytic and Regulatory Sites of Rat Liver Phenylalanine Hydroxylase

M. Parniak 327

Immunochemical Studies of Phenylalanine Hydroxylase

R.G.H. Cotton, I.G. Jennings, W. McAdam, S. Smith 339

Influence of Substrates and Cofactors on Tyrosine Hydroxylase

D. Kuhn 343

Mechanism of Activation of Tryptophan 5-Monooxygenase by Mercaptides

H. Hasegawa, A. Ichiyama 353

Effects of Tetrahydropterins on DOPA Production in vitro and in vivo
M. Bräutigam, W. Pfleiderer 359

Pterins and Monoamine Metabolites in Cerebrospinal Fluid
J.L. Dhondt, G. Forzy, J.M. Hayte, J.P. Farriaux, C. Largilliere .. 367

AUTHOR INDEX 375

SUBJECT INDEX 379

LIST OF PARTICIPANTS 393

In Memoriam

ALOIS NIEDERWIESER

1935 - 1987

On February 5, 1987 Professor Alois Niederwieser died after a long illness at the age of 52. We mourn the loss of an outstanding scientist of world renown, a dear colleague and friend, who shaped to a considerable extent international research in the field of inborn errors of metabolism.

Alois Niederwieser was not only my coworker, he also was a close friend for many years and we had been colleagues at the university, thus experiencing a quality of friendship which is not often found in the course of a lifetime.

Niederwieser came from a practicing catholic family in Laupheim/Württemberg and had originally planned to become a priest. But the life sciences had caught his interest early and one year before entering university he decided to study chemistry.

After two semesters at the University of Tübingen, Niederwieser continued his studies at the renowned Organic Chemistry Institute of Nobel prize winner Tadeus Reichstein in Basle and did his Ph.D. with Max Brenner on thin-layer chromatography of peptides and amino acids. That is where I first met Niederwieser. The years of post war reconstruction in Europe were happy and positive years for all of us.

At that time Niederwieser suggested to our future friend, Max Brenner, to separate peptides and amino acids with the help of thin-layer chromatography. In order to appreciate this proposal one must remember that in those days chromatography was still in its infancy: Egon Stahl had just held a lecture at Roche Company on the separation of pigments by thin-layer chromatography. W.H. Stein, from the USA, had held a lecture at the Organic Chemistry Institute on the separation of amino acids by ion exchange chromatography, and Max Brenner with his coworker, Rolf Weber, had constructed the first amino acid analyzer in Switzerland according to the specifications by Stein and S. Moore.

The first pioneering work on the separation of amino acids and peptides by thin-layer chromatography was published by Niederwieser, Brenner, and Pataki. Four books by Niederwieser and G. Pataki were published, followed by a number of publications. Among other inventions the "Brenner-Niederwieser chamber" was developed which was still available on the market until recently.

After successfully completing his dissertation, Niederwieser first started work as scientific assistant in the research laboratory of the Neurology Department at the University of Basle. In 1965 he moved to the University of Düsseldorf where he worked as research assistant until 1968. During those years we did not see much of each other. By accident I met Niederwieser again at the Analytica meeting in Munich and I asked him if he would like to join me at the University of Zurich Children's Hospital. At that time the new laboratory wing was being completed and we had just started with gas chromatography. Soon after the Analytica meeting, Niederwieser wrote that he would very much like to come to Zurich.

In Zurich the most important 18 years of his fruitful scientific work began. He published more than 130 papers and was invited to hold more than 60 lectures which underlined his worldwide reputation as scientist. He discovered and characterized six inborn errors of metabolism, such as β-mercaptopyruvate sulfur transferase deficiency, isolated formimino transferase deficiency, 4-hydroxyphenyl pyruvate dioxygenase deficiency, amino peptidase deficiency, and two enzyme defects of the tetrahydrobiopterin biosynthesis. In addition, he isolated and elucidated drug metabolites and artefacts of food technology, did analytic pioneering work, such as the separation of peptides as copper complexes, the selective isolation of peptides from biological material by chemical transformation to polyamino alcohols, GC-MS of oligopeptides in urine, quantitation of mercapturic acid by mass fragmentography, measurement of pterins in urine by GC-MS and HPLC, and much more.

In the last few years most of his energy was devoted to the pterins. He elucidated two inborn errors of metabolism, i.e., GTP cyclohydrolase I deficiency and 6-pyruvoyl tetrahydropterin synthase deficiency. He developed a treatment for tetrahydrobiopterin deficiency using tetrahydrobiopterin and neurotransmitter precursors. From Zurich he conducted the screening for atypical phenylketonuria in many central European countries and overseas. Just a short while ago it became possible to perform the prenatal diagnosis of a 6-pyruvoyl tetrahydropterin synthase deficiency. In the case of an Italian mother who had already given birth to a 6-pyruvoyl tetrahydropterin synthase deficient child, Niederwieser performed a prenatal diagnosis during her second pregnancy which showed normal enzyme activity. A healthy girl was born whom the grateful parents named Aloisia, after Alois Niederwieser.

Niederwieser played an important part in the characterization of the most important steps in the tetrahydrobiopterin biosynthesis. Recently, together with Wolf Endres in Munich he proposed that the so-called peripheral defects of 6-pyruvoyl tetrahydropterin synthase deficiency are to be found in patients with remaining activity and in heterozygotes.

A number of projects remain to be completed. Niederwieser's enormous knowledge and his lucid mind would have facilitated the solving of many more problems.

In June of 1986, already marked by his terminal illness, Niederwieser participated in the 8th International Symposium on Pteridines and Folic Acid Derivatives in Montreal. As you may have heard, he and I were at that time nominated to organize the 9th Symposium in Zurich in 1989.

His pioneering work brought him a number of honors. Many guest scientists from various parts of the world visited us in Zurich and Niederwieser always maintained a warm relationship with them. Representative for many other colleagues in various parts of the world, I would just like to mention the following names: Martin Gal, Paolo Giliberti, Judy Hammond, Bob Levine, Emilio Oradi, Alberto Pignero, Haruo Shintaku, Shin-Ichiro Takikawa, Patricia Tippett, and Wang Muti.

Niederwieser worked up to the last moment, manuscripts and a book were on his deathbed. Shortly before he died on February 5 at one o'clock in the afternoon, he had given instructions to his coworker, Ana Matasovic. He died the way he had lived.

The University of Zurich and the Children's Hospital, pediatricians throughout the world, and his colleagues have lost a dedicated scientist and a dear friend.

H.-Ch. Curtius

Section A: Chemistry and Analysis of Pterins

HOMOGENEOUS CATALYTIC HYDROGENATION OF L-BIOPTERIN: INFLUENCE OF THE CHIRALITY OF THE SIDE CHAIN AND THE CATALYST ON ITS DIASTEREOSELECTIVE INDUCTION

S. Antoulas, M. Viscontini

Organisch-chemisches Institut der Universität, Winterthurerstr. 190, CH-8057 Zürich, Schweiz

Introduction

Heterogeneous catalytic hydrogenation of L-biopterin (I) under acidic conditions yields a mixture of diastereomers 6α-II[1)] and 6β-III in a ratio of approximatively 1/2 [3].

In order to improve this ratio in favour of the 6β-III isomer (natural form, clinically employed for the treatment of atypical PKU [4]), we tried to hydrogenate the triacetylated derivative IV [5] of L-biopterin in presence ot the "Wilkinson catalyst" IX[6].

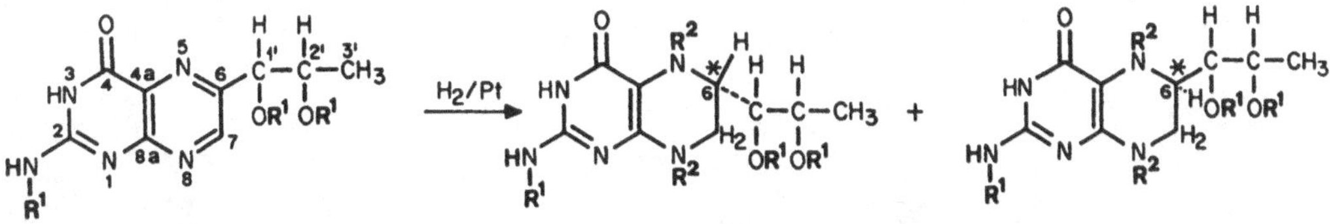

I : R^1 = H
IV : R^1 = CO-CH_3

II : R^1 = R^2 = H
V : R^1 = CO-CH_3 ; R^2 = H
VII : R^1 = R^2 = CO-CH_3

III : R^1 = R^2 = H
VI : R^1 = CO-CH_3 ; R^2= H
VIII : R^1 = R^2 = CO-CH_3

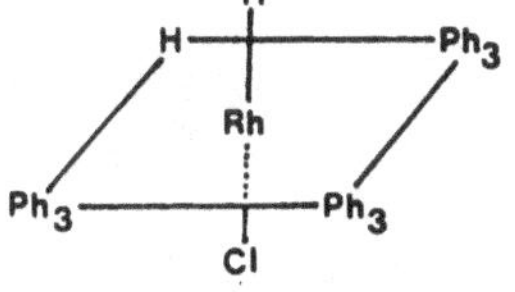

IX, Ph_3 : Triphenylphosphine

1) The terms β and α are used according the IUPAC rules for the nomenclature of steroids as proposed by one of us (M.V.)[1][2].

Unconjugated Pterins and Related Biogenic Amines

Even under drastic conditions (60°, 50 At.) no hydrogenation was observed and the substrate could not be recovered from the reaction mixture. On the contrary, the McQuillin catalyst (X)[7] proved to be very efficient. We postulate for this in situ generated catalyst, the following three structures: X, XI, XII.

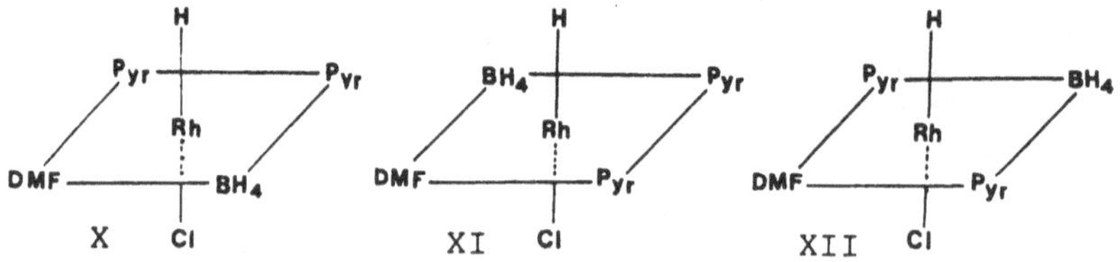

DMF = Dimethylformamide, Pyr = Pyridine, BH_4 = Sodium borohydride

X and XI are enantiomeric to each other, whereas the third strucsture XII represents a "meso"-form, with a plane of symmetry.

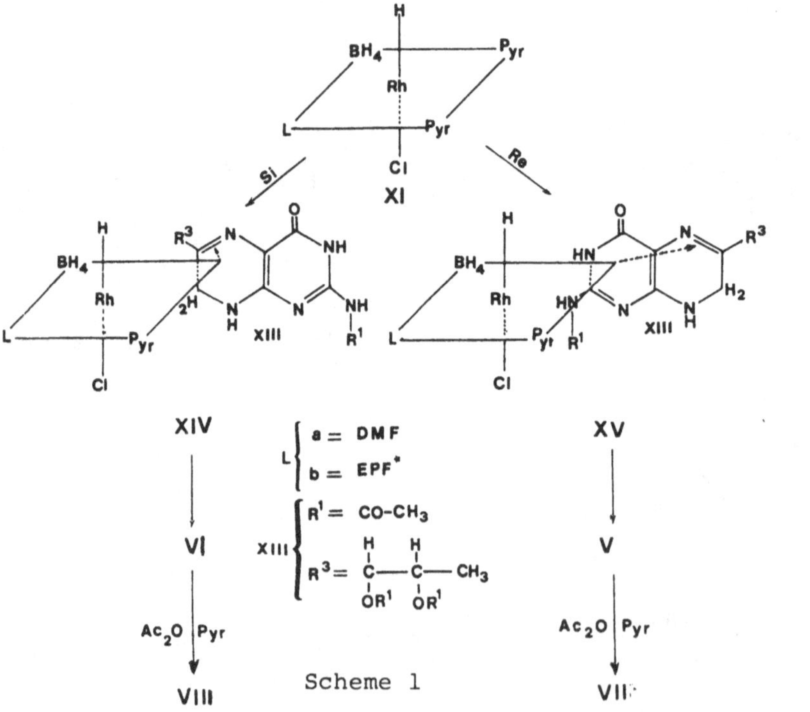

Scheme 1

We assume that each of the three postulated structures forms a complex with the substrate IV to yield the achiral 7,8-dihydro-derivative XIII, which undergoes a second complexation with each of the three structures of the catalyst. In scheme 1 we consider only the second complexation of XIII with the chiral catalyst XI. Depending on wether this new complex formation occurs on the Re-(XV) or the Si-side (XIV) of the prochiral C(6) center, hydrogen addition finally gives rise to 6β-triacetyl-5,6,7,8-tetrahydro-L-biopterin (V) and 6α-triacetyl-5,6,7,8.-tetrahydro-L-biopterin (VI) respectively, which then were converted to the corresponding pentaacetylated derivatives VII and VIII [8]. The approximative 6α/6β ratio was determinated by ^{13}C-NMR spectroscopy, because the spectrum and the absolute configuration of VIII are well known [8]. By comparing the relative intensities of the C(2), C(4) and C(8a) signals, we found that a ratio of ca 1/1 was obtained with McQuillin's catalyst (Figure 1).

In order to get an effective asymmetric induction, we modified the McQuillin catalyst by replacing dimetylformamide (DMF) with a chiral amide, 1R-1-phenylethylamino-formamide (XVIa) [9] (Figure 2 and 3).

The inverse ratio obtained by using the 1S-EPF* enantiomer (XVIb) is shown in figure 4.

With the described experiments we have tried to

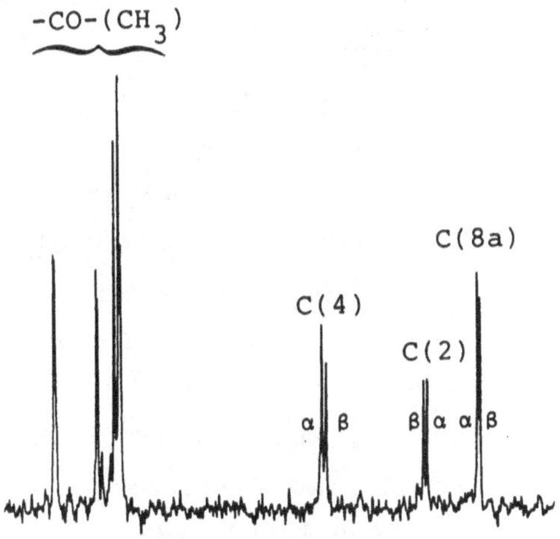

Figure 1. ^{13}C-NMR Spectrum of the mixture 6α- + 6β-pentaacetyl-tetrahydro-L-biopterin, in CD_3-SO-CD_3.

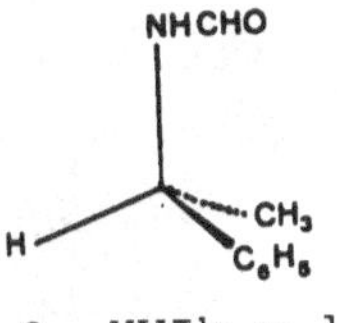

Figure 2. XVIb = 1S-EPF*;(XVIa = 1R-)

demonstrate the importance of homogeneous catalysis for the asymmetric hydrogenation of L-biopterin. In order to improve the diastereoselectiv excess of the natural 6β-tetrahydo-L-biopterin, other more efficient chirality inducing asymmetric ligands should be developed.

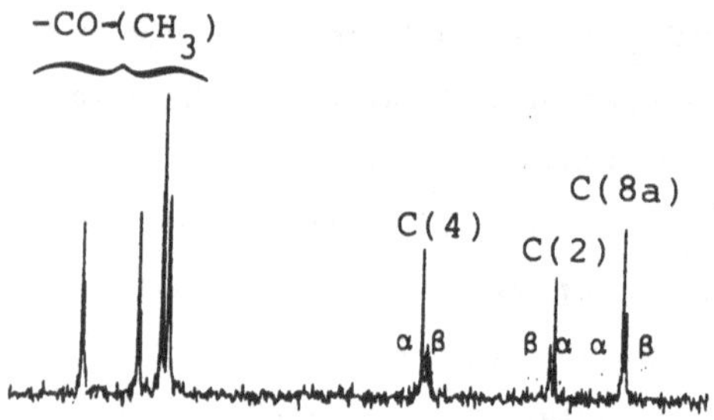

Figure 3. Part of the ^{13}C-NMR spectrum of the mixture VII + VIII after hydrogenation with XVIa (in CD_3SO-CD_3).

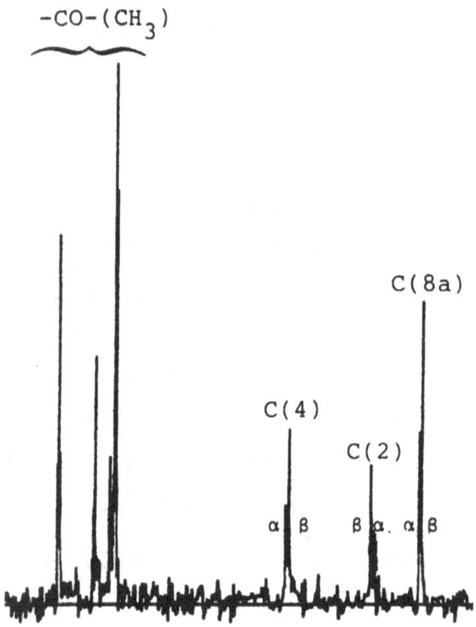

Figure 4. Part of the ^{13}C-NMR spectrum of the mixture VII + VIII after hydrogenation with XVIb (in CD_3SO-CH_3).

Experimental Part

McQuilling catalyst. According to [7].

1R- and 1S-Phenylethylaminoformamide. According to [9].

Homogeneous Hydrogenation. A solution of 1g (2,37 mmol) 2-N-acetyl-1',2'-O-diacetyl-L-biopterin (IV) in 50 ml CH_3OH was added to 50 ml CH_3OH containing 200 mg of in situ generated McQuillin catalyst, and reduced under H_2 for 3 hours (stirring). The reaction mixture was evaporated to dryness and 1,5 ml 12N HCl was added to the residue. The obtained concentred mixture was taken up in 10-20 ml water and the Rh-complex was extracted twice with 80 ml chloroform.

The aqueous phase was evaporated in vacuo and the remaining hydrochlorid was peracetylated under the conditions described in [8]. The crude product was then examined by ^{13}C-NMR spectroscopy in order to estimate its diastereomeric composition (cf. figures 1,3,4).

References

1. Ganguly, S.N., Viscontini, M. 1978. Helv. Chim. Acta 61,166.

2. Antoulas, S., M. Viscontini. 1986. In:Chemistry and Biology of Pteridines (B.A. Cooper, V.M. Whitehead eds.). Walter de Gruyter & Co. Berlin, New York, p. 95.

3. Schircks, B., J.H. Bieri, M. Viscontini. 1978. Helv. Chim. Acta 61, 2731; S.W. Bailey, J.E. Ayling. 1978. J. Biol. Chem. 253, 1598.

4. Schaub, J. S. Dämling, H.-Ch. Curtius, A. Niederwieser, K. Bartholomé, M. Viscontini, B. Schircks, J.H. Bieri. 1978. Arch. of Diseases in Child. 53, 674; H.-Ch. Curtius, A. Niederwieser, A. Otten, J. Schaub, M. Viscontini, S. Scheibenreiter, H. Schmidt. 1979. Clin. Chim. Acta 93, 251.

5. Furrer, H.J., J.H. Bieri, M. Viscontini. 1979, Helv. Chim. Acta 62, 2577.

6. Osborn, J.A., F.H. Jardine, J.F. Young, G. Wilkinson. 1966. J. Chem. Soc.(A) p. 1711.

7. Love, C.J., F.G. McQuillin. 1973. J. Chem. Soc. Perkins Trans. p. 2509.

8. Ganguly, S.N., M. Viscontini. 1982. Helv. Chim. Acta 65, 1090; R. Prewo, J.H. Bieri, S.N. Ganguly, M. Viscontini. 1982. Helv. Chim. Acta 65, 1094.

9. Kraus, M.A. 1973. Synthesis, p. 361.

SYNTHESIS OF D-NEOPTERIN-3'-PHOSPHATES

B. Zagalak, F. Neuheiser, R. Bosshard and U. Redweik

University of Zurich, Department of Pediatrics, Division of Clinical Chemistry, Steinwiesstrasse 75, CH-8032 Zurich, Switzerland

Introduction

Recent research in the field of pterins has been focused on the biosynthesis of (6R)-tetrahydrobiopterin (BH_4) from D-7,8-dihydroneopterin-3'-triphosphate (7,8-H_2-3'-NTP). Obviously, this investigation requires a suitable method for the preparation of pure 7,8-H_2-3'-NTP. Enzymatic preparations of 7,8-H_2-3'-NTP obtained from GTP and GTP-cyclohydrolase I are not pure and contain always free GTP and 7,8-dihydromonapterin-3'-triphosphate (7,8-H_2-MTP) and therefore they cannot be used in a precise spectroscopic or enzymatic study.

We have earlier reported that D-neopterin can be directly phosphorylated with $POCl_3$ and that D-neopterin-3'-monophosphate (3'-NMP) was formed (1). The 3'-NMP was further phosphorylated to 3'-NTP. Unfortunately, we observed that a cyclisation of D-neopterin-3'-monophosphoimidazolidate with a hydroxyl group of the chain takes place after the activation step of 3'-NMP with 1,1'-carbonyldiimidazole. Naturally, this cyclisation causes a very low yield of 3'-NTP. The obtained 3'-NTP after catalytic reduction to 7,8-H_2-NTP was active in the enzymatic synthesis of BH_4 (1).

Recently we improved the chemical synthesis of 3'-NMP and 3'-NTP and this is the subject of this communication.

Unconjugated Pterins and Related Biogenic Amines

Results

D-neopterin-3'-monophosphate

We found in the course of our further phosphorylation study that preparations of 3'-NMP obtained from direct $POCl_3$ phosphorylation of D-neopterin in triethylphosphate contained up to 15% of unknown D-neopterin monophosphate.
This was shown by performing an analytical HPLC on a RP-8 column, using 30 mM $Et_3N:H_3PO_4$ buffer, pH 7.0. The mixture of monophosphates was chromatographically unresolvable at the preparative scale when ion exchangers were used, therefore we gave up this method of phosphorylation of D-neopterin.

Recent improvements in the regiospecific synthesis of C-6 substituted pterins and easy synthesis of D-ribose-5-monophosphate gave us the idea to synthesize D-neopterin phosphates <u>via</u> condensation according to the method by Viscontini et al. (2).:

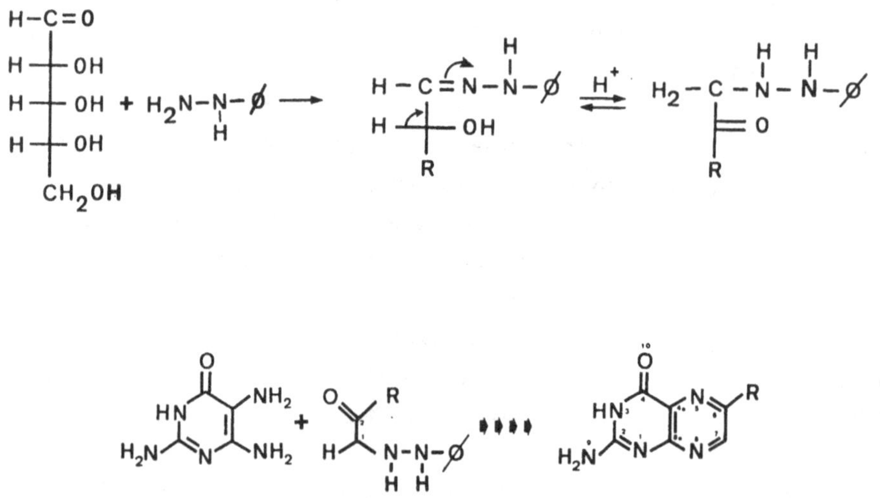

Fig. 1 C-6-Regiospecific condensation of phenylhydrazones of sugars with 4-hydroxy-2,5,6-triaminopyrimidine.

In order to know the pH range of the condensation reactions we began our investigation with precise determination of pK_a values of 4-hydroxy-2,5,6-triaminopyrimidine·2HCl·H_2O (HTAP) (Fig. 2). We determined the following pK_a constants: (1) 5-NH_2 pK_a = 5.214, (2) N-1 pK_a = 1.667 and (3) 4-OH pK_a = 10.076. According to Traub & Pfleiderer (3) the pK_a constants of 1, 2 and 3 should be positioned at N-5, N-1 and 4-OH respectively.

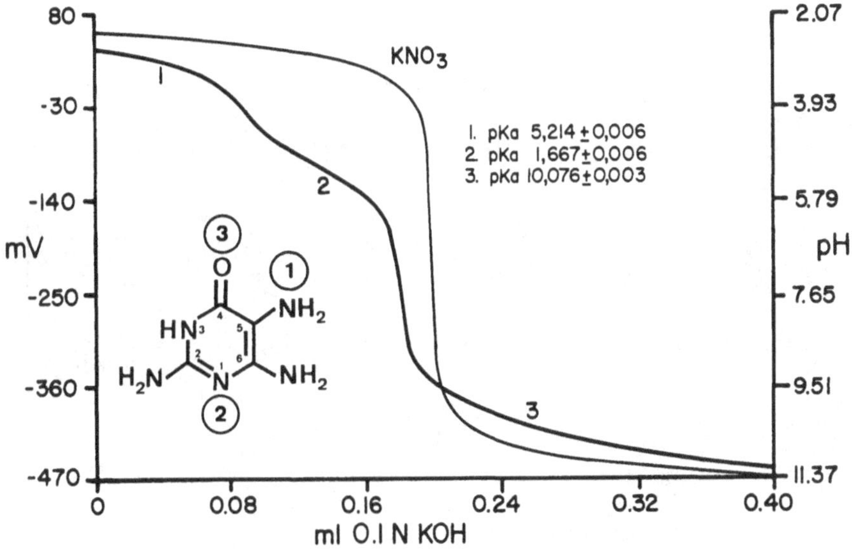

Fig. 2 Determination of pK_a values of 4-hydroxy-2,5,6-triaminopyrimidine·2HCl·H_2O (2.146mg) using KNO_3 (46.6mg) as a standard.

Synthesis <u>de novo via</u> condensation: Adenosine 5'-monophosphate was hydrolyzed with Dowex 50-H^+ (4) to D-ribose-5-monophosphate which was immediately reacted with phenylhydrazine at pH 6.2. Phenylhydrazone of D-ribose-5-monophosphate was isolated as a barium salt which was further condensed with HTAP in MeOH:H_2O (1:1, reflux, 60'). Finally the crude reaction mixture was oxidized with hydrogen peroxide and purified on ion exchangers (see Fig. 3). The overall yield of 3'-NMP triethylammonium salt was 10-17%, however according to the HPLC analysis the yield of the condensation reaction was at least 25-35% (HPLC-assay after the I_2:HCl oxidation). Since the triethylammonium salt is unstable, the end product can be con-

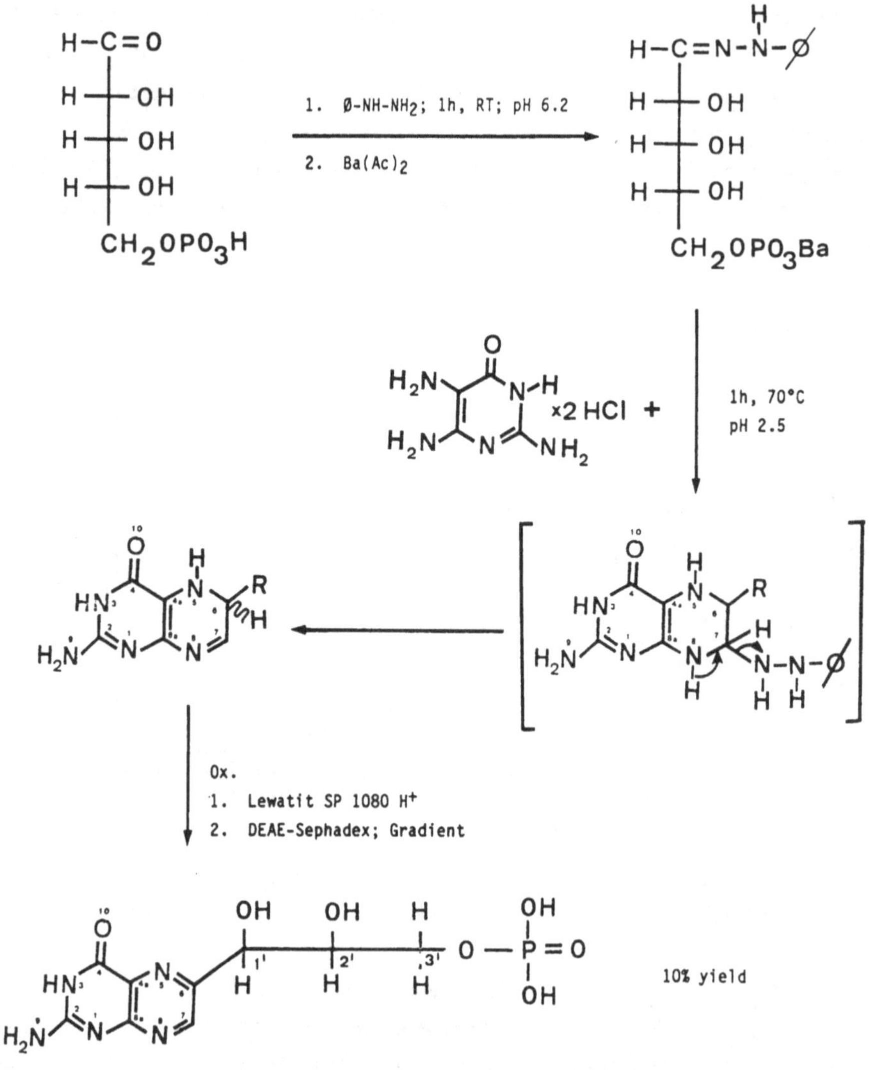

Fig. 3 Synthesis of D-Neopterin-3'-monophosphate via condensation. (Gradient: LiCl, 0.05-0.4M, pH 7.4 or $Et_3N:H_2CO_3$, 0.05-1M, pH 7.6)

verted to a stable sodium salt by treatment of its methanolic solution with an excess of NaI in acetone. The $KMnO_4$ oxidation of 3'-NMP obtained from this synthesis gave pterin-6-carboxylate. Structure of this 3'-NMP was also confirmed by spectroscopic (H-NMR, IR, UV, see Fig. 4) and chromatographic methods.

Partial reduction of this 3'NMP (H_2-Pd or sodium hyposulfite) gave 7,8-dihydro-3'-NMP which was identical with enzymatic one obtained from GTP (GTP-cyclohydrolase and alc. phosphatase).

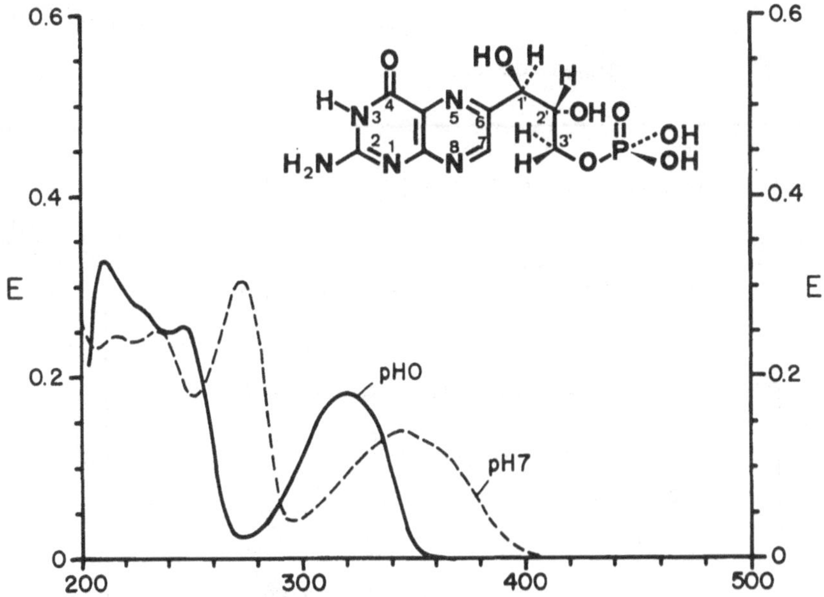

Fig. 4 UV-spectrum of 3'-NMP Na-salt obtained from the condensation.

Synthesis of D-neopterin-3'-triphosphate.

a) Enzymatic approach (see Figs. 5,6): Guanosine-5'-monophosphate kinase (EC 2.7.4.8) catalyses the phosphorylation of GMP using GTP as a phosphate substrate in this system. The produced GDP is further phosphorylated to GTP in the presence of phosphoenolpyruvate kinase and phosphoenolpyruvate (5). Unfortunately, in our hands this system did not work when we used 3'-NMP. Neither did we observe phosphorylation of 3'-NMP to 3'-NDP or 3'-NTP using a similar system with nucleoside-monophosphate kinase (EC 2.7.4.4) (6).

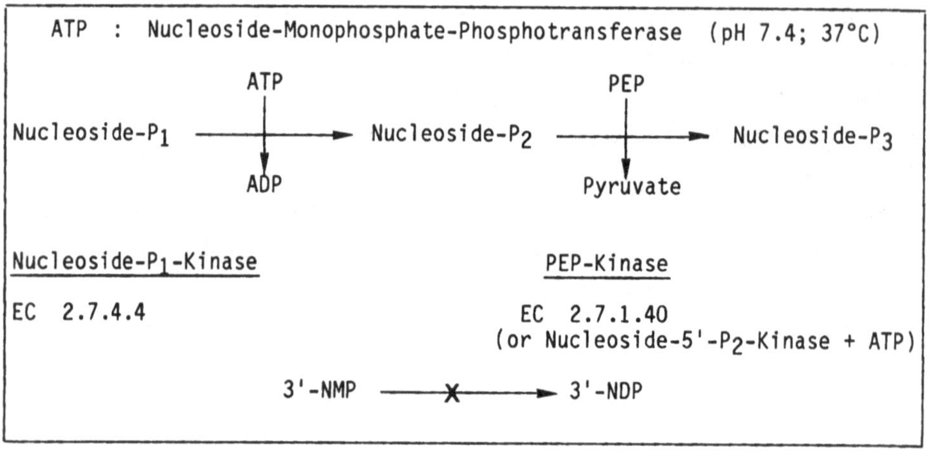

Fig. 5 Nucleoside-monophosphate-kinase (EC 2.7.4.4).

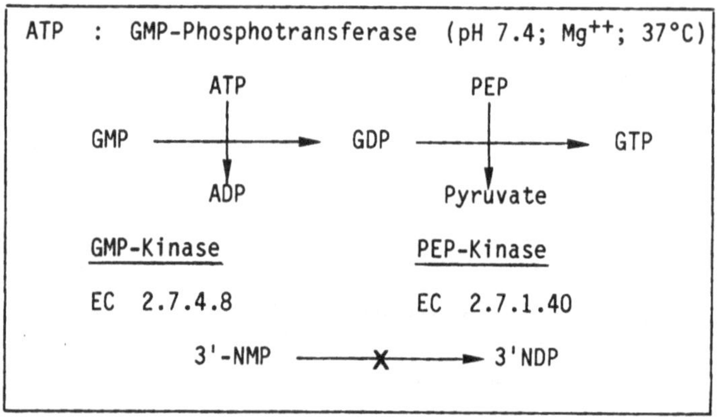

Fig. 6 Guanosine-5'-monophosphate-kinase (EC 2.7.4.8).

b) Synthesis *de novo via* the condensation of HTAP with phenylhydrazone of D-ribose-5-triphosphate: D-ribose-5-monophosphate monotriethylammonium salt was activated with 1,1'-carbonyldiimidazol (7,8) and further phosphorylated with pyrophosphate mono-n-tributylammonium salt. The yield of D-ribose-5-triphosphate was 60%. This triphosphate was converted to the phenylhydrazone and further condensed with HTAP. Unfortunately, no 3'-NTP formation was observed (see Fig. 7).

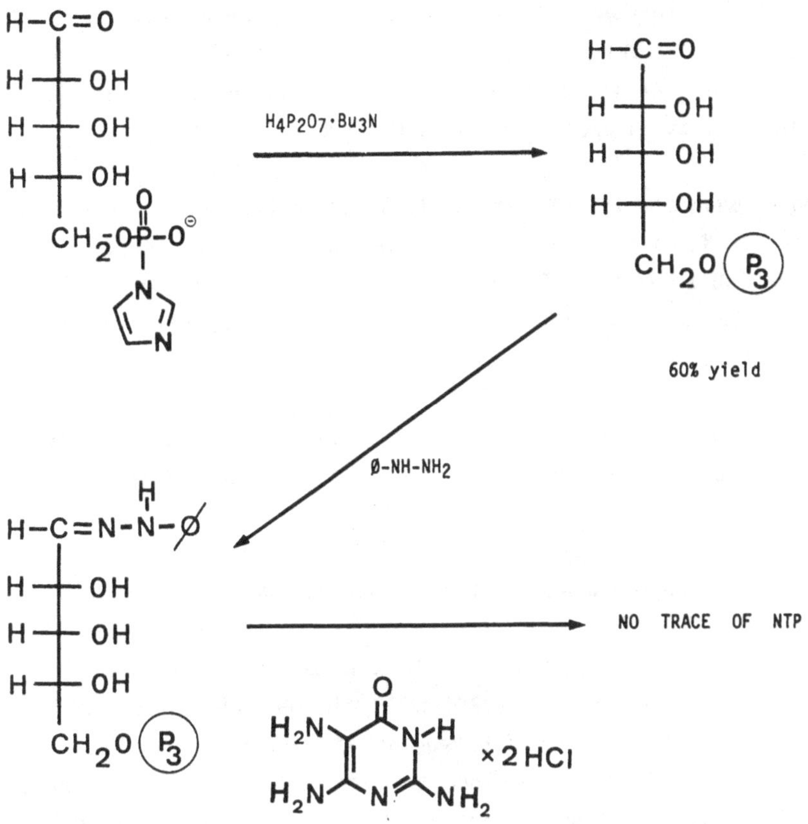

Fig. 7 Condensation of 4-hydroxy-2,5,6-triaminopyrimidine with phenylhydrazone of D-ribose-5-triphosphat.

c) Inserting of protective groups into 3'-NMP: In order to avoid the internal cyclisation of 3'-NMP-imidazolidate during the phosphorylation with pyrophosphate, we tried to insert several O-protective groups into the chain of 3'-NMP.

The 1'-2'-ketalisation of the 3'-NMP proceeds with a good yield when samples were refluxed with 3 equivalents of p-toluenesulfonic acid in 2,2-dimethoxypropane.

The 1',2'-ethoxyethylether of 3'-NMP was prepared at 0°-10°C using vinylethylether (excess) and 3 equivalents p-toluenesulfonic acid. Since the cleavage of this derivative proceeds relatively fast at pH 4.0 it could be used for the synthesis of 3'-NTP. On the other hand, we found the derivatisation procedure (8) to be irreproducible.

We have found that 3'-NMP can easily be esterified with 100% formic acid and the produced mono- and diesters can be cleaved under very mild conditions (pH 8.5-9.0, R.T., 2 h). This gave us the chance of further phosphorylation of 3'-NMP to 3'-NTP avoiding the cyclisation of 3'-NMP-imidazolidate.

d) Chemical phosphorylation of 3'-NMP formates to 3'-NTP: This synthesis is summarized in Fig. 8. The overall yield of 3'-NTP sodium salt was 10-18%.

Preparation of 7,8-dihydro-D-neopterin-3'-triphosphate lithium salt.

According to our experience 7,8-H_2-NTP can be easily prepared from much more stable 3'-NTP by partial H_2:PdO catalytic reduction at pH 7.0 or using sodium hyposulfite as a reducing reagent at pH 6.8. In both causes the reduction was terminated when the ratio of 7,8-H_2-derivatives to 3'-NTP reached 4:1. The reaction mixture was purified on a DEAE-Sephacel column (9) (see Fig. 9). The yield of this step depends on the aerobic conditions of the chromatographic purification procedure and may end up as low as approximatelly 12%.

The structure of both 7,8-H_2-3'-NTP preparations was identical and confirmed by HPLC (RP-8, Et_3N:H_3PO_4, pH 7.0 buffer), UV spectrum and kinetics of phosphate release by alcaline phosphatase, when compared with an enzymatically prepared 7,8-H_2-NTP. Also both preparations obtained by catalytic or hyposulfite reductions were fully active in two different enzymatic systems: the synthesis of BH_4 and in the epimerisation of 7,8-H_2-3'-NTP to 7,8-H_2-3'-MTP.

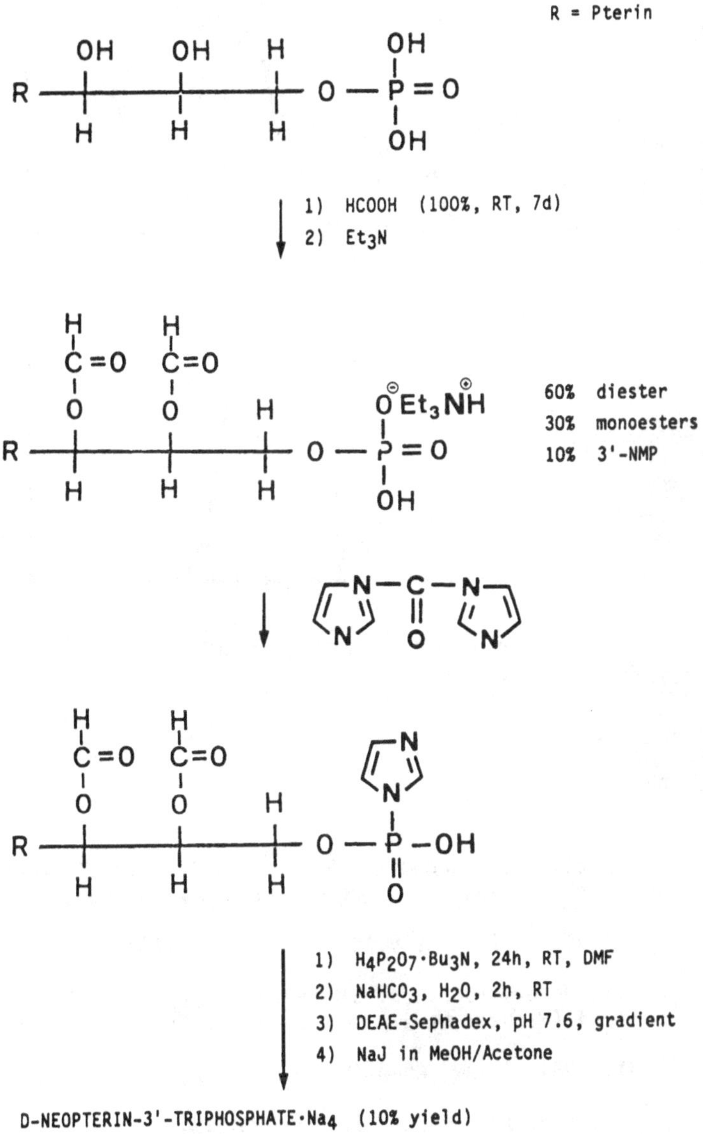

Fig. 8 One-pot synthesis of D-neopterin-3'-triphosphat.

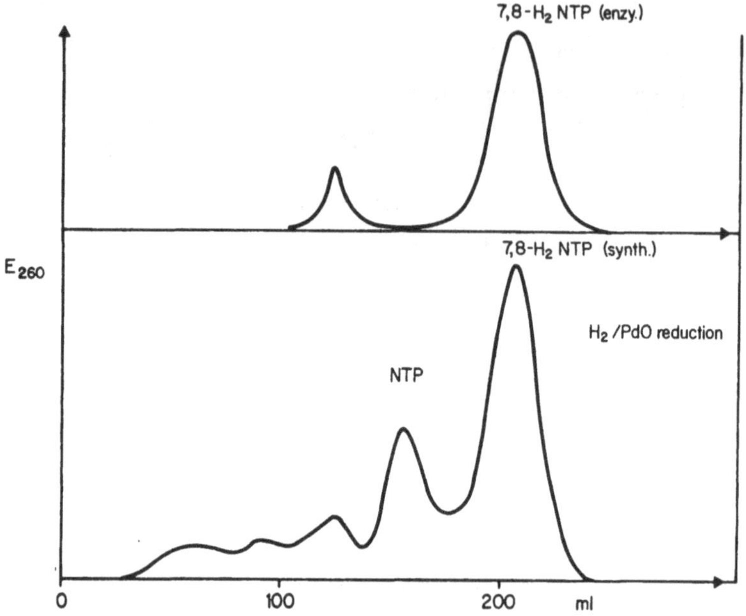

Fig. 9 Purification of 7,8-H_2-NTP on DEAE-Sephacel (1.6 x 25cm, 0.2M LiCl, flow 11ml/h, +4°C).

References

1. Zagalak et al. 1985. In: Biochemical & Clinical Aspects of pteridines Vol. 4, (Wachter et al. eds.). W. de Gruyter, Berlin p. 159.

2. Viscontini M. et al. 1970. Helv. Chim. Acta 53, 1202.

3. Traub H., W. Pfleiderer. 1985. In: Biochemical & Clinical Aspects of pteridines, Vol. 4 (Wachter et al. eds.). W. de Gruyter, Berlin p. 45.

4. Walt D.R. et al. 1984. J. Am. Chem. Soc. 106, 234.

5. Shimono H., Y. Sugino. 1971. Eur. J. Biochem. 19, 256.

6. Liebermann I. et al. 1955. J. Biol. Chem. 215, 429.

7. Hoard D.E., D.G. Ott. 1965. J. Am. Chem. Soc. 87, 1785.

8. Kozarich J.W. et al. 1975. Biochemistry 14, 981.

9. Smith G.R., Ch.A. Nichol. 1983. Arch. Biochem. Biophys. 227, 272.

5-ALKYL-5,6,7,8-TETRAHYDROBIOPTERINS

A. Kaiser

Pharmaceutical Research Department
F. Hoffmann-La Roche & Co. AG
CH-4002 Basel

Summary:

Catalytic hydrogenation of 2-N-acyl-1',2'-di-O-acyl-biopterins and of 1',2'-di-O-acylbiopterins leads to the corresponding 5-alkyl-5,6,7,8-tetrahydrobiopterins - a hitherto unknown class of compounds. The mechanism and the scope of this reductive alkylation and its similarity to the reduction of N(10)-formylfolic acid are being discussed.

During the work on tetrahydrobiopterins it was found that catalytic hydrogenation (RT, atmospheric pressure) of 2-N-acetyl- 1',2'-di-O-acetylbiopterin (<u>1</u>) and subsequent acetylation with acetic anhydride/pyridine yielded not only the expected mixture of 6α-[1)] and 6ß- 2-N, 1'-O, 2'-O, 5-tetraacetyl-5,6,7,8-tetra- hydrobiopterins (<u>3a</u> and <u>3b</u>) but in addition a mixture of 6α- and 6ß-5-ethyl-2-N, 1'-O, 2'-O, 8-tetraacetyl-5,6,7,8-tetrahydrobio- pterins (<u>4a</u> and <u>4b</u>). This result was practically independent of the solvents

1) The terms α and ß are used for the description of the stereochemistry at C-6 as proposed by Viscontini [1].

Unconjugated Pterins and Related Biogenic Amines

(CH_3COOH, CF_3COOH, MeOH) used in the catalytic hydrogenation. Hydrolysis of the mixture of 4a and 4b with aqueous 5N hydrochloric acid gave a mixture of the 6α- and 6ß-5-ethyl- 5,6,7,8-tetrahydrobiopterins (5a and 5b), from which one pure diastereoisomer (configuration at C6 not yet determined) could be isolated as its sulfate (Scheme 1). The structure of the 5-ethylated derivatives 4a, 4b, 5a and 5b has been established by ^{1}H- and ^{13}C-NMR-, MS-spectroscopy and elemental analysis. In addition, the chemical evidence shown below also points to a 5-alkylation:

a) Catalytic hydrogenation and subsequent acetylation of 1',2'-O-diacetylbiopterin (2) gave essentially the same products (Scheme 1). This result shows that the N-2-acetyl group in the triacetylbiopterin 1 is not involved in the 5-alkylation.
b) Catalytic hydrogenation of 2-N-propionyl-1',2'-di-O-propionyl-biopterin (6) followed by hydrolysis with aqueous 5N hydrochloric acid and subsequent acetylation with acetic anhydride/pyri- dine lead to a mixture of the 2-N, 1'-O, 2'-O, 5-tetra-acetyl-5,6,7,8-tetrahydrobiopterins 3a and 3b and the 5-propyl-2-N, 1'-O, 2'-O, 8-tetraacetyl-5,6,7,8-tetrahydrobiopterins 7a and 7b. Hydrolysis of the mixture of 7a and 7b then yielded a mixture of 6α- and 6ß-5-propyl-5,6,7,8-tetrahydrobiopterins (8a and 8b) from which again one pure diastereoisomer could be isolated as its sulfate (Scheme 2). This reaction sequence shows that the 5-alkyl substituent is derived from the 1'-O-(or 2'-O-) acyl group and is formed during the catalytic hydrogenation.

c) All attempts failed to hydrogenate the 6α- or 6ß-tetraacetyl- tetrahydrobiopterins 3a and 3b and the corresponding pentaacetylderivatives 9a and

SCHEME 1

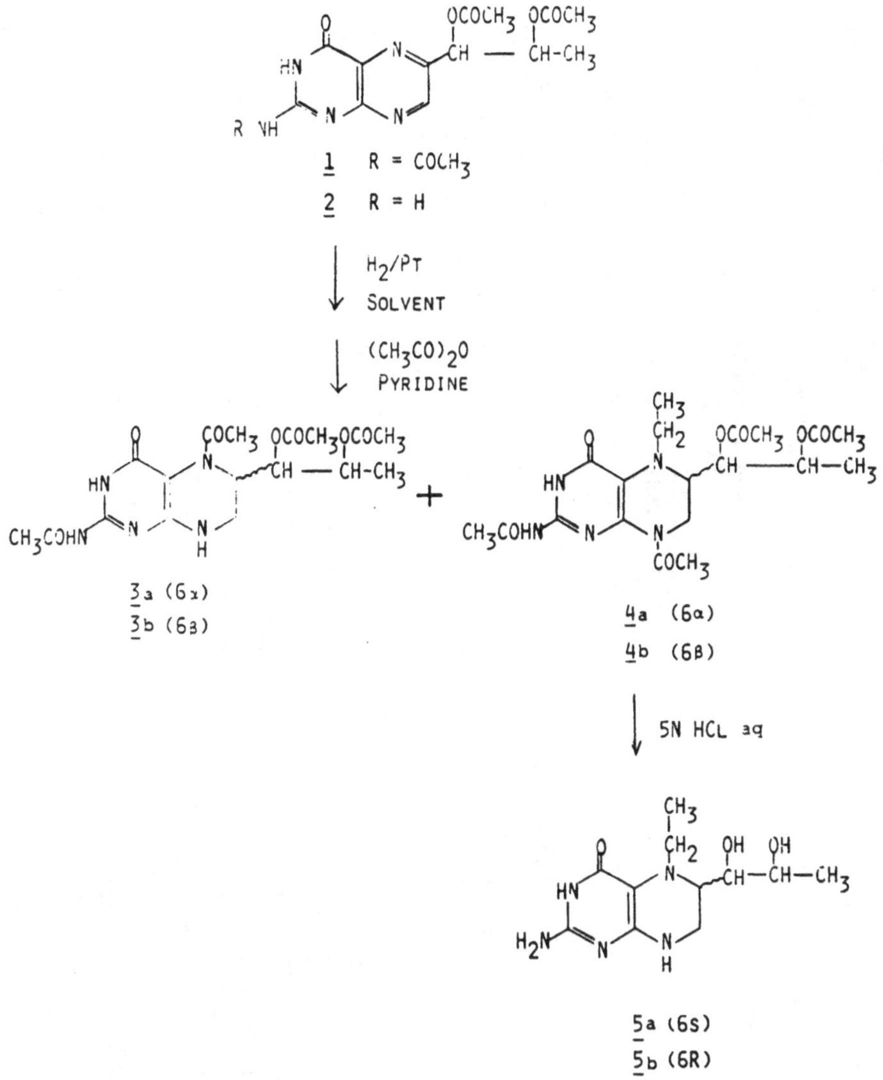

1 R = COCH3
2 R = H
H2/PT
SOLVENT
(CH3CO)2O
PYRIDINE
3a (6α)
3b (6β)
4a (6α)
4b (6β)
5N HCL aq
5a (6S)
5b (6R)

SCHEME 2

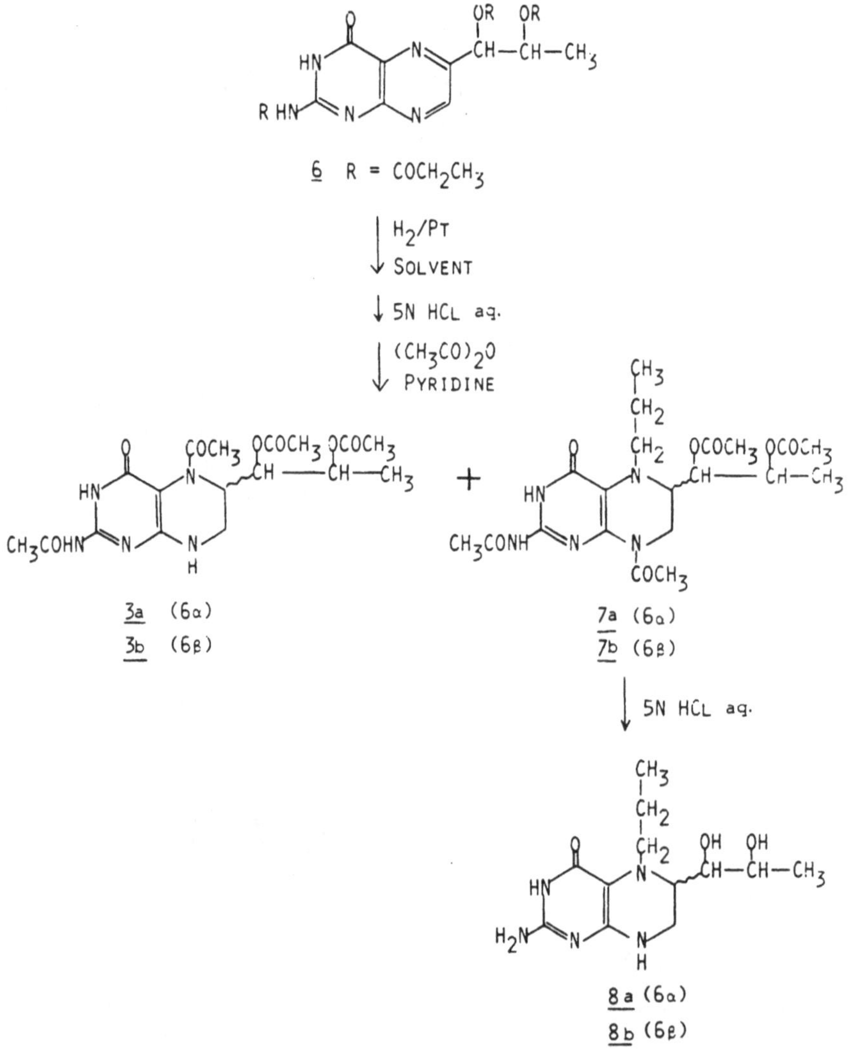
6 R = COCH2CH3
H2/PT
SOLVENT
5N HCL aq.
(CH3CO)2O
PYRIDINE
3a (6α)
3b (6β)
7a (6α)
7b (6β)
5N HCL aq.
8a (6α)
8b (6β)

9b, respectively (Scheme 3), indicating that the formation of these products during the hydrogenation of 1 and 2 is not the reason for the formation of 5-ethylated products.

SCHEME 3

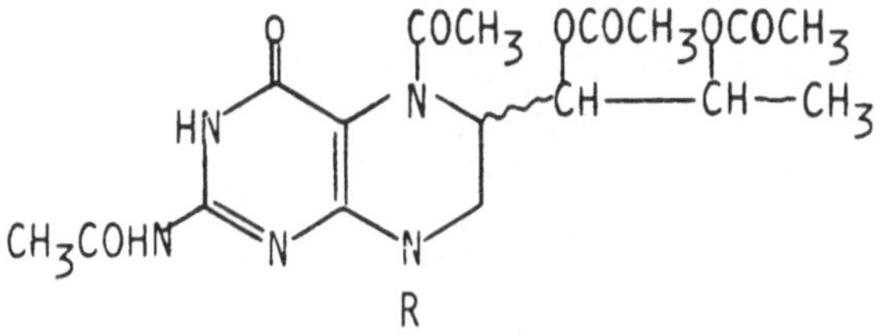

3a R = H 9a R = $COCH_3$ (6α)

3b R = H 9b R = $COCH_3$ (6β)

↓ H_2/PT, CF_3COOH

NO REACTION

The results described above point out that the formation of 5-alkylated products during the catalytic hydrogenation of 1',2'-di-O-acylbiopterins can best be explained by the intermediate formation of an oxazolinium compound (e.g. A or B) which is then hydrogenated to the 5-alkyl derivative. Theoretically, this oxazolinium compound can be formed either from N(5)- or from 1'-O-acylated tetrahydrobiopterins (e.g. C or D) (Scheme 4). All attempts to isolate acyl derivatives of these structures in pure form have not met with success.

SCHEME 4

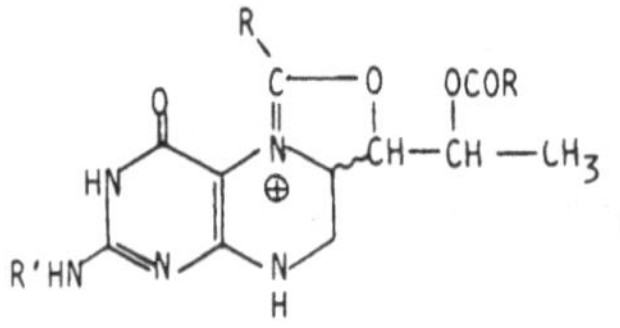

A

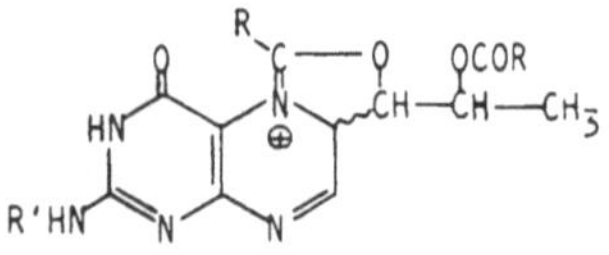

B

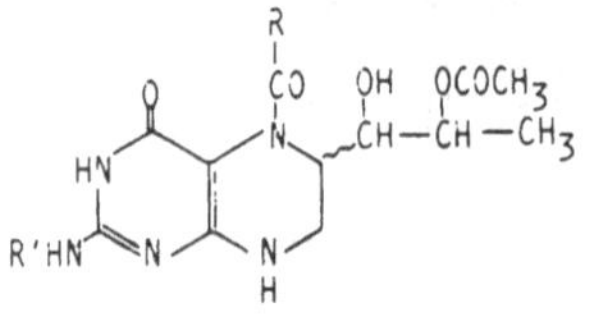

C

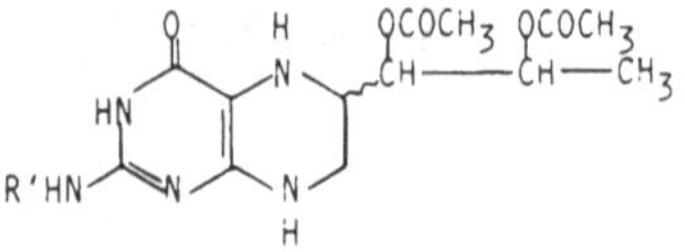

D

Preliminary results indicate that the catalytic hydrogenation described above is applicable to a broad range of 1',2'-di-O-acyl-biopterins and thus constitutes the first general synthesis of the hitherto unknown 5-alkyl-5,6,7,8-tetrahydrobiopterins. Moreover, the reaction sequence E-->F-->G (Scheme 5) seems to be applicable to other 6-hydroxyalkylpteridines as well.

Finally, the similarity of the hydrogenation of O-acylated biopterins to the reduction of N(10)-formylfolic acid should be mentioned. The question arises whether this reaction of O-acylated biopterin and other 6-hydroxyalkyl substituted pteridines or their tetrahydro derivatives is also of biological significance.

SCHEME 5

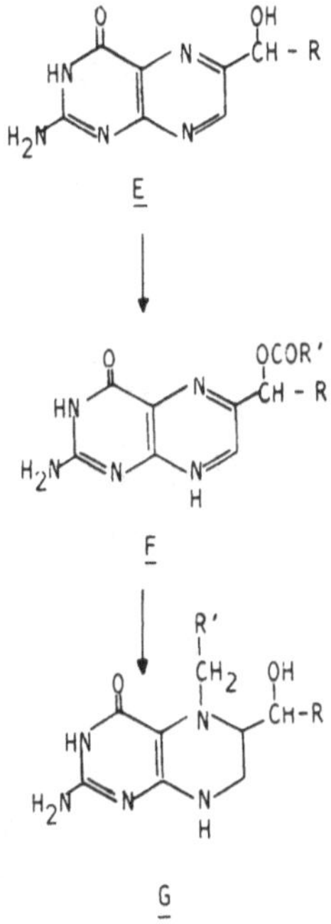

Acknowledgement

The help of Ms. N. Erhardt, Dr. U. Fischer, Ms. S. Herzig, Ms. C. Piers, Dr. W. Arnold, Dr. A. Dirscherl, Dr. E. Englert, Mr. W. Meister and Dr. W. Vetter is gratefully acknowledged.

Reference

[1] M. Viscontini 1985. In: Biological and Clinical Aspects of Pteridines (W. Wachter, H. Ch. Curtius, W. Pfleiderer, eds.) Walter de Gruyter u. Co., Berlin, New York, p. 57

MASS SPECTROMETRIC INVESTIGATIONS OF PTERIDINES

Th. Kuster, A. Niederwieser[t]
Department of Pediatrics, University of Zurich,
CH-8032 Zurich, Switzerland

INTRODUCTION

Due to the increasing interest pteridines gained in the field of biochemistry and medicine, various methods of analyzing this compound class have been developped. Whereas the microbiological and radioimmunoassays are mainly suited for routine analysis, high performance liquid chromatography (HPLC) yields very promising quantitative results (1,2). The inherent poor selectivity of chromatographic methods on the other hand however, prevents identification of unknown compounds in many cases. Mass spectrometry (MS) in combination or without gas chromatography (GC) arises to be very well suited to solve such problems (3-7) and we present here some examples from our laboratory. For experimental details see (6-8).

RESULTS

GC-MS

For the identification of unknown compounds as well as for trace analysis, GC-MS is the method of choice and we applied it in our laboratory for the monitoring of loading experiments, for the characterization of new metabolites and as a control for HPLC routine determinations.

Unconjugated Pterins and Related Biogenic Amines

a) Loading experiments

Atypical forms of PKU can be treated successfully with doses of BH_4, a cofactor which only slightly passes the blood brain barrier. During the efforts to find cofactor substitutes which enter the brain more easily than BH_4, we loaded a patient orally with 6-methyltetrahydrobiopterin and observed the excreted pterins before and after loading. Table 1 shows the mainly excreted and identified pterin metabolites found in urine samples during that laoading test.

Time after	Identified Pterins			
loading(hrs)	Biopterin	Neopterin	6-Methylpterin	6-Methylisoxanthopterin
before	+	+	-	-
2-4	-	+	+	+
4-6	+	+	+	+
12-24	+	+	+	+

Table 1: Excreted pterins after loading with 6-Methyltetrahydrobiopterin.

Fig. 1 shows the excretion pattern of 6-methylpterin and 6-methyl-isoxanthopterin relative to neopterin. The maximal value of 212 and 25 respectively is reached between 2and 4 hours after loading. The figure shows a significantly slower decrease of 6-methylisoxanthopterin (from 25 to 0.7) than of 6-methylpterin (from 212 to 0.3), a fact that is possibly due to the circumstance that the measured 6-methylpterin originates from 6-methyltetrahydropterin which was excreted mainly unchanged whereas 6-methylisoxanthopterin is produced by the action of xanthine oxidase.

b) Characterization of new metabolites

During BH_4 loading tests usually urine samples are investigated by HPLC for the occurence of pterin metabolites. Since the recovery lies in the region of 1-5% only, it was interesting if these metabolites are also excreted in the feces.

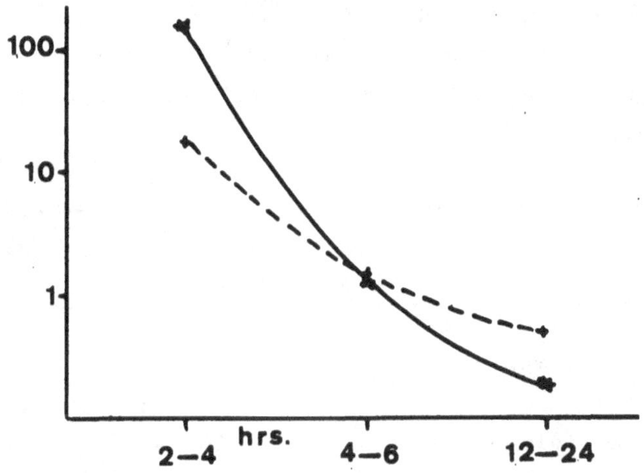

Figure 1: Total ion current peak area ratios of 6-Methylpterin* and 6-Methylisoxanthopterin+ to Neopterin versus time after loading with 6-Methyltetrahydropterin.

Figure 2 shows a total ion current chromatogram of such a feces fraction. For the determination of the methylene units, the sample has been spiked with even numbered n-alkanes C_{18}-C_{26}.

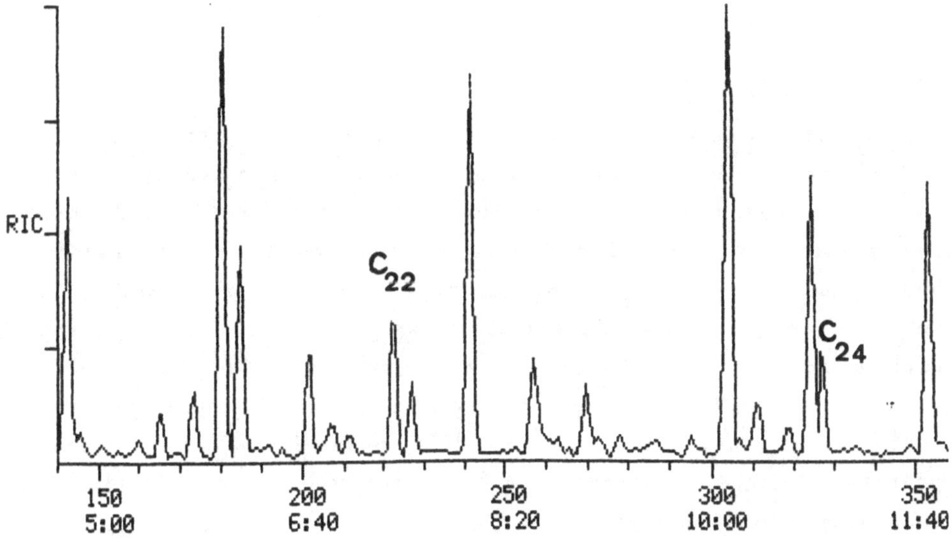

Figure 2: Total ion current chromatogram of a feces fraction

Table 2 lists the lumazines identified in feces after the BH_4 treatment together with their respective methylene units. All compounds listed have been verified by comparison with references and their methylene units (8).

Compound	No. TMS	MU	$M^{+\cdot}$(%)	Base peak
6-Hydroxylumazine	3	20.61	396(100)	
7,8-Dihydro-6-hydroxylumazine	4	21.55	470(100)	
2'-Deoxy-biolumazine	3	21.67	438(90)	409
7,8-Dihydro-2'-deoxybiolumazine	4	22.36	512(40)	422
2'-Deoxysepialumazine	3	22.72	438(100)	
Biolumazine	4	23.07	526(0.6)	410
7,8-Dihydrobiolumazine	5	23.85	600(100)	
Sepialumazine	4	23.93	526(100)	
2'-Deoxysepialumazine	4	24.67	510(100)	

Table 2: Lumazines identified in feces after ingestion of tetrahydrobiopterin. MU: methylene units, $M^{+\cdot}$: molecular ion,No.TMS:number of trimethylsilyl groups.

c) Control for HPLC

The amount of L-threo-neopterin (monapterin) excreted in urine normally is 10% of D-erythro-neopterin (neopterin). In some cases however the value exeeds this ratio and then the question arises if the HPLC method really yields reliable results An independent control then is indicated and we do this by GC-MS. Figure 3 shows two HPLC chromatograms with a normal excretion value of monapterin (A) and B with a ratio of 0.2 relative to neopterin. GC-MS here firstly could proof that the observed HPLC peak 2 really is monapterin through comparison of the methylene units and the mass spectrum with a reference compound.

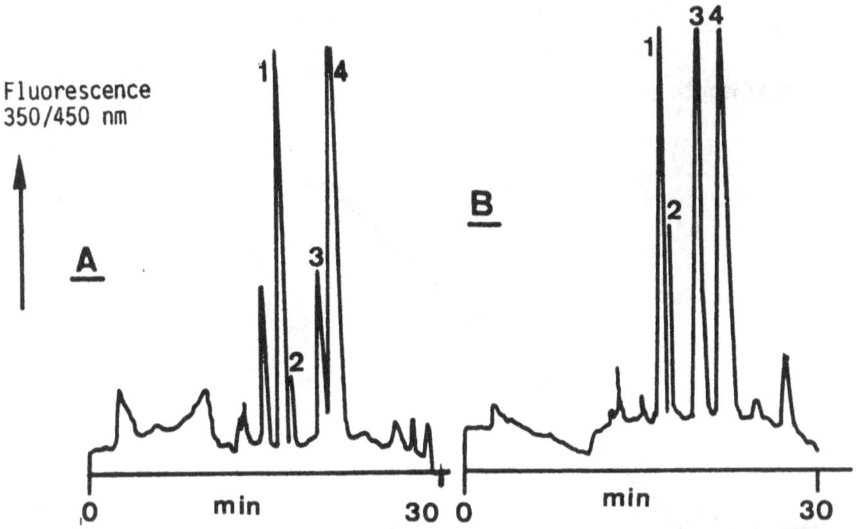

Figure 3: HPLC of pterins with (A) normal and (B) high monapterin to neopterin ratio. 1 = neopterin, 2 = monapterin, 3 = isoxanthopterin, 4 = biopterin.

The method of selected ion monitoring allowed a quantification of the excreted monapterin. Figure 4 shows the calibration plots for HPLC and GC-MS. The arrows mark the respective values determined for the urine sample B. Thus, HPLC yielded 0.21 for the mentioned ratio whereas GC-MS gives a value of 0.22 indicating that the HPLC reslults are "true".

Direct Chemical Ionization (DCI)

Whereas GC-MS has its advantages for mixture analysis of compounds with a relative high volatility, it fails with components less volatile which often cannot be vaporized without decomposition. As we did it here with the TMS-ethers, pterins can made more volatile by derivatization, but sometimes, this method results in more questions than answers and, especially for molecular weight determination a method which does not change the molecule is desirable. We investigated therefore some reference pterins by direct chemical

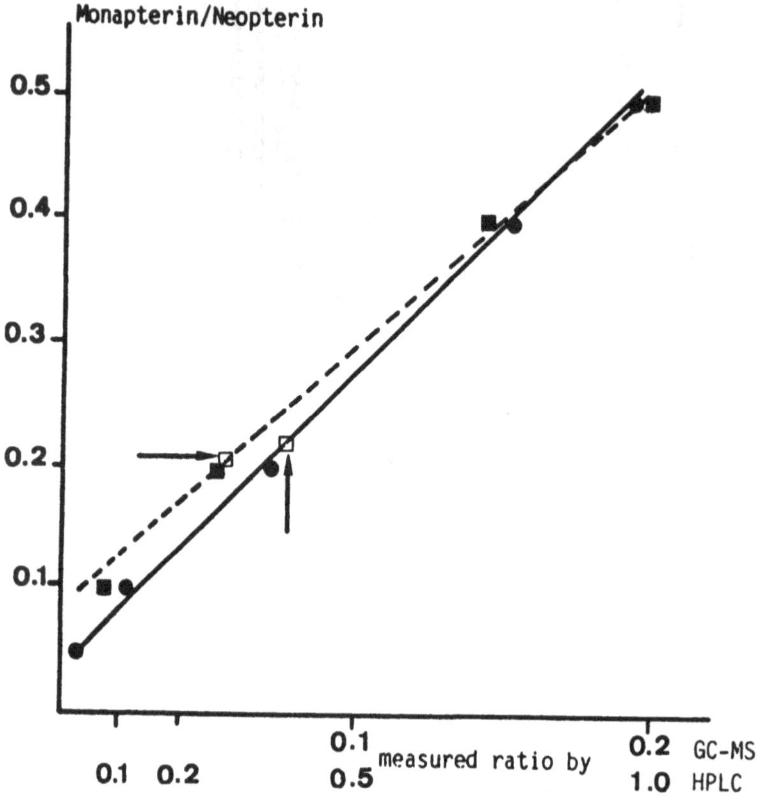

Figure 4: Ratio determination of Monapterin to Neopterin by HPLC(■----■) and GC-MS(●——●). Ratio values determined, (□) using the corresponding calibration graph: HPLC,0.21, GCMS,0.22.

ionization, a technique by which the sample is flash vaporized and so can be evaporated without decomposition. The result usually is a high intensity of ions in the region of the molecular weight. So, the protonated molecular ion often determines the respective molecular weight.
Table 3 lists the quasi molecular ions (MH^+) of the investigated pterins and Figure 5 shows the DCI mass spectrum of pterin-6-carboxylic acid. This example also shows some of the method's limits since the ion at m/z 164 arises to originate from decarboxylated acid rather than from the intact molecule

Compound	MG	MH^+ (%)	Base peak
Lumazine	164	165 (100)	
6-Hydroxylumazine	180	181 (100)	
Pterin	163	164 (100)	
Pterin-6-carboxylic acid	207	208 (27)	164
Xanthopterin	179	180 (100)	
Isoxanthopterin	179	180 (100)	
Sepiapterin	235	236 (48)	103
Biopterin	237	238 (100)	
Neopterin	253	254 (100)	
Monapterin	253	254 (75)	103

Table 3: Molecular weights (MG), Quasimolecular ions (MH^+) and base peaks of reference pterins under direct chemical ionization conditions with methane as reactant gas (9).

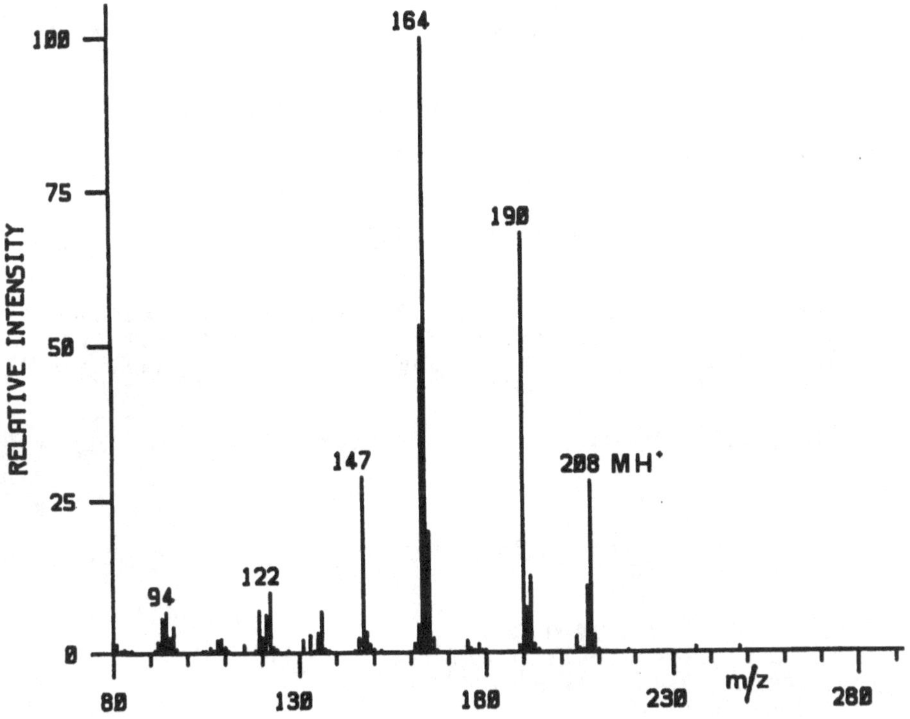

Figure 5: DCI (methane) spectrum of pterin-6-carboxylic acid (9).

ACKNOWLEDGEMENT

We express our thanks to Prof. E.C. Horning, University of Houston, Texas, USA for the DCI-analyses.

REFERENCES

1. Fukushima, T., J.C. Nixon. 1980. Anal. Biochem. 102, 176.

2. Niederwieser, a., W. Staudenmann, E. Wetzel. 1982. In: Biochemical and Clinical Aspects of Pteridines (H.Wachter, H.Ch. Curtius and W. Pfleiderer, eds.) de Gruyter, Berlin, 1, p. 81.

3. Kobayashi, K., M. Goto. 1970. In: Chemistry and Biology of Pteridines (M. Goto and Y. Iwanami, eds.) Int. Acad. Printing Co., Tokyo, p. 57.

4. Lloyd,T., S. Markey, N. Weinen. 1971. Anal. Biochem. 42, 108.

5. Röthler, F., M. Karobath. 1976. Clin. Chim. Acta 69, 457.

6. Kuster, Th., A. Niederwieser. 1983. J. Chromatogr. 278, 245.

7. Kuster, Th., A, Niederwieser. 1984. J. Chromatogr. 290, 303.

8. Niederwieser, A., A. Matasovič, Th. Kuster, W. Staudenmann W. Pfleiderer, S. Scheibenreiter. 1986. In: Chemistry and Biology of Pteridines (B.A. Cooper, V.M. Whitehead, eds.) de Gruyter, Berlin, p. 305.

9. Horning, E.C., H.-Ch. Curtius, A. Niederwieser: in preparation.

THE STRUCTURAL CHARACTERIZATION OF 6-PYRUVOYL TETRAHYDROPTERIN BY FAST ATOM BOMBARDMENT MASS SPECTROMETRY WITH ONE- AND TWO-DIMENSIONAL MASS ANALYSIS (FAB/MS AND FAB/MSMS)

W.J. Richter, F. Raschdorf, R. Dahinden
Central Function Research, Ciba-Geigy AG, CH-4002 Basel, Switzerland

A. Niederwieser+, H.-Ch. Curtius, W. Leimbacher, Th. Kuster
University of Zurich, Department of Pediatrics, Division of Clinical Chemistry, CH-8032 Zurich, Switzerland

Introduction

Tetrahydrobiopterin (BH_4), the cofactor of various hydroxylases, plays an important role in the biosynthesis of the neurotransmitters catecholamine and indoleamine (1), and inborn errors in the BH_4 biosynthesis cause atypical phenylketonuria. Clarifying the biosynthetic pathway (Scheme 1) therefore is of great importance for the elucidation of metabolic defects. The key intermediate hereby is a compound with a proposed 6-pyruvoyl tetrahydropterin (PPH_4) structure. This hypothesis has been deduced from its mode of formation, its chemical reactivity, and from trapping experiments (2-6). Whereas all these experiments led to indirect evidence for the PPH_4 structure, Ghisla et al. (7) presented direct ^1H-NMR evidence for the pyruvoyl tetrahydro structure. In order to have further direct evidence we present here mass spectrometric results toward the structure elucidation of the intermediate.

Results and Discussion

The intermediate in the BH_4 biosynthesis, for which a PPH_4 structure (A_1, in Scheme 2) had been inferred, was subjected to analysis by fast atom bombardment (FAB) mass spectrometry. A useful positive ion spectrum

Unconjugated Pterins and Related Biogenic Amines

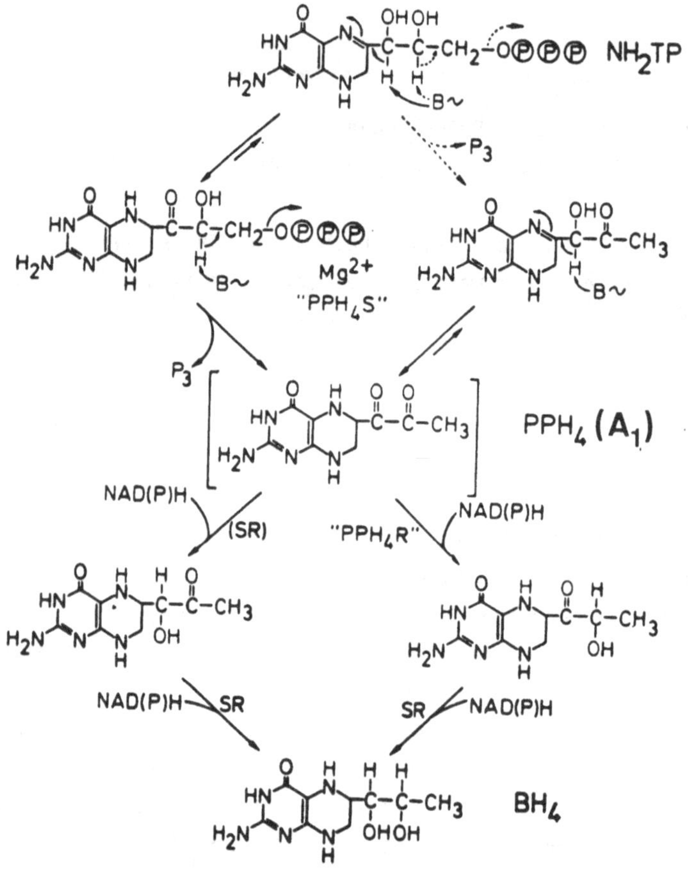

Scheme 1: Biochemical pathway of BH_4 biosynthesis from dihydroneopterin triphosphate.

(Fig. 1) was acquired using thioglycerol (glySH) as the liquid matrix, in which a $[M + H]^+$ ion of moderate abundance was observed at m/z 238 (ca. 20% rel. int.). This indicated the presence of a molecule with the correct elemental composition $C_9H_{11}N_5O_3$ (M=237). The intense signal at m/z 166 (base peak) was attributed to a $[M + H]^+$ ion of 7,8-dihydropterin (PH_2, $C_6H_7N_5O$, M=165) which was formed by decomposition of the moderately stable new compound either before or during analysis. An alternative route for the formation of the m/z 166 ion by genuine gas phase fragmentation of protonated PPH_4, e.g., via loss of $O{=}C{=}C(OH)CH_3$ from an intact $[M + H]^+$

precursor, appeared less probable since such a process is of only minor importance in the metastable ion (MI) decomposition which may be observed by FAB/MSMS (see below).

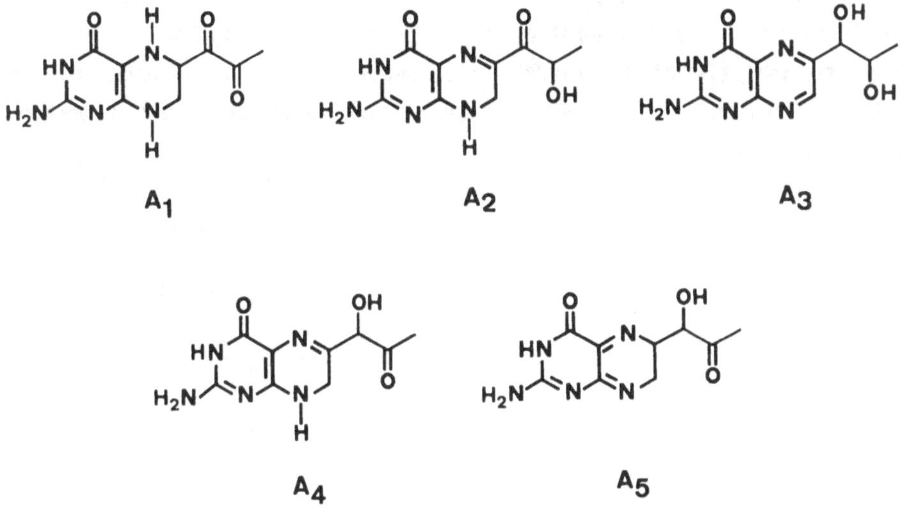

Scheme 2. Structures considered as the intermediate in tetrahydrobiopterin biosynthesis: A1, 6-pyruvoyl tetrahydropterin; A_2, sepiapterin; A_3, biopterin; A_4, A_5, see text.

The (+)FAB behavior of known isomers and hydrogenate analogues of PPH_4 in different matrices was of special interest in this context. Of these isomers, the most important ones, i.e., sepiapterin (A_2) and biopterin (A_3), were readily available for a preliminary study[1]. A_2 yielded a useful FAB spectrum in a glycerol (glyOH) matrix displaying a $[M + H]^+$ ion at m/z 238, ca. 15% rel. int., together with other major ions of the $(glyOH)_n$ cluster type. Difficulties were however encountered with A_3, as even modest $[M + H]^+$ formation required the use of m-nitrobenzylalcohol (NOBA) as a

---

[1]) A more detailed study of the FAB behavior of pteridines is currently in progress (W.J. Richter, F. Raschdorf, Th. Kuster, and H.-Ch. Curtius).

liquid matrix (8) since the 10% rel. int. signal at a "chemical noise" level of up to 5% appeared to be useless unless MS-MS was employed. While the A_4 and A_5 isomers were not obtainable, the highly unstable hydrogenated analogue o- or p-quinonoid dihydrobiopterin (quinonoid BH_2) was available for comparison. As expected, this analogue (M=239) also proved difficult to analyze (glySH matrix), as an efficient uptake of 2 H atoms was observed which probably yields BH_4 as the stable product (m/z 240/242 ratio about 1:5). The analysis of BH_4 (M=241) did however prove to be successful as the (+)FAB spectrum (glySH matrix) showed exclusive $[M + H]^+$ formation (m/z 242, base peak).

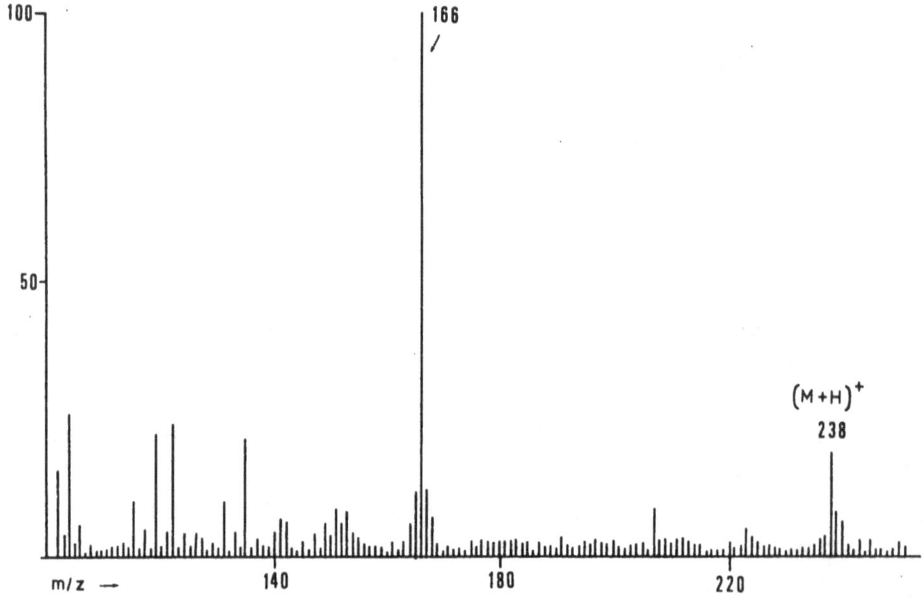

Fig. 1. (+)FAB spectrum of the intermediate A_1 (thioglycerol matrix, 10 keV xenon atoms).

From Fig. 1 and the preliminary results obtained with these isomers and analogues it is quite obvious that structural parameters, other than molecular size, cannot be directly deduced from FAB/MS, owing to a lack of pronounced fragmentation of a clear-cut origin. Under these circumstances, FAB in conjunction with two-dimensional mass spectrometry (FAB/MSMS) appeared highly promising since the ions of interest, i.e., the $[M + H]^+$ ions of intact PPH_4, can be singled out for further MS analysis as more or less pure species. The "dynamic", i.e., nonchromatographic "isolation" of the ions, that is, separation from other ions such as those of PH_2 and the prevailing "chemical noise", is accomplished in the first stage, MS-I of the employed tandem mass spectrometer. MS-II then performs the mass analysis of the daughter ions which are generated in a collision cell positioned between MS-I and MS-II. Operation of the cell with or without a target gas permits the options of collision induced dissociation (CID) and spontaneous, i.e, metastable ion (MI) decomposition, respectively.

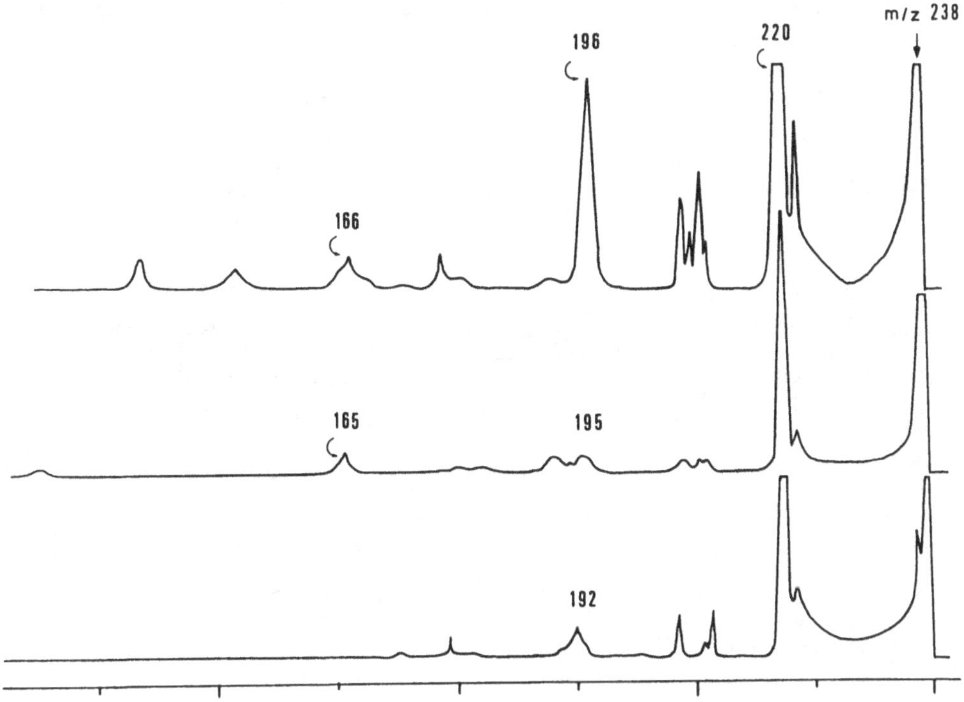

Fig. 2. Mass analyzed ion kinetic energy (MIKE), FAB/MSMS spectra of $[M + H]^+$ ions (m/z 238) of the intermediate A_1 (a, upper trace), sepiapterin (b, center trace) and biopterin (c, lower trace).

Fig. 2 shows the FAB/MSMS spectra of the relevant mass-selected $[M + H]^+$ ions, m/z 238, which were recorded in the MI mode. The $[M + H]^+$ ion of the "unknown", which is to be compared with those of the reference ions derived from A_2 and A_3, shares a pronounced loss of H_2O with its isomers. In the case of structure A_1, which lacks a hydroxyl substituent as a preformed leaving group, a rearrangement by hydride shift(s) is indicated (for loss of H_2O from protonated ketones see (9)). The pronounced elimination of a 42-Dalton neutral, which is most likely to be $CH_2=C=O$ from the terminus of the side-chain, seems to be highly specific for the new derivative. The genesis of the corresponding daughter ion at m/z 196 may also be rationalized by assuming that a fast 1,2-hydride shift in a carbonyl-protonated $[M + H]^+$ species occurs as the initial step (Scheme 3). The rearranged species, i.e., the product of an intramolecular "redox" process leading to a dihydropterin, should permit a "Grob-type" fragmentation which produces

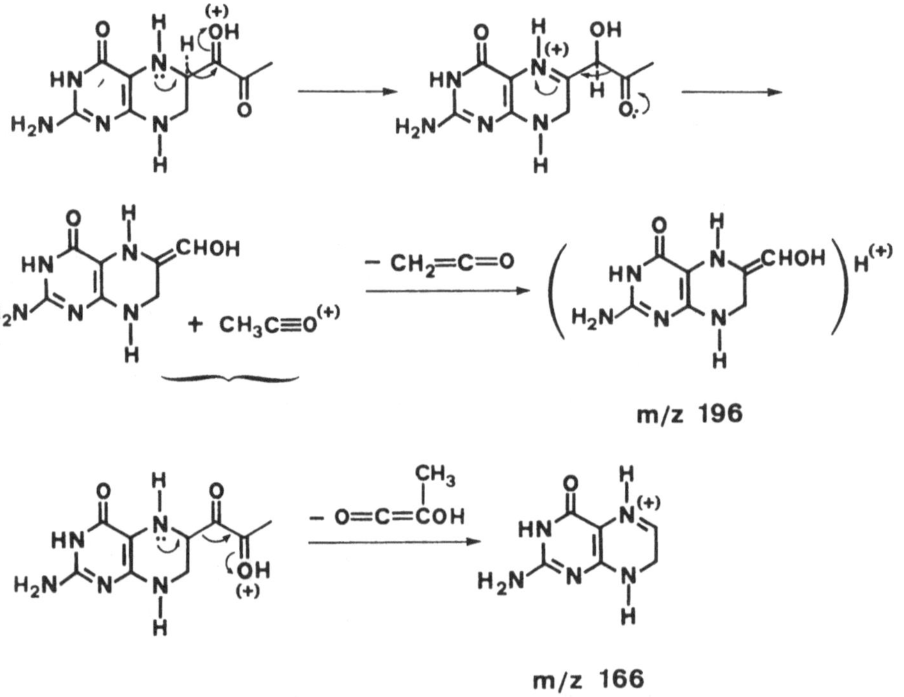

Scheme 3. Proposed mechanism for the elimination of $CH_2=C=O$ and $CH_3C(OH)=C=O$ from $[M + H]^+$ of PPH_4.

an acetyl cation $CH_3C{\equiv}O^+$ together with a neutral hydroxymethylenamine molecule. The high proton affinity of the neutral may favor subsequent proton abstraction from $CH_3C{\equiv}O^+$ (the potential Bronsted acid) prior to dissociation of the intermediate ion/dipole complex, thereby rendering $CH_2{=}C{=}O$ the final neutral ejected. It seems plausible that the di- and nonhydrogenated pterins such as A_2 and A_3 do not lend themselves readily to analogous reaction pathways. The "compound specificity" of the MI daughter ion spectrum of the $[M + H]^+$ of the new derivative supports not only the postulated existence of the new isomer, but also its isolation without large-scale decomposition by rearrangement into A_2 or A_3 during the workup and/or FAB analysis.

As mentioned above, the formation of a protonated PH_2 (m/z 166) by MI ejection of $CH_3C(OH){=}C{=}O$, rather than chemical decomposition, is inconspicuous. It should nevertheless be considered, as it places the additional oxygen atom clearly into the C(1')-carbonyl group of the assumed CH_3COCO substituent. The specificity of this latter process is unfortunately not sufficiently well established owing to the low relative intensity and broadness of the m/z 166 MI signal (see e.g. A_2). The MS evidence for a tetrahydrogenated pterin nucleus is therefore insufficient.

When specificity is only admitted for the MI elimination of ketene from the C_3 substituent in a di- or tetrahydrogenated pterin skeleton, structural assignment of A_1 has still two major competitors, i.e., A_4 and A_5. Both these alternatives represent dihydropterins, i.e., a 7,8-dihydro- and a quinonoid o- or p-tautomer 6,7-dihydro derivative carrying a common 1'-hydroxy-2'-oxopropyl substituent. A_4, which is tautomeric with sepiapterin via an endiol intermediate, is ruled out because of its chromophore which is incompatible with the observed UV absorption. The UV-compatible chromophore A_5 is a highly unlikely competitor owing to its FAB/MS behavior which sharply contrasts that of o- or p-quinonoid 6,7-dihydrobiopterin since the quinonoid BH_2 reference compound A_3 rapidly takes up 2 H atoms prior to or during FAB analysis. In the case of the new derivative no such hydrogen uptake was observed (Fig. 1). The PPH_4 structure A_1, which was derived from other spectroscopic evidence without actual isolation, is therefore complemented and fully supported by mass spectrometric means.

Section B: Biochemistry and Metabolism of Pterins

TETRAHYDROBIOPTERIN: OTHER PHYSIOLOGICAL ROLES IN ADDITION TO AROMATIC AMINO ACID HYDROXYLATION?[1]

Sheldon Milstien, Ph.D.

Laboratory of Neurochemistry
National Institute of Mental Health
Bethesda, MD 20892

Introduction

The two major unconjugated pterins present in human serum, urine and cerebrospinal fluid are dihydroneopterin and tetrahydrobiopterin (BH_4) (1). The only established physiological role for BH_4 is as the natural cofactor in the hydroxylation of the aromatic amino acids, phenylalanine, tyrosine, and tryptophan (2). BH_4 is used by the three different enzymes which catalyze the hydroxylation reactions in conjunction with enzyme-bound iron to reduce molecular oxygen (2). A genetic defect in phenylalanine hydroxylase blocks the metabolism of phenylalanine and causes phenylketonuria (3). Children with genetic defects in either de novo BH_4 biosynthesis or in dihydropteridine reductase, the enzyme responsible for maintaining BH_4 in its reduced form, have both an inability to metabolize phenylalanine, as well as a lack of the important neurotransmitters, dopamine, epinephrine, and serotonin, which are derived from the hydroxylated amino acids (4).

The origin of dihydroneopterin in human tissues has not yet been

---

[1] This paper is dedicated to the memory of Dr. Alois Niederwieser

Unconjugated Pterins and Related Biogenic Amines

unequivocally established. It probably arises as a result of dephosphorylation of dihydroneopterin·triphosphate, the first pterin derivative in the de novo BH_4 pathway (5). The measurement of dihydroneopterin levels in blood and urine has proven to be a useful clinical parameter in cancer (6), AIDS (7), and organ transplantation (8). Dramatic increases in neopterin excretion are seen whenever lymphoid cells, such as T cells or macrophages, are stimulated by interferon or IL-2 (9). The role of dihydroneopterin in this or any other physiological process has as yet not been established. Clearly, dihydroneopterin cannot be an indispensable factor, since: many species have essentially no neopterin in their body fluids or tissues (1); many human cells, such as fibroblasts, do not produce any neopterin; and children with a genetic block in GTP-cyclohydrolase, the enzyme which catalyzes the formation of dihydroneopterin triphosphate, have no apparent anomalies in their immune systems (10).

The presence of BH_4 in many tissues and in brain regions nearly devoid of any aromatic amino acid hydroxylase activity (1), raises the possibility that BH_4 might also be necessary for the activity of a different kind of enzyme system. Several other oxidation reactions have been proposed to utilize a tetrahydropterin as a cofactor. These include: ether lipid oxidation (11); proline hydroxylation (12); and mitochondrial electron transport (13). The evidence for the requirement for a tetrahydropterin in these enzyme systems is reviewed below. In addition, as a guide to the discovery of other roles for BH_4 and it's biosynthetic enzymes, the significance of the discovery that dihydropteridine reductase, an enzyme which had previously been thought to only be involved in maintaining BH_4 in the fully reduced state, also plays an important role in tetrahydrofolate metabolism, is discussed.

Distribution of BH_4 and Its Biosynthetic Enzymes in Rat Brain

BH_4 is present in every region of the rat brain (14) and the human brain (15) with a fairly narrow concentration distribution. Since the only known function of BH_4 in the CNS is to serve as a cofactor for tryptophan and tyrosine hydroxylases, it is not surprising that a correlation has been found between the activity of these enzymes in various regions of the rat brain and the concentration of BH_4 (14). Using an enzymatic cofactor assay, Levine, et al (14), showed that there was a high degree of correlation ($r = 0.8$) between reduced cofactor content and total tryptophan plus tyrosine hydroxylase activities in the rat brain. In Figure 1, are plotted BH_4 concentrations, determined by a high performance liquid chromatographic assay (1), against the total tryptophan and tyrosine hydroxylase activities in seven areas of the rat brain. There is a high degree of correlation between BH_4 content and hydroxylase activities, even though the HPLC method gives somewhat greater values than the cofactor assay. It should be noted that there is more than a 120 fold range in the hydroxylase activities between the cerebellum and the striatum, whereas there is only a 5 fold range in BH_4 concentration. This difference in the ranges suggests that if the BH_4 in the striatum is concentrated almost solely in aminergic neurons as has been suggested (16), this cannot be the case in other areas of the brain since the concentration of BH_4 would be extremely high in those areas with only relatively few aminergic neurons, such as the cerebellum and cortex. Also, extrapolation of the data in Figure 1 to a hypothetical region of the brain with no hydroxylase activities indicates that there would still be significant amounts of BH_4 remaining.

Another approach which can be used to evaluate the localization of BH_4 and its biosynthetic enzymes in specific aminergic neurons is the application of neurotoxins, such as 6-hydroxydopamine, 5,7-dihydroxytryptamine, kainic acid, and MPTP. These toxins are taken up by receptor-mediated processes and then kill those cells which concentrate them by still not well characterized mechanisms. 6-Hydroxydopamine and MPTP are taken up selectively by cells in the substantia nigra dopaminergic pathway and cause symptoms in animals similar to those of Parkinson's Disease in

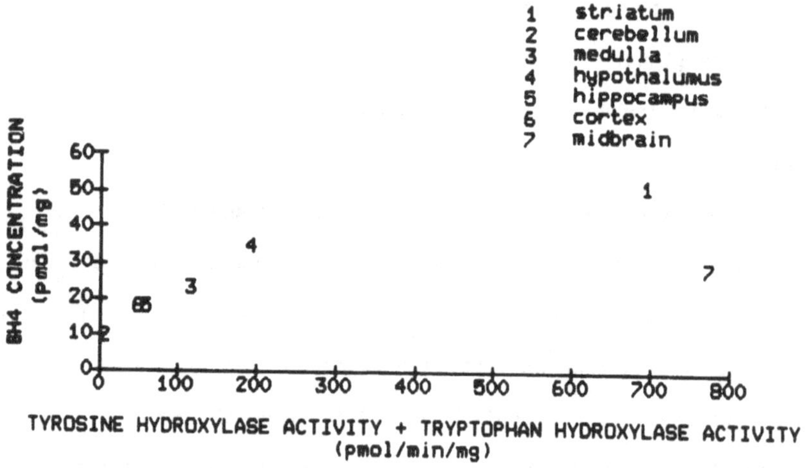

Figure 1. Correlation of tetrahydrobiopterin concentration and hydroxylase activities in the rat brain.

humans (17). Recently, Levine, et al showed that treatment of rats with 6-hydroxydopamine, caused a 93% loss of striatal tyrosine hydroxylase activity (16). This was accompanied by a 73% decrease in BH_4 concentration and a 68% loss of GTP-cyclohydrolase activity, the first, and rate-limiting enzyme in de novo BH_4 biosynthesis. This finding, which suggests a high degree of co-localization of tyrosine hydroxylase activity, a marker of dopaminergic neurons, and BH_4 and its biosynthetic enzymes in the striatum, leaves open the possibility that even in the striatum, some BH_4 may be present in other cells. It should be noted that there are a substantial number of serotonergic neurons in the striatum which also presumably contain BH_4. Thus, while it is difficult to accurately determine the concentration of BH_4 in all cell types in the striatum, it is possible that BH_4 occurs in non-aminergic neurons.

The de novo biosynthesis of BH_4 is still not completely characterized. The first enzyme in the pathway, GTP-cyclohydrolase, catalyzes the formation of dihydroneopterin triphosphate from GTP (5). Dihydroneopterin triphosphate is converted to the first tetrahydropterin on the pathway, 6-pyruvoyl-tetrahydropterin, by the enzyme 6-pyruvoyl-tetrahydropterin synthase (18). The conversion of this diketo tetrahydropterin to BH_4 requires two keto reductases (19), 6-pyruvoyl-tetrahydropterin reductase, and finally, sepiapterin reductase. The distribution of these enzymes in the rat brain is shown in Figure 2. There is a significant correlation between GTP-cyclohydrolase activity and BH_4 concentration in these seven gross regions of the rat brain. However, the activities of the remaining enzymes of the de novo BH_4 biosynthetic pathway are at relatively the same concentration in the entire rat brain and do not show any correlations with BH_4 content. This unusual distribution would be more logical if these enzymes had other functions besides those of BH_4 biosynthesis. Recent experiments with MPTP-treated mice, showed that this toxin caused a marked decrease in striatal tyrosine hydroxylase, striatal BH_4, striatal GTP-cyclohydrolase, and had no effect on the other enzymes of BH_4 biosynthesis (20). Clearly, at least in the striatum, the major fraction of the terminal enzymes in the de novo BH_4 biosynthetic pathway after GTP-cyclohydrolase must be located in non-dopaminergic cells. Therefore, either there is only a small fraction of the total amount of these enzymes co-localized in the same striatal neurons with GTP-cyclohydrolase and whose loss could not be detected in these experiments, or BH_4 biosynthesis is completed outside of these dopaminergic neurons and then is taken up. If there really is a small amount of all of the enzymes of de novo BH_4 biosynthesis co-localized with tyrosine hydroxylase, then it is likely that the other enzymes besides GTP-cyclohydrolase have as yet other undiscovered functions.

Ether Lipid Oxidation

Ether lipids are derivatives of glycerol which have a fatty acid attached through an ether bond, rather than the usual ester linkage found in most phospholipids. Compounds of this type, such as the plasmalogens, are components of many membranses including the myelin sheath. Platelet activating factor, 1-alkyl-3-acetyl-sn-glycero-3-phosphocholine, is a naturally occuring ether lipid with potent activity that modulates hypotensive and platelet-aggregating responses. In 1964, Tietz, Lindberg, and Kennedy reported the discovery of a microsomal hydroxylating system

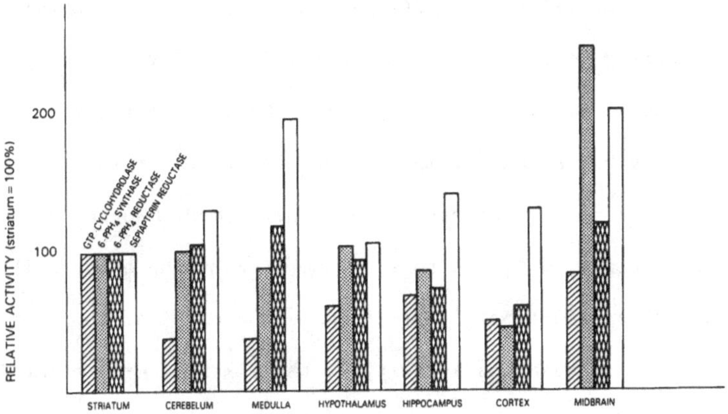

Figure 2. Distribution of BH_4 biosynthetic enzymes in the rat brain. The activities of the enzymes in the striatum were: GTP-cyclohydrolase (10), 300 pmol/min/g; 6-pyruvoyl-tetrahydropterin synthase (19), 390 pmol/min/g; 6-pyruvoyl-tetrahydropterin reductase (19), 450 pmol/min/g; sepiapterin reductase (32), 17 nmol sepiapterin reduced/min/g.

that catalyzed a molecular oxygen and tetrahydropterin-dependent cleavage of glyceryl ethers (11). They found that this activity was localized to mitochondria and microsomes of rat liver and was stimulated by the addition of a tetrahydropterin to washed subcellular preparations. The tetrahydropterin substrate specificity for this reaction was examined in some detail since Tietz, et al were aware that phenylalanine hydroxylase activity shows markedly different properties with pterin analogs than it does with the natural cofactor, tetrahydrobiopterin. They compared the ether cleavage activity of liver microsomes with 6,7-dimethyltetrahydropterin ($DMPH_4$), 6-methyltetrahydropterin ($6MPH_4$), and BH_4. There was almost no activity without added tetrahydropterin. All three cofactors gave almost the same maximal activity, but the apparent K_m for $6MPH_4$ was lower than that for $DMPH_4$. Unfortunately, at that time they only had a small sample of BH_4 and were not sure of its purity, so that no kinetic analyses could be performed. No further studies have been reported on the role of the tetrahydropterin in this ether lipid oxidation reaction. It was found, however, that there is an inverse correlation between the activity of the alkylglyceryl ether monooxygenase system and the concentration of ether lipids in that tissue (11). Thus, tissues such as bone marrow and spleen, which have a high content of glyceryl ether lipids have very low activity. Whereas, liver, which has a very low concentration of ether lipids has the highest activity of ether lipid oxidation of any organ. The physiological importance of this ether lipid oxidation reaction has not yet been elucidated. If, BH_4 is the natural cofactor for this reaction, a point not yet established, it would be expected that the reaction would be blocked in patients with a genetic deficiency in the de novo biosynthesis of BH_4. An absence of BH_4 would presumably lead to an accumulation of those ether lipids that were substrates for this reaction. Since no actual measurements of ether lipids in any patient with a BH_4 deficiency have been reported, it is not possible to determine if this is a factor in the neuropathology of untreated patients. In fact, the consquences of inhibition of the ether lipid oxidation pathway are unknown, but merit further investigation. In a disease with severe neurological symptoms known as the Zellweger syndrome, there is a complete absence of peroxysomes, a subfraction of microsomes. These patients show dramatic alterations in many aspects of long chain fatty acid metabolism since many of the enzymes for normal lipid metabolism are also located in the peroxysomes. It has not been determined if the ether cleavage activity is also located in peroxysomes, but it is a likely possibility.

Proline Hydroxylation

The processing of collagen molecules involves the hydroxylation of proline residues of the nascent polypeptide chains while still attached to ribosomes. The hydroxylation reaction is catalyzed by proline hydroxylase and requires molecular oxygen, a-ketoglutarate, and a reducing agent, usually ascorbate. While ascorbate is generally considered to be the physiological reductant, other compounds, such as tetrahydropterins, cysteine, and dithiothreitol are also active (12). Several murine fibroblastic cell lines have been described which have lost the ability to synthesize ascorbate and which contain no measurable ascorbate and yet are still able to hydroxylate proline in a normal manner. In a study of the proline hydroxylase from L-929 cells, Peterkofsky, et al, showed that 6,7-dimethyltetrahydropterin was equally as effective as ascorbate as cofactor and both were somewhat better than cysteine (12). The L-929 cells did not contain measurable ascorbate but contained a reduced proline hydroxylase cofactor that has still not been identified (12). Using a phenylalanine hydroxylase cofactor assay, they did not find enough tetrahydropterin euivalents to suggest that the endogenous reducing substance could be a tetrahydropterin. However, since they did not directly determine the concentration of pterins in their cells by HPLC, nor did they even compare the activity of the natural cofactor, BH_4, in their assays, the question of any participation by BH_4 in in vivo proline hydroxylation reactions remains open. BH_4-deficient children do not show any symptoms similar to those seen in ascorbic acid deficiency, which are clearly due to an impaired collagen synthesis. However, since collagen is part of the extracellular matrix, subtle alterations of the matrix in the CNS could lead to profound neurlological alterations, especially during development.

Mitochondrial Electron Transport

It is has long been known that the reduction potential for the quinonoid dihydrobiopterin-tetrahydrobiopterin couple is lower than that for the oxidized cytochrome c-reduced cytochrome c couple (22). In fact, the reduction of oxidized cytochrome c by tetrahydropterins forms the basis of a very convenient assay of dihydropteridine reductase activity (22). In 1972, Rembold and Buff demonstrated that tetrahydropterins could feed electrons into the mitochondrial respiratory chain at the level of cytochrome c, resulting in a substrate-dependent increase in respiration (13). High concentrations of tetrahydropterins inhibited respiration, probably by preventing reoxidation of cytochrome oxidase. Thus, they found that tetrahydropterins could potentially have a dual effect on respiration depending upon the concentration. They found that dihydropterins were inactive and that tetrahydroneopterin, $DMPH_4$, and BH_4 were equally as effective at stimulating mitochondrial respiration. For reasons which still remain unclear, they did not find any increase in ATP formation with any tetrahydropterin which accompanied the increased respiration. Rembold and Buff also showed for the first time that the concentration of biopterin in mitochondria was higher than in the cytosol, lending some credence to the possibility that BH_4 could participate in electron transport.

In 1975, Taylor and Hochstein repeated studies of the tetrahydropterin-dependent respiration reaction in rat liver mitochondria (23). In contrast to the results of Rembold and Buff, they were able to demonstrate that ATP production was coupled to the increase in respiration produced by $DMPH_4$. They found the P:O ratio with $DMPH_4$ was 0.85 as expected for reduction at the level of cytochrome c. The possible contributions of BH_4 to mitochondrial electron transport in vivo still remain to be established.

Both the adrenal medulla (1) and the anterior pituitary (24) contain very high concentrations of BH_4. In the adrenal medulla it is thought that the BH_4 is necessary for tyrosine hydroxylation, the first step in the synthesis of norepinephrine. However, the concentrations of tryptophan and tyrosine hydroxylase are very low in the pituitary gland even though there is a higher concentration of BH_4 than in the striatum (14). The secretory granules of the anterior pituitary contain an oxygenase that catalyzes the c terminal amidation of many peptides, such as vasopresin, oxytocin, and MSH. This oxygenase requires ascorbic acid which is present in the granules at high concentrations (24). While it has been reported that tetrahydropterins were not cofactors for this reaction, it is clear that in secretory granules of both the adrenal medulla and the anterior pituitary, there are cytochromes in the outer membranes which function as electron carriers to maintain ascorbate in its reduced form. A likely role for tetrahydrobiopterin could be as an electron donor to ultimately reduce ascorbate, although the potential of the cytochrome in the secretory granules (+140 mV) would require a very high local concentration of BH_4 (+150 mV) to provide a driving force for this reaction.

Dihydropteridine Reductase

During the hydroxylation of phenylalanine, tyrosine, and tryptophan, there is a stoichiometric oxidation of BH_4 to quinonoid-7,8-dihydrobiopterin (qBH_2)(2). In the absence of any reducing agents, the qBH_2 spontaneously rearranges to the more stable 7,8-dihydro isomer which is no longer active as a cofactor. The enzyme, dihydropteridine reductase (DHPR), catalyzes the reduced pyridine nucleotide-dependent reduction of qBH_2 back to BH_4, allowing the cofactor to function catalytically. When this role of DHPR was elucidated, it was clear that not only would phenylalanine hydroxylation be impaired if DHPR were not active or missing, but also the synthesis of the neurotransmitters derived from tyrosine and tryptophan would be affected, since DHPR is also necessary for the formation of the hydroxylated precursors of the neurotransmitters.

DHPR is a ubiquitous enzyme occurring in almost every cell (25). The distribution of DHPR activity in the brain does not follow the distribution of either tyrosine or tryptophan hydroxylase or the distribution of BH_4 (Figure 3). Thus, even though the only role that DHPR was supposed to play was that of maintaining BH_4 in the reduced state, it was found at high levels in tissues where there either was not any BH_4 or hydroxylase activity, such as in fibroblasts. These results made it likely that DHPR also had some other function, but it was not until a

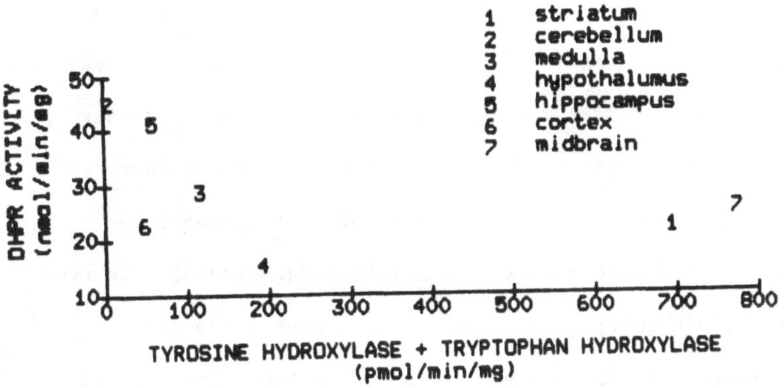

Figure 3. Correlation of dihydropteridine reductase activity and hydroxylase activities in the rat brain.

genetic disease was discovered in which DHPR activity was missing was another role found for DHPR.

In 1975, the first case of hyperphenylalaninemia due to DHPR deficiency was described (26). The child had severe neurological problems despite adequate dietary therapy to maintain low serum phenylalanine levels. The neurological degeneration in this case was originally attributed to low neurotransmitter levels in the brain (26). However, in this case and in several others, neurotransmitter replacement therapy with dihydroxyphenylalanine and 5-hydroxytryptophan was not very effective at preventing the neurological symptoms. During the original clinical workup of the first case, it was noticed that the patient had very low serum folate levels. Treatment with folic acid did not result in any improvement of the neurological symptoms. It had previously been shown that the oxidation of tetrahydrofolates proceeds through quinonoid dihydro intermediates (27), and that quinonoid dihydrofolates were good substrates for DHPR (28). Thus, if DHPR was also actually the enzyme necessary for the maintenance of tetrahydrofolates in their fully reduced form, the absence of DHPR would result in a deficiency of tetrahydrofolates. It might also be expected that a tetrahydrofolate deficiency of this type would be more evident in the brain since the concentration of dihydrofolate reductase, the enzyme that normally keeps folates reduced, is very low in the brain (29). Pollock and Kaufman showed by direct measurements on brain biopsy material from a DHPR deficient child that there indeed was a deficiency of tetrahydrofolates (30). In addition, they demonstrated that DHPR activity stimulated reactions where tetrahydrofolate was a substrate, such as the serine transhydroxymethylase reaction (30). In homogenates of frontal cortex biopsies, additions of DHPR stimulated, in an NADH-dependent manner, the conversion of serine to glycine in the DHPR-deficient sample but not in a normal sample (30). This was the first demonstration of another role for DHPR other than that of reducing qBH_2 formed during the hydroxylation of the aromatic amino acids. These findings clearly pointed out the fact that DHPR-deficient children should also receive some form of tetrahydrofolate therapy.

Recently, the effect of folinic acid therapy, in addition to a low phenylalanine diet and neurotransmitter replacement therapy, in 4 children with DHPR deficiency was studied (31). In three out of four cases where treatment was begun before profound neurological impairment, there was an improvement in neurological symptoms. In one case, treatment was started at 2 months of age and at least up to 2 years of age, there is normal neurolgical development. These results provide evidence that the irreversible damage due to DHPR-dependent tetrahydrofolate deficiency may be prevented by early treatment with folinic acid, and confirm the hypothesis that DHPR is responsible for maintaining both conjugated and unconjugated tetrahydropterins in their fully reduced, active state.

Summary

BH_4 and the enzymes which are thought to catalyze the conversion of dihydroneopterin triphosphate to BH_4 occur in relatively high concentrations in tissues for which no role for BH_4 has been found. Several other possible candidate enzyme reactions have been described which could use BH_4 as a cofactor. Further studies are needed to clarify this issue. The wide distribution of the enzymes of BH_4 biosynthesis and the lack of correlation with BH_4 concentration, suggests that they may catalyze other reactions in these tissues. Certainly, sepiapterin reductase, the terminal enzyme in the pathway, has been shown to be a non-specific keto reductase (32). The occurrence and distribution of dihydropteridine reductase in virtually all cells in the body has recently been shown to be a logical consequence of it's recently discovered role in tetrahydrofolate metabolism.

Acknowledgement

The author is grateful to Ms. Ann Orr for her help in the dissection of rat brains.

References

1. Fukushima, T., Nixon, J.C. 1980. Anal. Biochem. 102, 176-188.

2. Kaufman, S., Fisher, D.B. 1974. In: Molecular Mechanisms of Oxygen Activation (O. Haiyaishi, ed.) Academic Press, New York, pp. 285-369.

3. Kaufman, S. 1958. Science 128, 1506-1508.

4. Kaufman, S. 1976. Adv. Neurochem. 2, 1-132.

5. Burg, A.W., Brown, G.M. 1968. J. Biol. Chem. 243, 2349-2358.

6. Stea, B., Halpern, R.M., Halpern, B.C., Smith, R.A. 1981. Clinica Chimica Acta 113, 231-242.

7. Abita, J.P., Cost, H., Milstien, S., Kaufman, S., Saimot, G. 1985. Lancet, 51-52.

8. Margreiter, R., Fuchs, D., Hausen, A., Huber, C., Reibnegger, G., Spielberger, M., Wachter, H. 1983. Transplant. 36, 650-653.

9. Huber, C., Batchelor, J.R., Fuchs, D., Hausen, A., Lang, A., Niederwieser, D., Reibnegger, G., Swetly, P., Troppmaier, S., Wachter, H. 1984. J. Exp. Med. 160, 310-316.

10. Niederwieser, A., Blau, N., Wang, M., Joller, P., Atares, M., Cardesa-Garcia, J. 1984. Eur. J. Pediatr. 141, 208-214.

11. Tietz, A., Lindberg, M., Kennedy, E.P. 1964. J. Biol. Chem. 239, 4081-4090.

12. Peterkofsky, B., Kalwinsky, D., Assad, R. 1980. Arch. Biochem. Biophys. 199, 362-373.

13. Rembold, H., Buff, K. 1972. Eur. J. Biochem. 28, 579-585.

14. Levine, R.A., Kuhn, D.M., Lovenberg, W. 1979. J. Neurochem. 28, 1575-1578.

15. Sawada, M., Hirata, Y., Arai, Y., Iizuka, R., Nagatsu, T. 1987. J. Neurochem. 48, 760-764.

16. Levine, R.A., Miller, L.P., Lovenberg, W. 1981. Science 214, 919-921.

17. Langston, J.W., Ballard, P., Tetrud, W., Irwin, I. 1983. Science 219, 979-980.

18. Switchenko, A.C., Primus, J.P., Brown, G.M. 1984. Biochem.

Biophys. Res. Comm. 120, 754-760.

19. Milstien, S., Kaufman, S. 1986. In: Chemistry and Biology of Pteridines (B.A. Cooper, V.M. Whitehead, eds.) Walter deGruyter, Berlin, pp. 169-181.

20. Milstien, S., Kuhn, D.M., Levine, R.A. 1987. Abstract presented at the 6th International Catecholamine Symposium, Jerusalem, Israel.

21. Kaufman, S. 1959. J. Biol. Chem. 234, 2677-2682.

22. Hasegawa, H. 1977. J. Biochem. (Tokyo) 81, 169-177.

23. Taylor, D., Hochstein, P. 1975. Biochem. Biophys. Res. Comm. 67, 156-162.

24. May, B., Eipper, B.A. 1985. J. Biol. Chem. 260, 16224-16231.

25. Armarego, W.L.F., Randles, D., Waring, P. 1984. Medicinal Res. Rev. 4, 267-321.

26. Kaufman, S., Holtzman, N.A., Milstien, S., Butler, I.J., Krumholz, A. 1975. New Engl. J. Med. 293, 785-790.

27. Kaufman, S. 1961. J. Biol. Chem. 236, 804-810.

28. Lind, K.E. 1972. Eur. J. Biochem. 25, 560-562.

29. Pollock, R.J., Kaufman, S. 1978. J. Neurochem. 30, 253-256.

30. Pollock, R.J., Kaufman, S. 1978. J. Neurochem. 31, 115-123.

31. Irons, M., Levy, H.L., O'Flynn, M.E., Stack, C.V., Langlais, P.J., Butler, I.J., Milstien, S., Kaufman, S. 1987. J. Pediatr. 110, 61-67.

32. Katoh, S., Sueoka, T. 1984. Biochem. Biophys. Res. Comm. 118, 859-866.

^{1}H-NMR STUDIES WITH TETRAHYDROBIOPTERINS, EVIDENCE FOR THE STRUCTURE OF 6-PYRUVOYLTETRAHYDROPTERIN, AN INTERMEDIATE IN THE BIOSYNTHESIS OF TETRAHYDROBIOPTERIN

S. Ghisla
Faculty of Biology, University of Constance, D-7750 Constance, FRG

P. Steinerstauch, Th. Hasler, N. Blau and H.-Ch. Curtius
Division of Clinical Chemistry, Department of Pediatrics, University of Zurich, CH-8032 Zurich, Switzerland

Introduction

Recent reports on the biosynthesis of tetrahydrobiopterin (BH_4) have led to the recognition of 6-pyruvoyl tetrahydropterin (PPH_4) as an intermediate of key importance (1-7). Its exact chemical structure has long resisted elucidation; this was partly due to its instability and to the difficulties in obtaining substantial quantities of material. In fact, the only known procedure for the preparation of PPH_4 requires purified PPH_4-synthase (PPH_4S) and purified dihydroneopterin triphosphate (NH_2TP), the latter also having to be prepared enzymatically from GTP (8). Although the body of evidence presented in favour of the 6-pyruvoyl tetrahydropterin structure is convincing, it is mainly of secondary nature. In view of this, we consider it important to obtain further information about this intermediate's chemical properties and constitution. In a preliminary report at the 1986-meeting on "Chemistry and Biology of Pteridines", we presented ^{1}H-NMR experiments designed to follow directly the PPH_4S catalyzed conversion of NH_2TP in the NMR-tube (7). These data could correlate the disappearance of the NH_2TP signals with the appearance of a new signal which was attributed to the (3')-CH_3 group of PPH_4 (7).
Unfortunately, the signals of the hydrogens at position 7 of the pterin nucleus of PPH_4 could neither be assigned nor identified. This was due to the relatively low signal to noise ratio, as well as to the interference of

Unconjugated Pterins and Related Biogenic Amines

impurities. The C(7)-H_2 signals are, however, important for determining the mode of attachment of the side chain and will reflect the oxidation state of the pterin nucleus: In the case of a dihydropterin, a singlet at 4 to 4.5 ppm is expected, while a tetrahydropterin carrying a substituent at C(6) should yield an AB system centered at 3 to 3.7 ppm. In order to clarify these points, we have prepared some model tetrahydropterins, and compared their ^{1}H-NMR spectra with those of the intermediate in question. A suitable model for PPH_4 would be a tetrahydropterin with a carbonyl at C(1') such as 6-lactoyl tetrahydropterin (dihydrosepiapterin). In addition, the conversion of NH_2TP was reexamined under improved conditions, which allowed the identification of the C(7)-methylene signals.

Results and Discussion

^{1}H-NMR spectra of pterin model compounds

An ample literature exists on ^{1}H-NMR spectra of pterins and biopterins at different redox states (9-15). Unfortunately, however, with a single exception (16), we are not aware of reports on ^{1}H-NMR investigations of tetrahydrobiopterins in aqeous medium in the neutral pH region, the latter conditions being essential for kinetic and structural studies involving the use of enzymes. The choice of other solvents in the previous studies might have been dictated by solubility problems, and/or by the relative instability of neutral tetrahydropterins towards oxygen. As pointed out in the introduction, an AB signal should be expected for the C(7)-H_2 of tetrahydropterins due to the asymmetric C(6) center. This signal might be further complicated when the side chain also carries further centers of asymmetry, or if cyclisation with the nucleus occured.
The spectrum of (6R)-BH_4 at pH ~7 illustrates these problems (Fig. 1). Its interpretation is not trivial since all but the (3')-CH_3 signals of BH_4 appear in the region 3 to 4 ppm. The multiplet assigned to C(2')-H (octet) can easily be identified at 3.9 ppm by decoupling the (3')-CH_3 signal. The remaining signals are difficult to be attributed directly, although the signal at 3.2 ppm might be that of C(6)-H. We have therefore resorted to generation of selectively deuterated tetrahydropterins.

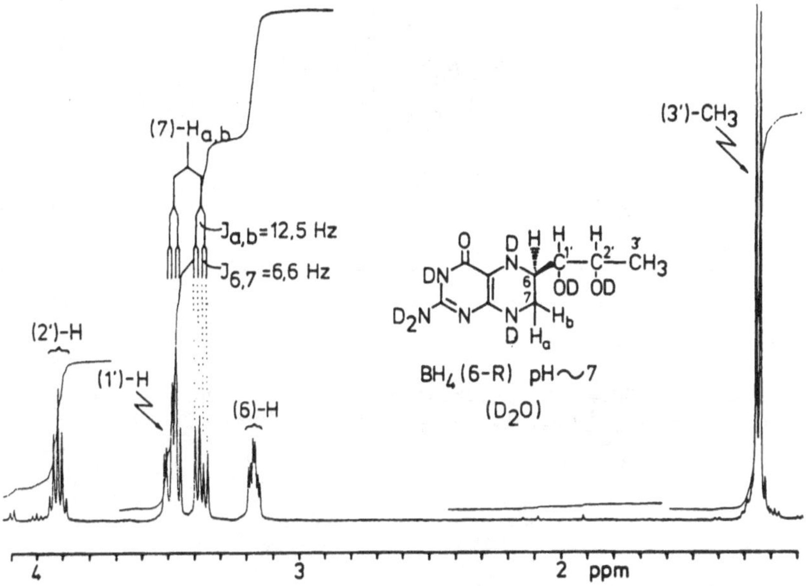

Fig. 1: ^{1}H-NMR spectrum of 6-R-tetrahydrobiopterin (BH_4). Approx. 3 mg of BH_4 hydrochloride were dissolved in ~0.4 ml D_2O containing 0.01 M phosphate buffer, and the pH was adjusted to ~7 with solid potassium carbonate (phosphate buffer prepared by dissolving P_2O_5 in D_2O, adjusted with NaOD to the pH value indicated, not corrected for H/D effect). The spectra were recorded with a Bruker 250 MHz fourier instrument (100 acquisitions) at 25°C, and the ppm scala was calibrated upon addition of a trace ethanol to the sample.

The spectrum of dihydrobiopterin (BH_2) is shown in Fig. 2 for comparative purposes.

Here, the attribution of the signals is straightforward since C(7)-H_2 is a singlet and the signals of C(1')-H and C(2')-H can be located by successive decoupling of the neighboring hydrogens.

Catalytic reduction with $^2H_2$ (Platin on asbestos in 0.1 M DCl) leads to (6-R,S),6-D-BH_4, the spectrum of which is shown in Fig. 3.

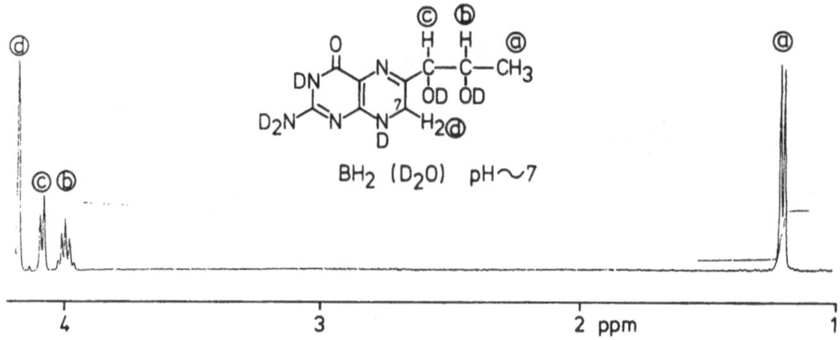

Fig. 2: ^{1}H-NMR spectrum of dihydrobiopterin. The conditions were as described in the legend of Fig. 1.

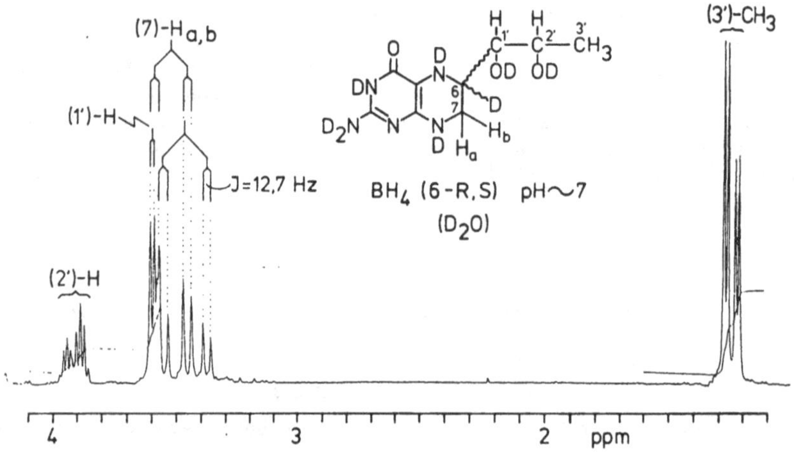

Fig. 3: ^{1}H-NMR spectrum of 6-deutero-6-(R,S)-tetrahydrobiopterin. Conditions as in the legend of Fig. 1. The compound was prepared by reduc-

tion of BH_2 with $^2H_2$ and Pt/asbestos in 0.1 N DCl. The hydrochloride obtained upon lyophilisation of the reaction mixture was then dissolved in D_2O and the pH was adjusted to a reading of 7 with NaOD. Note that comparison with the spectrum of R-BH_4 (Fig. 1) allows the facile attribution of the signal of C(6)-H of the latter and of the C(7)-H_2 AB system of S-BH_4. The signals of C(1')-H and C(2')-H can easily be assigned to the respective neighboring hydrogens by decoupling.

The first striking feature of this spectrum is the relatively large difference between the signal of the 6-R and of the 6-S-BH_4 forms (cf. two doublets of the (3')-CH_3 groups). In this case the signal of C(1')-H and of C(2')-H can be easily attributed by selective decoupling. The remaining signals must thus be those of two C(7)-H_2 AB systems belonging to the R,S pair. Again, the difference in chemical shift between the H_a and the H_b hydrogens seems to be important and is considerably different for the R form, as compared to the S form; this suggests significantly different conformations of the side chain.

The next spectrum of relevance for our purpose would be that of 6-lactoyl tetrahydropterin. This compound has been described by various groups, and can be prepared by enzymatic reduction of sepiapterin. It is, however, quite instable, and in our hands enzymatic preparation did not yield a product of sufficient quality for good ^1H-NMR spectra. Most recently, Katoh and Sueoka (16) have reported the ^1H-NMR of 6-lactoyl tetrahydropterin in neutral aqueous buffer; yet, they have neglected to show or discuss the spectral region 3 to 4 ppm, the most interesting one for our problem. We therefore decided to attempt the catalytical reduction of sepiapterin using $^2H_2$, since this would also have the advantage of leading to a 6-deuterated product. The spectrum of the starting material, sepiapterin, is simple (Fig. 4). In contrast to this, that of the product(s) was unexpectedly complex as shown in Fig. 5. In addition to some residual sepiapterin (doublet at 1.38 ppm), several signals appear below 1.0 ppm (4 doublets of different intensity). The chemical shift of these signals is typical of -(CH_2)-CH_3 groups as opposed to a -(CHOH)-CH_3 function (cf. sepiapterin, Fig. 4, or BH_4, Fig. 1, with doublets at 1.3 ppm).

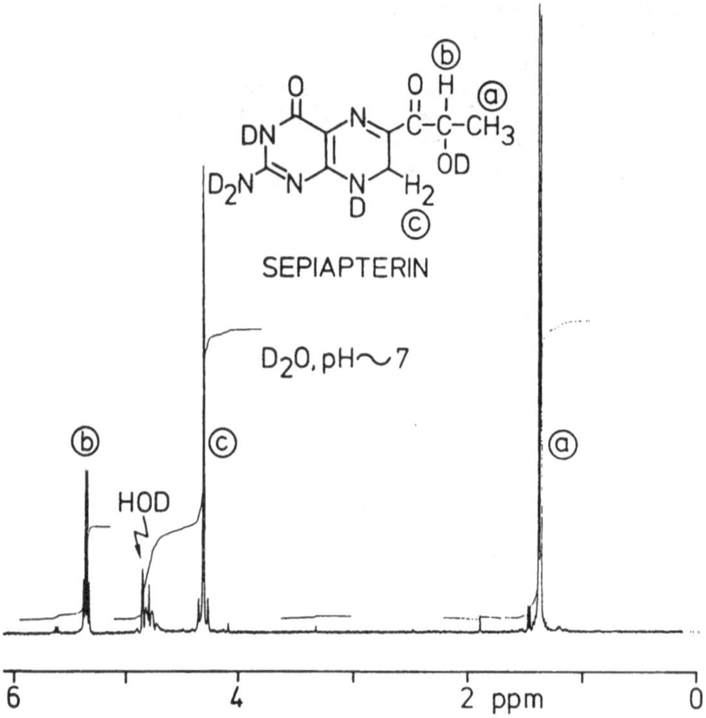

Fig. 4: ^{1}H-NMR spectrum of neutral sepiapterin. Conditions as in the experiment of Fig. 1.

Since the observed signals are doublets and not triplets, the presence of $-CDH-CH_3$ is most probable. In agreement with this interpretaion, the integration of the signals at 1.3 to 1.6 ppm yields 1 hydrogen, and they collapse to singlets upon decoupling the (3')-CH_3 group (Fig. 5). The chemical shift of these signals is compatible with C-H α to a carbonyl or to a hydroxymethyl function, the latter not carrying further (electronegative) substituents. The integration of the signals at 3.0 to 3.7 ppm (Fig. 5) yields 2 hydrogens (compared to 3 hydrogens for the signals at 0.9 to 1.0 ppm) which are thus attributed to the C(7)-H_2. From this follows that the C(6)-side chain of the product(s) contains 4 hydrogens and the pterin nucleus two.

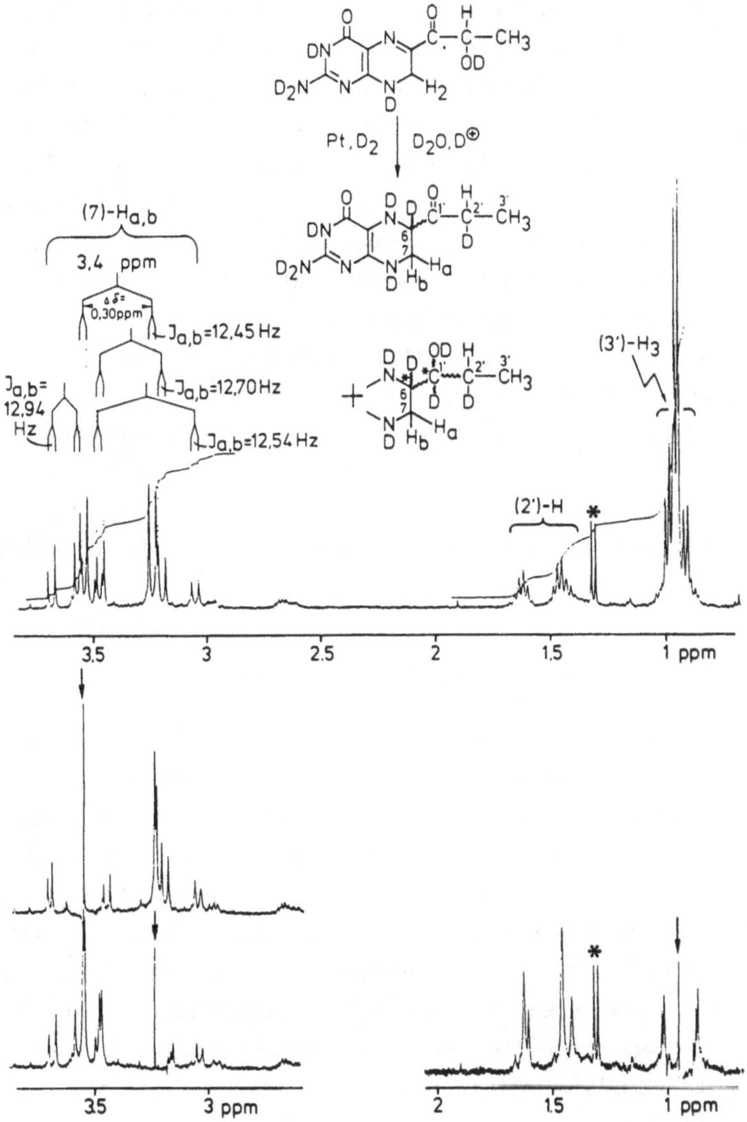

Fig. 5: ^{1}H-NMR spectrum of the mixture of products obtained upon hydrogenation of sepiapterin with $^2H_2$. Reduction conditions as described in the legend of Fig. 3. (*) denotes the signal of a trace unmodified sepiapterin. The lower part shows some decoupling experiments carried out in order to assign some of the signals. See text for further details.

As will be detailed below, the chemical properties and the gas chromatographic-mass spectrometric analysis of these products are consistent with this interpretation. All signals located at 3.0 to 3.7 ppm can be attributed to AB signals having different coupling constants, chemical shifts and differences in their chemical shifts of the A and B parts as shown in Fig. 5. This requires that upon or during hydrogenation, elimination of the (2')-OH and partial reduction of C(1') have occured. An attribution of the AB signals to individual components of the mixture (6-propionyl tetrahydropterin, (R/S) 6-(1'-hydroxy)propyl tetrahydropterin, cf. structures, Fig. 5) was not attempted. This would require the isolation of the pure components, a rather difficult task considering their instability. The conclusions which can be drawn from the experiments of Figs. 3 and 5 are nevertheless clear: They show the range in which to expect AB signals of tetrahydropterins C(7)-H_2 (3.0 to 3.7 ppm), the magnitude of the H_a, H_b coupling constants, and that the difference in chemical shift of H_a and H_b can vary considerably.

Analysis of products of ^{2}H-hydrogenation of sepiapterin

When the colourless (anaerobic) solution containing the products of hydrogenolysis of sepiapterin is exposed to air, a deep yellow colour rapidly evolves, and after several minutes a yellow precipitate formes. The UV visible absorption spectrum of this solution is similar in shape and λ_{max} to that of sepiapterin itself. This indicates that a (major) portion of the mixture is 6-propionyl tetrahydropterin which is oxidized by oxygen to (yellow) 6-propionyl-7,8-dihydropterin (2'-deoxy sepiapterin). HPLC analysis of the precipitate and of the supernatant confirm the presence of these two compounds. When the reaction mixture is analyzed by GC-MS after trimethylsilylation, three main components are found as shown in Fig. 6A. These are monodeutero-2'-deoxysepiapterin, dideutero-2'-deoxybiopterin and monodeuteropterin, the latter probably arising from (oxidative) cleavage of the side chain.

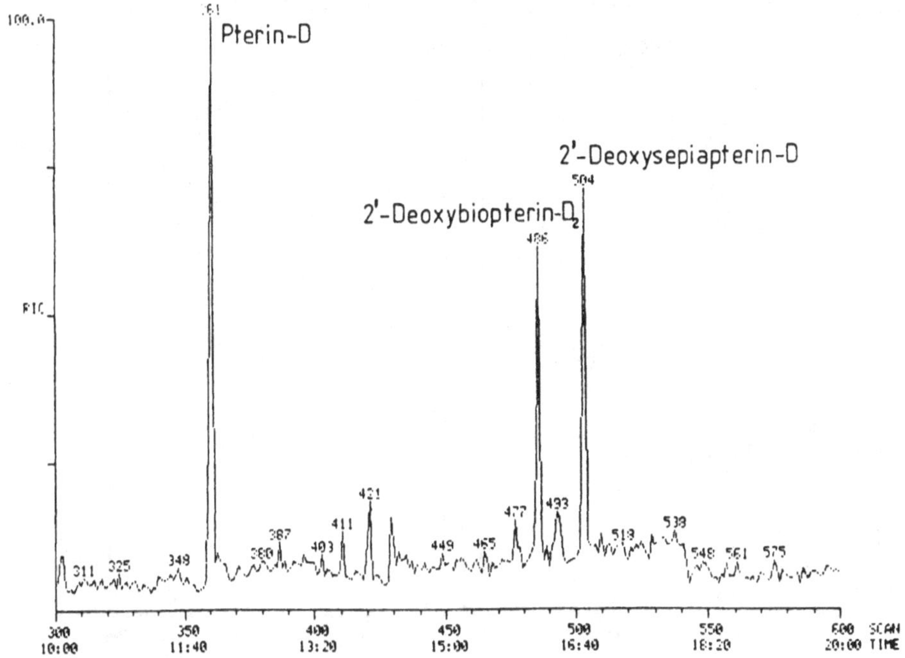

Fig. 6A: Total ion current obtained by GC-MS analysis of the silylated reaction mixture after hydrogenolysis of sepiapterin and exposure to air. These are the same products as shown in Fig. 5.

The fragmentation pattern of the 2'-deoxy biopterin (Fig. 6B) is compatible with the incorporation of a deuterium each at its position 1' and 2' (see structures on Figs. 5 and 6B). These results clearly support the structural assignments deduced from the 2^1H-NMR spectra discussed above.

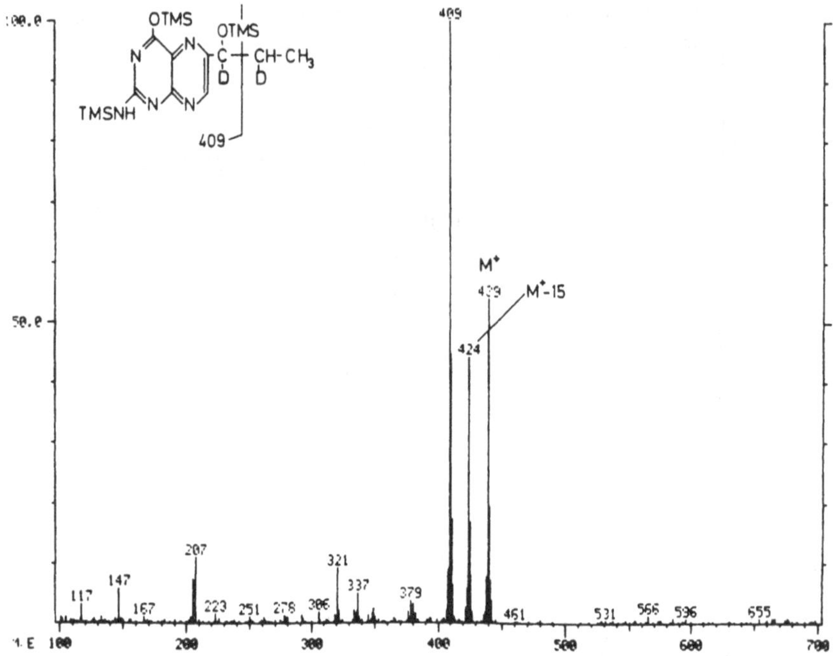

Fig. 6B: Electron impact mass spectra of dideutero-2'-deoxy biopterin TMS 3

^1H-NMR-spectrum of 6-pyruvoyl-tetrahydropterin (PPH_4)

Starting from NH_2TP, which was prepared enzymatically from GTP with cyclohydrolase (8) (Fig. 7, A), PPH_4 was prepared in situ using purified PPH_4-synthase (PPH_4S). The conditions were analogous to those described earlier (7), i.e. D_2O, pH 7.6 (adjusted to pH 7.6 without correction for H/D effect) and T = 27°C.

Fig. 7, A, shows that the starting material contains, in addition to NH_2TP, some GTP and ethanol. The latter is carried over from the preparation of NH_2TP, and was used for the calibration of the spectrum. Upon addition of PPH_4S, and after 210 min spectrum B was obtained.

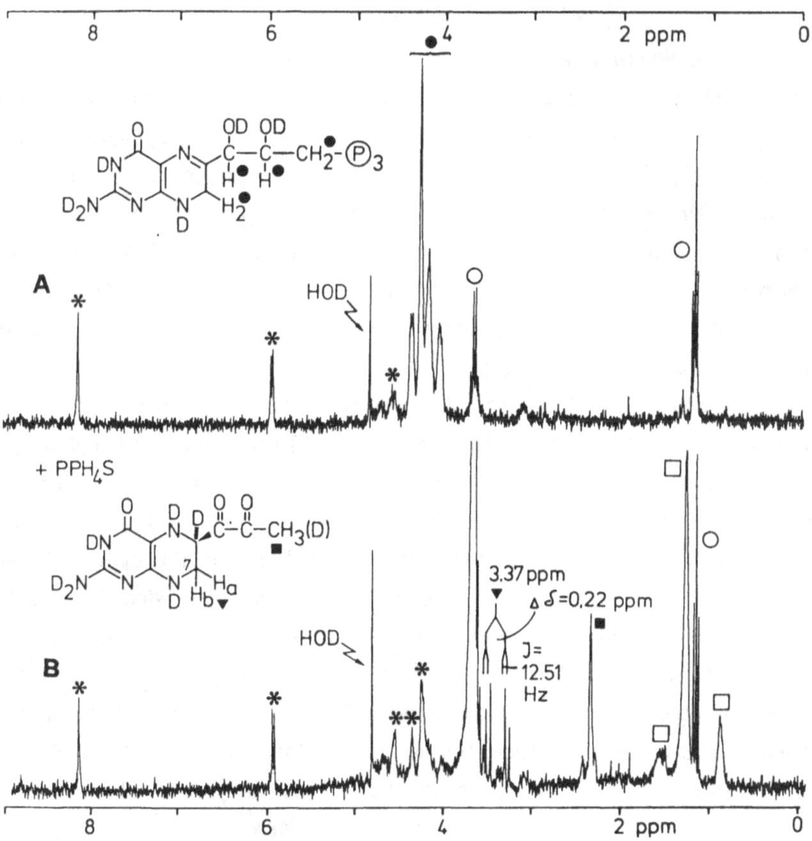

Fig. 7: ^1H-NMR spectrum of NH_2TP and of the product (PPH_4) obtained upon incubation with PPH_4S.
Spectrum A: solution of ~0.5 mg NH_2TP in 0.3 ml buffer containing 8 mM MgCl (200 acquisitions). Further conditions as described in the legend of Fig. 1. (*): signals of traces GTP (used for the enzymatic preparation of NH_2TP). (●): signals of the nonexchangeable hydrogens of NH_2TP (see structure insert). (○): trace ethanol (impurity from preparation of NH_2TP).
Spectrum B: same as A, but upon addition of 0.36 units PPH_4S and incubation for 210 min at 27°C. (*): as above. (□): protein signals and impurities introduced with PPH_4S. (○): as above. (■): signal attributed to the (3')-CH_3 group of PPH_4, and (▼): AB system attributed to C(7)-H_2. For further details see text.

Spectrum B gives information on several important points: GTP is essentially unchanged while new (broad) signals due to PPH_4S and to an impurity are present, which do not undergo changes during the incubation. The signals of NH_2TP disappear completely, concomitant with the appearance of two new ones (Fig. 7, B). The first of these, at 2.3 ppm, corresponds to that already observed earlier, its approximate intensity corresponds to 1-2 protons and the chemical shift is compatible with that of a methyl group α to a carbonyl. It should be noted, that this signal is relatively broad; this can be taken as suggesting the presence of different conformers or of internal cyclisation products e.g. of the side chain carbonyl group(s) with the 5- or 8-amino functions of the pterin nucleus. The relative signal intensity was estimated using the signals of GTP as standard and attributing 6 protons to the signals of NH_2TP (Fig. 7, A). Thus, the lower then expected intensity of the 2.3 ppm peak is attributed to incorporation of deuterium at position 3' during the incubation (7). In the spectral region 3.2 to 3.5 ppm a new signal is now apparent which has the shape typical for an AB system. The signal intensity, the coupling constant, and the difference in chemical shift between H_a and H_b is similar to that observed for the model compounds discussed above. It is thus reasonable to assign this signal to the C(7)-H_2 function of PPH_4. The presence of an AB signal is of importance since it proves that the neighboring C(6) atom is chiral, i.e. that the pterin must be in its tetrahydro form.

Conclusions

The present ^{1}H-NMR results have allowed to identifiy the signal of the C(7)-methylene function of PPH_4 as an AB system. This is direct and conclusive evidence for the existence of a tetrahydropterin nucleus in 6-pyruvoyl tetrahydropterin. They are compatible with formation of a (3')-CH_3 function and with removal of the C(1')-, and C(2')-hydrogens of dihydroneopterin triphosphate. The study of the NMR properties of various model pterins extends and completes those published by others. They suggest marked differences of the chemical properties between the Re and the Si side of tetrahydropterins carrying optically active substituents at C(6).

Acknowledgements

This work was supported by the Swiss National Science Foundation, project no. 3.395-0.86. We are grateful to Mr E. Krienitz for his skillful technical assistance and to Professor M. Viscontini for his valuable discussion.

References

1. Takikawa, S., H.-Ch. Curtius, R. Redweik, W. Leimbacher, S. Ghisla. 1986. Eur. J. Biochem. 161, 295.

2. Curtius, H.-Ch., D. Heintel, S. Ghisla, T. Kuster, W. Leimbacher, A. Niederwieser. 1985. Eur. J. Biochem. 148, 413.

3. Masada, M., M. Akino, T. Sueoka, S. Katoh. 1985. Biochem. Biophys. Acta 840, 235.

4. Switchenko, A.C., G.M. Brown. 1985. J. Biol. Chem. 260, 2945.

5. Smith, G.K., C.A. Nichol. 1986. J. Biol. Chem. 261, 2725.

6. Milstien, S., S. Kaufman. 1985. Biochem. Biophys. Res. Commun., 1099.

7. Ghisla, S., S. Takikawa, P. Steinerstauch, Th. Hasler, H.-Ch. Curtius. 1986. In: Chemistry and Biology of Pteridines (B.A. Cooper and V.M. Whitehead, eds.). de Gruyter, Berlin, pp. 299.

8. Niederwieser et al.: in preparation.

9. Viscontini, M., L. Merlini, W. von Philipsborn. 1963. Helv. Chim. Acta 46, 1181.

10. Viscontini, M., R. Provenzale, S. Ohlgart, J. Mallevialle. 1970. Helv. Chim. Acta 53, 1202.

11. Weber, R., M. Viscontini. 1975. Helv. Chim. Acta 58, 1772.

12. Schircks, B., J.H. Bieri, M. Viscontini. 1976. Helv. Chim. Acta 59, 248.

13. Schircks, B., J.H. Bieri, M. Viscontini. 1977. Helv. Chim. Acta 60, 211.

14. Schircks, B., J.H. Bieri, M. Viscontini. 1978. Helv. Chim. Acta 61, 2731.

15. Amarego, W.L.F., H. Shou. 1977. J. Chem. Soc. Perkin I, 2529.

16. Katoh, S., T. Sueoka. 1987. J. Biochem. 101, 275-278.

PURIFICATION OF 6-PYRUVOYL TETRAHYDROPTERIN SYNTHASE FROM SALMON LIVER

Th. Hasler, A. Niederwieser+, and H.-Ch. Curtius
Division of Clinical Chemistry, Department of Pediatrics, University of Zurich, CH-8032 Zurich, Switzerland

Introduction

In atypical phenylketonuria, 6-pyruvoyl tetrahydropterin synthase (PPH_4S) deficiency is one of the known defects which could be clearly identified (1). Big efforts have been made in the analysis of PPH_4S activity in liver biopsies and erythrocytes of normal persons and patients (1,2). However, as normal values are very low, such an assay is still very troublesome. The use of an antibody against this enzyme might be one possibility to increase sensitivity, in case that an appropriate antibody could be produced.

Another point is, that PPH_4S is one of the key enzymes in the biosynthetic pathway of tetrahydrobiopterin (BH_4). Its product of catalysis, 6-pyruvoyl tetrahydropterin (PPH_4), is a very labile intermediate, and it showed to be very difficult to proof its postulated structure (3-7). To produce higher amounts of this intermediate a lot of PPH_4S activity is necessary. Takikawa et al. (8) were able to purify this enzyme to apparent homogeneity from human liver. However, only low amounts of enzyme activity (30mU: 1U=1μmol PPH_4/min.) and of protein (300 μg) were isolated from 500 g of human liver.

Therefore we decided to look for another source for this enzyme. In a comparative study we determined the activities of the enzymes involved in BH_4 biosynthesis in the livers of 11 different species (9). Activity of PPH_4S was lowest in man and about 200-fold higher in trout, where it showed to be highest. As trout is a rather small fish, PPH_4S activity was also measured in carp and salmon. The activity in the carp liver was much lower than in trout but in salmon it was even twice as high.

So we hoped that PPH_4S isolated from salmon liver would provide enough

Unconjugated Pterins and Related Biogenic Amines

enzyme activity and enzyme protein for further studies of PPH_4 as well as for the production of antibodies.

Materials and Methods

Enzyme activity determinations were carried out as described previously (1) except that dithioerythritol (DTE) was added at a concentration of 10 mM. The enzyme purification procedure and the assays for the characterization of the enzyme will be described elsewhere in detail.

Results and Discussion

Purification: Starting to purify PPH_4S from salmon liver, we tried to follow the purification scheme of Takikawa et al. for the human liver enzyme (8). Soon it became obvious that the two enzymes although very similar show some important differences following their purification. We noticed that the loss of activity especially during the last purification steps was dramatically higher for the salmon liver enzyme. This loss of activity seems to be due to adsorption of the enzyme to plastic, glass or column material, as addition of either bovine serum albumin (BSA) or the nonionic detergent Tween 20 can prevent this at least in part. Another important finding was that the salmon liver enzyme has a much higher specific activity than that of human liver. Conventional hydroxyapatite column chromatography turned out to be not well suited for the salmon liver enzyme; recoveries were low and purification was poor. In this respect the use of hydrophobic interaction chromatography was more suitable (Table 1). For the gel filtration step BSA was added to the enzyme solution to increase recovery. This was possible because BSA has the same elution volume as PPH_4S. In order to minimize sample volumes and to reduce purification time, the last two chromatographic steps were performed using a FPLC system (Table 1). A 9500-fold purification with an overall recovery of about 12 % was obtained. The purified enzyme was submitted to SDS-PAGE (Fig. 1). Two bands stain with approximately the same intensity representing molecular weights of 16 and 17 kDa. Up till now, we were not able to separate these two protein bands except in SDS-PAGE.

Table 1: Purification of salmon liver PPH_4S

	total activity (mU)	total protein (mg)	specific activity (mU/mg)	recovery (%)	purification (-fold)
- homogenate	38100	79000	0.482	100	1.0
- ammonium sulfate (40-55%)	27500	25000	1.10	72	2.3
- heat treatment (80°C)	19700	1700	11.6	52	24
- HIC (Butyl-Fractogel)	14600	78	187	38	389
- gel filtration (AcA 44 / BSA)	13300	(19)	(700)	35	(1450)
- ion exchange * (Mono Q / Tween 20)	6100	1.6	3800	24	7880
- hydroxyapatite ** (HPHT / Tween 20)	2700	0.62	4600	12	9500

One unit of enzyme activity produces 1µmole of PPH_4/min from NH_2TP at 37°C. Only 66 % (*) and 89 % (**) respectively of the enzyme have been applied to the last two chromatographic steps.

Characterization: The properties of the purified salmon liver PPH_4S are summarized in Table 2 and compared with the data from the purified human liver enzyme. First of all, there is a difference in the molecular weight of the native enzymes as determined by gel filtration. Both enzymes consist of four subunits of similar molecular weight as judged by SDS-PAGE. The isoelectric point is rather different for both of the enzymes. It is pH 6.0 to 6.3 for the salmon enzyme whereas the pI for the human enzyme was determined to be 4.5 by Heintel et al. (10). The pH optimum for both of the enzymes is the same with a value of 7.5. Magnesium ions have to be added for catalytic activity, as EDTA stops the enzymatic reaction immediately. The Michaelis Menten constant of the enzyme from salmon liver for dihydroneopterin triphosphate (NH_2TP) is with 2.2 µM about 5-fold lower than that of the human liver. Most surprising was that the specific activity is about 50-fold higher for the salmon enzyme. Both enzymes seem to have thiol groups essential for activity. Without the addition of the disulfide

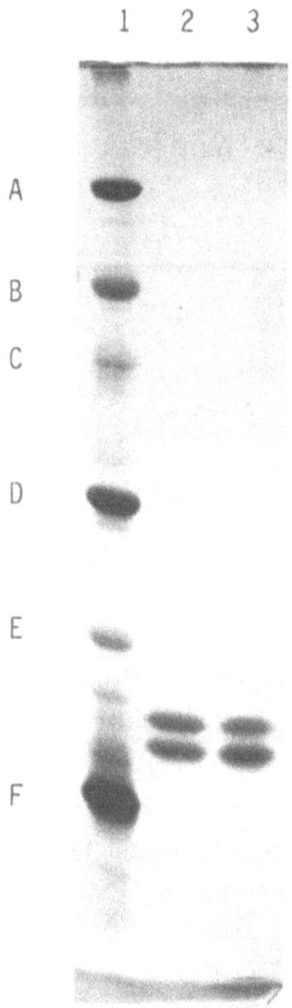

Figure 1:

SDS-PAGE with 12.5% polyacrylamide: (1) molecular weight standards: A bovine serum albumin (66 kDa), B egg albumin (45 kDa), C glyceraldehyde-3-phosphate dehydrogenase (36 kDa), D trypsinogen (29 kDa), E trypsin inhibitor (20 kDa), F α-lactalbumin (14 kDa); (2,3) purified salmon liver PPH_4S after hydroxyapatite HPHT.

reducing agent DTE to the purified salmon enzyme, no activity can be observed at all. Maximum activity of both enzymes can be obtained by preincubation with DTE, and subsequent dilution of the DTE for the enzyme assay. Both enzymes are very heat stable. Heating to 80°C for 5 minutes without substantial loss of activity is possible and can be used for the purification of both enzymes.

Glycoprotein stain: Takikawa et al. (11) suspected that the human enzyme is a glycoprotein. They performed a periodic acid Schiff stain on a nondenaturating Davis PAGE gel and found a band at the same place where the PPH_4S activity was measured. As this method using the Schiff's reagent is rather insensitive we tried another approach. We used the concanavalin A - horseradish peroxidase method described by Hawkes in 1982 (12) with slight modifications. Concanavalin A (Con A) is a lectin with a very broad binding

Table 2: Properties of salmon and human liver PPH_4S

	salmon liver	human liver
- molecular weight	68 kDa	83 kDa
- composition	4 x 16/17 kDa	4 x 19 kDa
- isoelectric point	6.0 - 6.3	4.5 **
- pH optimum	7.5 (Pipes)	7.5 (Tris)
- Mg^{2+} dependence	+	+
- K_m (NH_2TP)	2.2 μM	10 μM
- specific activity (*)	4600 mU/mg	100 mU/mg
- essential thiol groups	+	+
- heat stability (5 min.)	80°C	80°C

(*) One unit of enzyme activity produces 1μmole of PPH_4/min from NH_2TP at 37°C. Results for the human liver enzyme are from Takikawa et al. (8) except (**) from Heintel et al. (10).

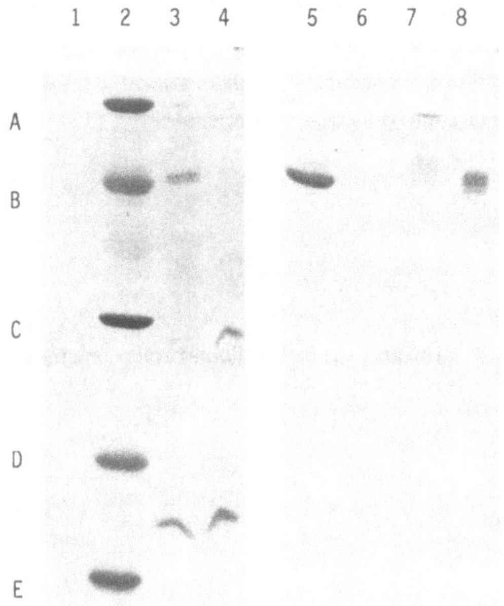

Figure 2:

Western blot on nitrocellulose after 12.5% SDS-PAGE: (5,6,7,8) Concanavalin A - horseradish peroxidase stain for glycoproteins; (1,2,3,4) Amidoblack stain for proteins; (2,5) molecular weight standards: A bovine albumin (66 kDa), B egg albumin (45 kDa), C carboanhydrase (29 kDa), D trypsin inhibitor (20 kDa), E α-lactalbumin (14 kDa); (3,7) and (4,8) two different preparations of human liver PPH_4S; (1,6) salmon liver PPH_4S.

specificity for glycoproteins. In a spot test on nitrocellulose we found that Con A was binding to purified PPH_4S from human liver but not to the enzyme prepared from salmon liver. Next, these enzymes were run on a SDS-PAGE gel and transferred afterwards to nitrocellulose by electroblotting. The Con A - horseradish peroxidase stain as well as a conventional protein stain using amidoblack were performed on these blots (Fig.2). The stained glycoproteins in two purified human liver PPH_4S samples are some high molecular weight contaminants. The low molecular weight subunit is not stained at all. As expected, no glycoprotein was detected in the salmon enzyme sample, however, blotting of its subunits was not sufficient as can be seen in the protein stain (Fig.2).

Conclusions: PPH_4S was purified to a high degree from salmon liver, but the question of the inhomogeneity of this enzyme remains to be answered. Different explanations for the two bands on SDS-PAGE are possible: The enzyme may consist of different subunits, the lower molecular weight band might be a proteolytic degradation product, or one of the two bands might belong to a contaminating protein.

An about 100-fold higher yield of PPH_4S activity was obtained from salmon liver than from human liver; but since the salmon enzyme has a high specific activity, only slightly higher amounts of protein were isolated.

The high activity of pure PPH_4S was used to produce higher amounts of PPH_4 for subsequent elucidation of its structure by mass spectrometry (13) and nuclear magnetic resonance (14) techniques.

Acknowledgement

This work was supported by the Swiss National Science Foundation, project no. 3.395-0.86.

References

1. Niederwieser, A., W. Leimbacher, H.-Ch. Curtius, A. Ponzone, F. Rey, and D. Leupold (1985) Eur. J. Pediatr. 144, 13.

2. Niederwieser, A., H. Shintaku, Th. Hasler, H.-Ch. Curtius, H. Lehmann, O. Guardamagna, and H. Schmidt (1986) Eur. J. Pediatr. 145, 176.

3. Switchenko, A.C. and G.M. Brown (1985) J. Biol. Chem. 260, 2945.

4. Curtius, H.-Ch., D. Heintel, S. Ghisla, T. Kuster, W. Leimbacher, and A. Niederwieser (1985) Eur. J. Biochem. 148, 413.

5. Masada, M., M. Akino, T. Sueoka, and S. Katoh (1985) Biochim. Biophys. Acta. 840, 235.

6. Smith, G.K. and C.A. Nichol (1986) J. Biol. Chem. 261, 2725.

7. Milstien, S. and S. Kaufman (1985) Biochem. Biophys. Res. Commun. 128, 1099.

8. Takikawa, S., H.-Ch. Curtius, U. Redweik, W. Leimbacher, and S. Ghisla (1986) Eur. J. Biochem. 161, 295.

9. Hasler, Th. and A. Niederwieser (1986) In: Chemistry and Biology of Pteridines 1986 (Cooper, B.A. and Whitehead, V.M., eds.) Walter de Gruyter, Berlin, p. 319.

10. Heintel, D., W. Leimbacher, U. Redweik, B. Zagalak, and H.-Ch. Curtius (1985) Biochem. Biophys. Res. Commun. 127, 213.

11. Takikawa, S., H.-Ch. Curtius, U. Redweik, and S. Ghisla (1986) Biochem. Biophys. Res. Commun. 134, 646.

12. Hawkes, R. (1982) Anal. Biochem. 123, 143.

13. Richter, W.J., F. Raschdorf, R. Dahinden, Th. Kuster, A. Niederwieser, W. Leimbacher, and H.-Ch. Curtius (1987) In: Unconjugated Pterins and Related Biogenic Amines (Curtius, H.-Ch. and Blau, N., eds.) Walter de Gruyter, Berlin, this volume.

14. Ghisla, S., S. Takikawa, P. Steinerstauch, Th. Hasler, and H.-Ch. Curtius (1986) In: Chemistry and Biology of Pteridines 1986 (Cooper, B.A. and Whitehead, V.M., eds.) Walter de Gruyter, Berlin, p. 299.

PURIFICATION OF 6-PYRUVOYL TETRAHYDROPTERIN 2'-KETO REDUCTASE FROM HUMAN LIVER

H.-Ch. Curtius, P. Steinerstauch, W. Leimbacher, U. Redweik
Division of Clinical Chemistry, Department of Pediatrics,
University of Zurich, CH-8032 Zurich, Switzerland

Sh.-I. Takikawa
Biological Laboratory, Kitasato University, Sagamihara City,
Kanagawa 228, Japan

S. Ghisla
Department of Biology, University of Konstanz, D-7750 Konstanz, FRG

Introduction

The biosynthetic pathway of tetrahydrobiopterin (BH_4) (Fig. 1) has been studied intensively by several groups in the last years and most steps have been elucidated (1-6).

The first reaction is the conversion of GTP to dihydroneopterin triphosphate (NH_2TP) catalyzed by a single enzyme, GTP cyclohydrolase I (GTPCH) (7). The nature of the next step, a critical one since it is probably rate-limiting in man, has long remained unclear due to the high instability of the product and the resulting problems of structure elucidation. This key intermediate is 6-pyruvoyl tetrahydropterin (PPH_4) whose structure has by now been well defined using NMR and MS (8,9), as well as other techniques (10,11,12). The elimination of triphosphate and the intramolecular rearrangement are catalyzed by one single enzyme, 6-pyruvoyl tetrahydropterin synthase (PPH_4S), which was purified to homogeneity by Takikawa et al. (13,14). The enzymes which catalyze the two-step reduction of PPH_4 to BH_4 are sepiapterin reductase (SR) and 6-pyruvoyl tetrahydropterin reduc-

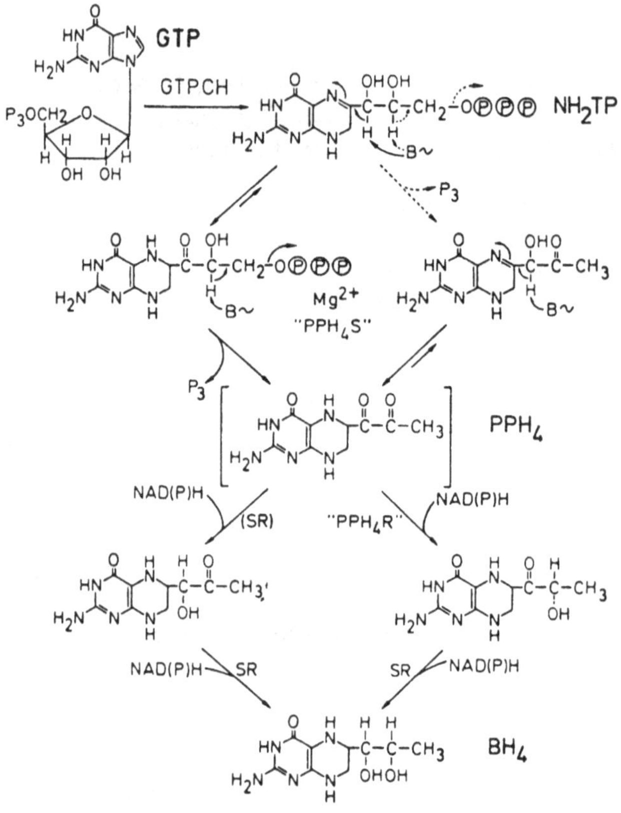

Fig. 1. Proposed pathway for the biosynthesis of tetrahydrobiopterin from guanosine triphosphate in human liver.

tase (PPH_4R), respectively. The occurrence of the second reductase, PPH_4R, was described by several groups, e.g., Milstien et al. in rat and bovine adrenal medulla (15), Switchenko et al. in Drosophila heads (10), Masada et al. in silk worm (5), and Takikawa et al. from our group in human liver (14). This enzyme was named 6-lactoyl-PtH_4 synthase (15), sepiapterin synthase B (10), and sepiapterin synthetizing enzyme 2 (SSE_2) (5).

It is not yet clear whether PPH_4R is on the BH_4 biosynthetic pathway *in vivo*. Neither has the sequence of the two-step reduction, i.e., whether reduction occurs first at C-1' or at C-2', been elucidated yet. For the study of this question N-acetyl serotonin, a potent inhibitor of SR, has been used (16). However, we have observed that in contrast to SR from rat tissues which is completely inhibited (16), this reagent inhibits human liver SR to about 70% only, and, in addition, PPH_4R is partially inhibited.

In collaboration with M. Masada from Tokyo Metropolitan University we have studied the SR-mediated reduction of synthetic aromatic substrates carrying a 1',2'-dicarbonyl side chain. It appears that the reduction of the 1'-carbonyl group precedes that of the second carbonyl group. As was shown by Masada et al. (5) and also by our group, SR has to be present in large excess in order to guarantee efficient BH_4 biosynthesis (1,17).

B. Wermuth (18) has reported the isolation of an NADPH-dependent carbonyl reductase from human brain and liver, which reduces a number of biologically and pharmacologically active carbonyl compounds. Using Wermuth's pure enzyme, we have confirmed that this carbonyl reductase is different from PPH_4R and is not involved in the biosynthesis of BH_4 (Masada et al., in preparation).

In this communication we outline the purification procedure of PPH_4R from human liver and describe some of its properties. The availability of pure enzyme is important in order to clarify the role of PPH_4R in the biosynthetic pathway _in vivo_.

Materials and Methods

Human liver was obtained from traffic accident victims. BH_4 and sepiapterin were purchased from Dr. Schircks Laboratories (Buechstrasse 17a, 8645 Jona, Switzerland). NH_2TP, 6-(1'-oxo-2'-hydroxypropyl)-5,6,7,8-tetrahydropterin and 6-(1'-hydroxy-2'-oxopropyl)-5,6,7,8-tetrahydropterin were prepared as described by Takikawa (13). N-acetyl serotonin and dihydrofolate reductase were from Sigma, NADPH and $NADP^+$ from Boehringer Mannheim. Other chemicals were obtained from commercial sources. PPH_4S and SR were purified as described by Hasler (19) and Takikawa (13). The detailed purification procedure of PPH_4R from human liver will be described elsewhere (Curtius et al., in preparation).

Protein was determined with the Bio-Rad protein assay kit using bovine gamma globulin as standard. PAGE was according to Laemmli (20) (12.5% acrylamide, 0.1% SDS, visualization by silver staining). The molecular mass of the enzyme was also estimated by gel filtration (Ultrogel ACA 54, equilibrated with 20 mM KP buffer, pH 7.0, containing 200 mM KCl; markers: bovine serum albumin (66 kD), egg albumin (45 kD), bovine erythrocyte carbonyl reductase (29 kD), wale skeletal muscle myoglobin (14 kD), and horse heart muscle cytochrome c (12.5 kD), elution monitored by A_{280nm}). Glycoprotein detection was according to Clegg (21), however, using 4-chloro-1-naphtol as horseradish peroxidase color development reagent. The isoelectric point was determined as described by Radola (22).

<u>Assays for PPH_4R and SR.</u> The assay mixture (100 μl) for PPH_4R activity contained 20 μM NH_2TP, 100 mM Tris/HCl, pH 7.4, 8 mM $MgCl_2$, 1mM DTE, 1 mM NADPH, 2 mU PPH_4S from salmon liver obtained after gel filtration (19), and the enzyme fraction. The formed product, 6-(1'-oxo-2'hydroxypropyl)-5,6,7,8-tetrahydropterin, was measured by electrochemical detection using HPLC (24). In the presence of SR traces identified by 6-(1'-hydroxy-2'-oxopropyl)-5,6,7,8-tetrahydropterin and/or BH_4 production (electrochemical detection), PPH_4R was assayed with 1 mM N-acetyl serotonin. SR activity was essentially determined as described by Katoh (23).

Results and Discussion

The procedure for the purification of PPH_4R is outlined in Scheme 1. Table 1 gives the activity and recovery data of a typical purification. Because of the presence of SR PPH_4R activity can be measured only after the hydroxyapatite step when SR is largely removed. The assay for PPH_4R activity is not possible in the presence of SR since the latter also reduces PPH_4. On the other hand, N-acetyl serotonin inhibits human SR only par-

Human Liver (300 g)

↓

Homogenization, 4 vol of 0.01 M KP buffer, pH 7.0

↓

$(NH_4)_2SO_4$ fractionation, 40-60% saturation

↓

Sephadex G 25, pH 6.0

↓

Hydroxyapatite, pH 6.0

↓

Gel filtration, Ultrogel AcA-54, pH 7.0

↓

Butyl-Fractogel, pH 7.0

↓

Blue Sepharose CL6B, pH 7.0

↓

Sephadex G 25, pH 7.5

↓

Mono Q FPLC, pH 7.5

↓

Mono Q FPLC, pH 7.0

↓

Mono Q FPLC, pH 8.0

Scheme 1. Procedure for purification of 6-pyruvoyl tetrahydropterin reductase from human liver.

tially. In order to approximate the actual PPH_4R activity before SR is completely removed, N-acetyl serotonin is used in assay mixtures with only slight SR contamination.

The hydroxyapatite column chromatography was found to be essential since it provides separation from SR and also good recovery of PPH_4S. After the AcA 54 gel chromatography SR was essentially removed. Gel filtration on Ultrogel AcA 54 allows estimation of the molecular weight of PPH_4R which is about 30'000 D.

Table 1: Purification of PPH_4R from Human Liver

Procedure	Volume (ml)	Total activity (mU)	Total protein (mg)	Specific activity (mU/mg)	Recovery (%)	Purification (-fold)
Crude extract	1'065	*	44'400		*	*
Ammonium sulfate 40-60% fraction	100		9'400			
Hydroxyapatite	53	135.0	600	0.23	100	1
Ultrogel AcA 54	16	122.4	150	0.82	91	3.6
Butyl Fractogel	2.6	62.0	17.7	3.50	46	15.2
Blue Sepharose	1.1	38.0	0.3	126.60	28	550.4

*PPH_4R activity cannot be determined reliably before the hydroxyapatite column step due to the presence of SR which catalyzes the same reaction.

After Blue Sepharose affinity chromatography the increase of specific activity was about 550-fold. This purification factor does not take into account the first two purification steps which lead to a 100-fold purification, approximately. The total amount of protein was 0.3 mg from 300 g of human liver corresponding to an overall recovery of 28%. After the Blue Sepharose step some contaminating proteins were still present (see SDS-PAGE in Fig. 1). These were eliminated by threefold FPLC chromatography on Mono Q ionic exchanger. This resulted in two fractions (detection at 280 nm), both of them showing PPH_4R activity. While the first peak was still contaminated with minor protein impurities, the second yielded a single band corresponding to a molecular weight of approx. 35'000 D on

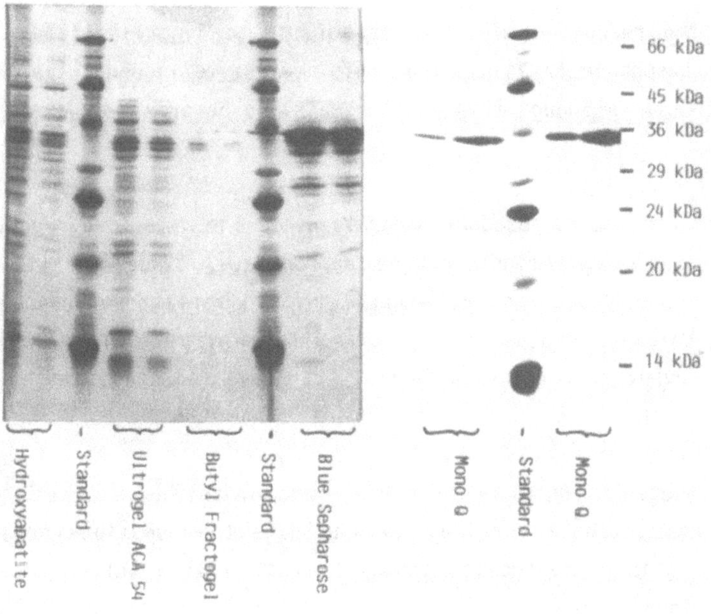

Fig. 2. SDS-Page with 12.5% polyacrylamide. Left gel: enzyme preparation from different purification steps, as mentioned in the figure; right gel: active fractions from the third Mono Q run. The mixture of standard proteins contained: bovine albumine (66 kD), egg albumine (435 kD), glycerinaldehyde-3-phosphate dehydrogenase (36 kD), carbonic anhydrase (29 kD), trypsinogen (24 kD), trypsin inhibitor (20 kD), α-lactalbumin (14 kD).

SDS-PAGE (Fig. 2). Since this corresponds to the molecular weight determined by using gel filtration, it must be concluded that the enzyme is a monomer.

Analytical isoelectric focusing (pH range 4-9, Servalyt T 4-9) on G-200 superfine revealed an isoelectric point of pH 6.0 ±0.2. Glycoprotein detection using Concanavalin A and peroxidase was negative. In the presence of 10 mM DTE less activity was found than with a 1 mM solution, suggesting that PPH_4R is not particularly sensitive to -SH group oxidation. Only PPH_4 and not 6-(1'-oxo-2'-hydroxypropyl)-5,6,7,8-tetrahydropterin or sepiapterin is a substrate for this enzyme. Most probably 6-(1'-hydroxy-2'-oxopropyl)-5,6,7,8-tetrahydropterin will also turn out to be a substrate for this enzyme, however, so far this has not been investigated in detail. This indicates a much higher specificity of PPH_4R compared to SR, the latter catalyzing the reduction of a variety of diketo functions (25). It is, therefore, reasonable to assume that this new enzyme indeed plays a role in BH_4 biosynthesis, although this still remains to be proven by direct experiments.

A particular problem is posed by nomenclature. The name "6-pyruvoyl tetrahydropterin reductase" might not be appropriate, since SR also reduces 6-pyruvoyl tetrahydropterin, although with a different preference for the side chain carbonyl groups. "6-pyruvoyl tetrahydropterin 2'-keto reductase" or "6-pyruvoyl tetrahydropterin 2'-carbonyl reductase" might be better choices.

The use of specific inhibitors for PPH_4R and of antibodies which are currently in preparation might help to elucidate the question whether this enzyme is on the biosynthetic pathway or not, and to solve the sequence problem mentioned above.

Conclusion

An enzyme which catalyzes the reduction of the 2'-keto group of 6-pyruvoyl tetrahydropterin, the intermediate in tetrahydrobiopterin biosynthesis, has been purified from human liver to apparent homogeneity. The enzyme has a native molecular weight of approx. 35'000 D and is a monomer. It apparently contains no carbohydrates and has an isoelectric point of pH 6.0. In the presence of NADPH this enzyme efficiently catalyzes the formation of 6-(1'-oxo-2'-hydroxypropyl)-5,6,7,8-tetrahydropterin from 6-pyruvoyl tetrahydropterin. 6-(1'-oxo-2'-hydroxypropyl)-5,6,7,8-tetrahydropterin and sepiapterin are no substrates.

Acknowledgements

This work was supported by the Swiss National Science Foundation, project no. 3.395-0.86. We are grateful to Dr. B. Wermuth for providing us with the carbonyl reductase.

References

1. Curtius, H.-Ch., S. Takikawa, A. Niederwieser, S. Ghisla. 1986. In: Chemistry and Biology of Pteridines 1986 (B.A. Cooper and V.M. Whitehead, eds.). de Gruyter, Berlin, p. 141.

2. Curtius, H.-Ch., D. Heintel, S. Ghisla, Th. Kuster, W. Leimbacher, A. Niederwieser. 1985. J. Inher. Metab. Dis. 8 Suppl. 1, 28.

3. Switchenko, A.C., G.M. Brown. 1985. J. Biol. Chem. 260, 2945.

4. Milstien, S., S. Kaufman. 1985. Biochem. Biophys. Res. Commun. 128, 1099.

5. Masada, M., M. Akino, T. Sueoka, S. Katoh. 1985. Biochim. Biophys. Acta 840, 235.

6. Nichol, C.A., G.K. Smith, D.S. Duch. 1985. Ann. Rev. Biochem. 54, 729.

7. Blau, N., A. Niederwieser. 1983. Anal. Biochem. 128, 446.

8. Ghisla, S., P. Steinerstauch, Th. Hasler, N. Blau, H.-Ch. Curtius. 1987. In: Unconjugated Pterins and Related Biogenic Amines. (H.-Ch. Curtius and N. Blau, eds.). de Gruyter, Berlin (this volume).

9. Richter, W.J., F. Raschdorf, R. Dahinden, A. Niederwieser, H.-Ch. Curtius, W. Leimbacher, Th. Kuster. 1987. In: Unconjugated Pterins and Related Biogenic Amines. (H.-Ch. Curtius and N. Blau, eds.). de Gruyter, Berlin (this volume).

10. Switchenko, A.C., J.P. Primus, G.M. Brown. 1984. Biochem. Biophys. Res. Commun. 120, 754.

11. Smith, G.K., C.A. Nichol. 1984. Biochem. Biophys. Res. Commun. 120, 761.

12. Smith, G.K., C.A. Nichol. 1985. J. Biol. Chem. 261, 2725.

13. Takikawa, S., H.-Ch. Curtius, U. Redweik, S. Ghisla. 1986. Biochem. Biophys. Res. Commun. 134, 646.

14. Takikawa, S., H.-Ch. Curtius, U. Redweik, W. Leimbacher, S. Ghisla. 1986. Eur. J. Biochem. 161, 295.

15. Milstien, S., S. Kaufman. 1983. Biochem. Biophys. Res. Commun. 115, 888.

16. Katoh, S., T. Sueoka, S. Yamada. 1982. Biochem. Biophys. Res. Commun. 105, 75.

17. Heintel, D., W. Leimbacher, U. Redweik, B. Zagalak, H.-Ch. Curtius. 1985. Biochem. Biophys. Res. Commun. 127, 213.

18. Wermuth, B. 1981. J. Biol. Chem. 256, 1206.

19. Hasler, Th., A. Niederwieser, H.-Ch. Curtius. 1987. In: Unconjugated Pterins and Related Biogenic Amines. (H.-Ch. Curtius and N. Blau, eds.). de Gruyter, Berlin (this volume).

20. Laemmli, U.K. 1970. Nature, London 227, 680.

21. Clegg, J.C.S. 1982. Anal. Biochem. 127, 389.

22. Radola, B.J. 1974. Biochim. Biophys. Acta 386, 181.

23. Katoh, S. 1971. Arch. Biochem. Biophys. 146, 202.

24. Niederwieser, A., W. Staudenmann, E. Wetzel. 1984. J. Chromatogr. 290, 237.

25. Sueoka, T., S. Katoh. 1985. Biochim. Biophys. Acta 843, 193.

HUMAN GTP CYCLOHYDROLASE I: PURIFICATION AND PREPARATION OF MONOCLONAL ANTIBODIES *

G. Schoedon, A. Niederwieser+ , H.-Ch. Curtius
Division of Clinical Chemistry, Department of Pediatrics, University of Zurich, CH-8032 Zurich, Switzerland

Introduction

The first step in the biosynthesis of tetrahydrobiopterin (BH_4) is the conversion of GTP to dihydroneopterin triphosphate (NH_2TP) and formic acid, which is catalyzed by the enzyme GTP cyclohydrolase I (GTPCH; EC 3.5.4.16). This enzyme has first been characterized from Escherichia coli (1,2), later its purification from several other microorganisms (3,4,5), insects (6), chicken (7), rat (8) and, partially, from human (9) has been described.
GTP cyclohydrolase I deficiency in man causes atypical phenylketonuria due to the lack of BH_4 biosynthesis and deficiency of the neurotransmitters dopamine and serotonin (10). Immune stimulation causes a drastical increase of GTPCH activity in the responsible cells of both man and mouse (11). These findings gave the reason for further studies of the human GTPCH on the molecular and gene level, including purification and stabilization of the enzyme as well as production of monoclonal antibodies.
Two characteristics of human GTPCH made its purification especially difficult: One is the failure of affinity chromatography for the mammalian enzyme and the other is its extreme instability when beeing purified. The first has been overcome by development of a new affinity material using 8-aminoguanosine triphosphate and sepharose 4B, which resulted in a 300-fold purification (12). But due to the instability of the enzyme, only 5% of activity has been recovered.

* This paper is dedicated to the memory of Prof. Alois Niederwieser

Unconjugated Pterins and Related Biogenic Amines

The present paper describes a purification procedure of GTPCH from human liver modified with respect to the stabilization of enzyme activity.

Materials and Methods

All chemicals used are commercially available. The detailed purification procedure for GTPCH will be reported elsewhere (13). GTPCH activity was measured as described in (9). Proteins were determined by the dye binding method (Bio Rad). Production of monoclonal antibodies has been done according to Köhler and Milstien (14). Monoclonal antibodies were selected by ELISA and Western blotting. Isotyping was performed by use of Ig-subclass specific double sandwich ELISA (Bio Rad, subisotyping kit).

Results and Discussion

Based on the repeated observation that GTPCH activity is unstable in dilute solution, studies of the conditions under which the activity is either unstable or stable have been made. For this, GTPCH activity was measured after storage for 24 h at 4°C, -20°C or -70°C in liver homogenate as well as in diluted samples in different buffers with and without stabilizing agents.

Generally, all diluted fractions tested were very unstable, freezing of these samples at -20°C eliminated even more activity than storage at 4°C. Lyophilization of the samples was not successful. The loss of activity in the diluted samples was partially avoided only by the addition of large amounts of bovine serum albumin, but not by DTE or 2-mercaptoethanol to the buffer. In homogenate, the enzyme was found to be more stable in phosphate buffer than in Tris, therefore a 50 mM potassium phosphate buffer at pH 7.8 was chosen as basis solution for GTPCH extraction. Addition of glycerol to this buffer even increased the stability of the enzyme, but did not cause an alteration of activity.

Almost no loss of activity was found in the ammonium sulfate fraction, and, when it was necessary to store the enzyme between purification steps, the best way was precipitation and storage of the pellet at 0-4°C.

Additionally, highly diluted samples were found to be fully active after storage as a suspension in Sephadex G-200 superfine (sf) at 4°C.The activity from aliquots of this suspension could easily be eluted over small G-25 columns. Slurries of purified enzyme in G-200 sf and 10% glycerol can be stored several days at 0-4°C without loss of activity.

For longer time, rapid concentration and storage in 50% glycerol at -70°C was found to counter the loss of activity at least for 3 months.

Since it became clear that GTPCH stability is dependent on concentration and protein environment, and that loss of enzyme activity is a factor of time, a purification procedure had to be developped, which is fast, keeps the enzyme as concentrated as possible, avoids freezing steps or longer storage periods, and in which the main stabilizers, ammonium sulfate, glycerol and superfine dextran particles, can be used.

Thus, human liver GTPCH was purified by 1. extraction with 50 mM phosphate buffer, pH 7,8, containing 5% glycerol, 2. ammonium sulfate precipitation between 30% and 40% saturation, 3. chromatography on Ultrogel AcA 34 and finally, preparative isoelectric focussing on Sephadex G-200 superfine. The first two steps can be done in 1 day from a large batch of liver, and the 40% ammonium sulfate pellet can be stored at 4°C overnight or, if not the whole batch is processed further, aliquots can be stored at -20°C. Step 3 has the advantage that no desalting step (dialysis, G-25) which causes loss of activity is needed for the processing of the ammonium sulfate fraction. The enzyme is eluted from the AcA 34 column in phosphate buffer containing 25% glycerol and is running as a sharp peak just behind the void volume. This makes separation fast and removes proteins of lower molecular weight nearly totally. The AcA 34 pool containing GTPCH is concentrated rapidly on an Amicon PM-10 ultrafilter and a slurry is made of 50 mg Sephadex G-200 sf/ml protein solution. The slurry can be stored at 4°C until it is aplied on a horizontal preparative isoelectric focussing gel made of Sephadex G-200 sf, 10% (v/v) glycerol, and 2% (v/v) ampholine T 4-9. The dimension of the gel layer is dependent on the amount of protein in the mixture, the loading capacity beeing 5-10 mg protein per ml gel suspension. GTPCH focusses at pH 5,6 =pI with high recovery of activity.

A purification summary of the GTPCH is given in Table 1. Analysis of the fractions by SDS-polyacrylamid gel electrophoresis showed reduction of the complex protein mixture to a few bands. The major band is enriched during the purification procedure, beeing a protein of 50'000 daltons. By gel filtration, an apparent molecular mass of 440'000 daltons was estimated for the active enzyme. Thus, human GTPCH is a complex of 8 subunits.

Table 1 Purification of GTP-Cyclohydrolase I from Human Liver

Procedure	Total protein mg	Total units uU	Specific activity uU/mg prot.	Yield %	Purification fold
Crude extract	57 050	44 000	0.7	100	1
Ammonium sulfate 30-40 %	5 240	38 800	7.4	88	10
Ultrogel AcA 34	1 100	32 600	30.0	74	43
IEF G-200 sf/ G-25 m pool	2.1	18 980	9 040.0	43	13 300

GTPCH I is purified from 300 g of human liver. After isoelectric focussing, the enzyme is eluted from the G-200 gel over Sephadex G-25 m, concentrated, substituted with glycerol to 50 % (v/v), and stored at -70°C. The activity is stable for 3 months.
1 uU = 1pmol NTP/min at 37°C.

Recently, we were able to produce monoclonal antibodies against human GTPCH. For immunization the protein isolated by isoelectric focussing was used. Mixed with complete Freund's adjuvant, 20 ug were injected subcutaneously in each of 4 Balb/c mice. One month after the primary immunization, antigen in incomplete Freund's adjuvant was injected. Serum obtained after two additional immunizations was tested for activity of specific antibodies by Ouchterlony double diffusion and ELISA. The two mice with highest titers were boosted with antigen in 0.9% NaCl intraperitoneally one week and 3 days before fusion. The fusion was achieved in PEG, and HAT-selection as well as cloning in HT-medium by use of the limited dilution method were performed. Eleven monoclonal antibodies were obtained, two of IgM subclass, the others IgG_1.

In ELISA against homogenous GTPCH from E.coli, 4 of the monoclonals showed activity. In western blotting, they bound against the 50'000 dalton protein of the human enzyme. With these antibodies, an immunocytochemical staining method is under investigation to localize GTPCH in the stimulated human macrophages and T-cells. They may also be useful for enzyme studies on the molecular and gene level.

Acknowledgement

This work was supported by the Swiss National Science Foundation, project No. 3.601-0.84 and No. 3.395-0.86. We are grateful to Dr. H. Hengartner, Zurich, for the possibility to perform monoclonal antibody production in his laboratory and Mr. U. Redweik for excellent technical assistance during isoelectric focussing experiments.

References

1. Burg, A.W., G.M. Brown. 1968. J. Biol. Chem. 243, 2349.
2. Yim, I.I., G.M. Brown. 1976. J. Biol. Chem. 251, 5087.
3. Cone, I.E., I. Plowman, G. Gurolf. 1974. J. Biol. Chem. 249, 5551.
4. Jackson, R.J., T. Shiota. 1975. Biochim. Biophys. Acta 403, 232.
5. Kohashi, M., T. Itadani, K. Iwai. 1980. Agric. Biol. Chem. 44, 271.
6. Weisberg, E.P., J.M. O'Donnell. 1986. J. Biol. Chem. 261, 1453.
7. Fukushima, K., W.E. Richter, T. Shiota. 1977. J. Biol. Chem. 252, 5750.
8. Bellahsere, Z., J.L. Dhondt, J.P. Farriaux. 1984. Biochem. J. 217, 59.
9. Blau, N., A. Niederwieser. 1983. Anal. Biochem. 128, 446.
10. Niederwieser, A., N. Blau, M. Wang, P. Joller, M. Atares, J. Gardesce-Garcia. 1984. Anal. Biochem. 141, 208.

11. Schoedon, G., J. Troppmair, A. Fontana, Ch. Huber, H.-Ch. Curtius, A. Niederwieser: Eur. J. Biochem. (in press)

12. Blau, N., A. Niederwieser. 1986. Biochim. Biophys. Acta 880, 26.

13. Schoedon, G., A. Niederwieser, H.-Ch. Curtius: et al.: in preparation

14. Köhler, G., C. Milstien. 1975. Nature (Cond.) 256, 495.

DIHYDROMONAPTERIN TRIPHOSPHATE: OCCURRENCE, ANALYSIS, AND EFFECT ON TETRAHYDROBIOPTERIN BIOSYNTHESIS *in vivo* AND *in vitro*

N.Blau, P.Steinerstauch, U.Redweik, K.Pfister, G.Schoedon, L.Kierat and A.Niederwieser+

Division of Clinical Chemistry, Department of Pediatrics, University of Zürich, CH-8032 Zürich, Switzerland

SUMMARY

6-Pyruvoyl-tetrahydropterin synthase (PPH_4S) from human and salmon liver is inhibited by L-monapterin (L-threo-neopterin). D-Monapterin as well as dihydro- and tetrahydro-L-monapterin have no effect on PPH_4S. Lowered biopterin excretion as well as induction of neopterin excretion in urine of mice was observed after treatment with L-monapterin. Incubation of myeloma cells with L-monapterin indicates that a pterin deficiency can be induced *in vivo* by inhibition of PPH_4S. L-Monapterin is formed from dihydromonapterin triphosphate (MH_2TP) after dephosphorylation and oxidation. MH_2TP produced from dihydroneopterin triphosphate (NH_2TP) by dihydroneopterin triphosphate 2'-epimerase was separated as monapterin triphosphate (MTP) by HPLC. Dihydroneopterin triphosphate 2'-epimerase from *E. coli* was purified by ammonium sulfate precipitation, DEAE-Sephadex A25 chromatography, Sephadex G150 gel filtration and isoelectrofocusing on Sephadex G200 SF.

INTRODUCTION

L-Monapterin is the major pterin found in some microorganisms. Rembold and Heinz (1) isolated this compound from cell lysates of *Escherichia coli* in 1971 but its

Unconjugated Pterins and Related Biogenic Amines

biological function is still unknown. The function of monapterin in Comamonas sp. was proposed by Guroff and Rhodas to be that of a cofactor of phenylalanine hydroxylase (2). They also described the occurrence of a dihydroform of monapterin in this bacteria. Urushibara et al. (3) suggested that the cyclic monophosphate of monapterin might be involved in methanol oxidation by a non-specific primary alcohol dehydrogenase in Bacterium methylococcus capsulatus. Monapterin has been reported to occur also in Serratia indica (4) and in Pseudomonas roseus fluorescens (5), however in neither report the biological significance was stated.

A number of authors (6,7,8) suggested that the cyclization of D-erythro-neopterin monophosphate to 2',3'-cyclic-neopterin monophosphate in some microorganisms may result in an inversion of the 2'-hydroxyl group of the propyl side chain of neopterin to yield the L-threo epimer. However, later on it was verified that dihydroneopterin triphosphate is the initial pterin formed from GTP and that the initially reported cyclic monophosphate of neopterin appears to be an artifact produced during purification procedures when an alkaline solvent is used.

An interesting point is the occurrence of monapterin in human tissues and urine but neither the biosynthesis nor its biological function in mammalians was explained. Monapterin like neopterin in urine is mostly present in its dihydro form and the ratio of monapterin to neopterin is usually about 1:10. In different blood cells this ratio varies from 1:10 to 1:3 (9). However, we have no proof that its formation occurs nonenzymatically from dihydroneopterin or enzymatically from dihydroneopterin triphosphate. We just have some observations that monapterin may possibly be formed from dihydroneopterin if untreated urines are stored for a longer period of time. Furthermore, we never noticed that monapterin was formed enzymatically in any mammalian material.

Ten years ago, Heine and Brown (10) described in E. coli an enzyme, named D-erythro-dihydroneopterin triphosphate 2'-epimerase, that catalyzes the conversion of NH_2TP to

MH_2TP. They were able to show that all three phosphate residues of the substrate are retained in the product and that the product is of the L-threo configuration but they were not able to separate the phosphorylated product from the substrate chromatographically.

Recently, we noticed that different preparations of NH_2TP which we use as a substrate in the PPH_4S assay show different activities. We were surprised that some preparations of NH_2TP, which we synthesized enzymatically from GTP by GTP cyclohydrolase I, contained up to 50% MH_2TP. In this paper we describe the separation of MH_2TP from NH_2TP by HPLC after oxidation with iodine, as well as the effect of MH_2TP and its metabolites on BH_4 biosynthesis <u>in vitro</u> and <u>in vivo</u>. The purification of NH_2TP 2'-epimerase from <u>E. coli</u> is also described.

MATERIALS AND METHODS

Most of the pterins were purchased from Dr. B. Schircks Laboratories, 8645 Jona, Switzerland. D-Monapterin was synthesized by Prof. M. Viscontini, and dihydro- and tetrahydro monapterin were prepared by reduction with Pd/H_2. NH_2TP was synthesized with GTP cyclohydrolase I immobilized on Sepharose 4B (11). Measurement of PPH_4S and sepiapterin reductase activities were performed as described earlier (12). The HPLC system for measurement of pterins was the same as described elsewere (13). Isoelectrofocusing was performed on Desaga flat bed electrophoresis unit (Heidelberg, GFR). Gel (8 mm thickness) consists of Sephadex G200 SF with 10% glycerol and 2% Servalyt T4-9. Electrofocusing was performed at 5 W for 100 min and at 7 W for 170 min at 3°C. 5-mm slices of the gel were assayed for enzyme activity.

HPLC of neopterin and monapterin phosphates

The HPLC analytical procedure was essentially as described earlier (14), except that the RCM system with Guard Pak C18 + Nova Pak C18 5µ (5 x 100 mm) from Waters was used.

Isopropanol in mobile phase was omitted and fluorescence was measured on a Hitachi F 1000 fluorometer.

Assay for NH_2TP 2'-epimerase

The assay for NH_2TP 2'-epimerase activity was based on the HPLC measurement of MTP (oxidation product of MH_2TP) formed from NH_2TP. The procedure of Heine and Brown (10) was used to generate MH_2TP from NH_2TP. The reaction mixture (100 µl) contained 25 µM NH_2TP, 5 mM $MgCl_2$, 1.4 mM EDTA, 0.1 g/l BSA, and 50 µl enzyme preparation in 25 mM 2-(N-morpholine) ethanesulfonic acid buffer pH 6.25. The reaction proceeded at 37°C for 15 min in the dark and was terminated by the addition of 10 µl of a 1 M HCl solution containing 1% iodine and 2% KI. After standing at room temperature for 15 min, proteins were removed by centrifugation and excess iodine was reduced by addition of 10 µl 1% ascorbic acid. The content of MTP was analyzed in 5 µl of supernatant by HPLC. The volume of the enzyme preparation in the assay mixture was chosen so that the ratio MTP/NTP at the end of incubation never exceeded 1.0.

Purification of NH_2TP 2'-epimerase

The enzyme was purified from *Escherichia coli* using modifications of the procedure of Heine and Brown (10). The protein precipitate obtained after fractionation between 50 and 80% ammonium sulfate was dissolved in 60 ml 50 mM potassium phosphate buffer, pH 8.0. The concentrated and dialyzed enzyme preparation was applied to a DEAE-Sephadex A25 column (5 x 30 cm) equilibrated with 0.1 M Tris-HCl buffer, pH 8.0 (starting buffer). NH_2TP 2'-epimerase was eluted with 2000 ml of a linear gradient to 0.5 M LiCl in the same buffer at a flow rate of 100 ml/h. Fractions between 520 and 960 ml containing NH_2TP 2'-epimerase activity were pooled, concentrated to 16 ml on an Amicon YM-10 ultrafilter, and applied to a Sephadex G150 column (2.6 x 100 cm) equilibrated with 0.1 M potassium phosphate buffer, pH 8.0. The proteins were eluted with the same buffer at a flow rate of 10 ml/h and 5 ml fractions were collected. The fractions

53 to 66 with NH_2TP 2'-epimerase activity were combined and concentrated to a volume of 5.8 ml. An aliquot of the fraction (2.5 ml) was concentrated to a volume of 0.9 ml on an Amicon YM-10 ultrafilter and applied to a Sephadex G200 SF gel prepared for isoelectrofocusing as described above. The gel slide at 8 to 9 cm (pH 7.0 - 6.4) was taken out and 2 ml of 1 M NaCl were added. The enzyme was eluted from gel with 0.5 M NaCl in 10 mM potassium phosphate buffer, pH 7.0 and washed from ampholyte on a Sephadex G25m column (1.6 x 26 cm) equilibrated with 10 mM potassium phosphate buffer, pH 7.0. The eluate was concentrated on an Amicon YM-10 ultrafilter to a volume of 2.8 ml and stored at -20°C.

RESULTS AND DISCUSSION

We modified our HPLC system for the separation of neopterin phosphates and guanine nucleotides so that we were able to separate neopterin phosphates from monapterin phosphates (Fig. 1).

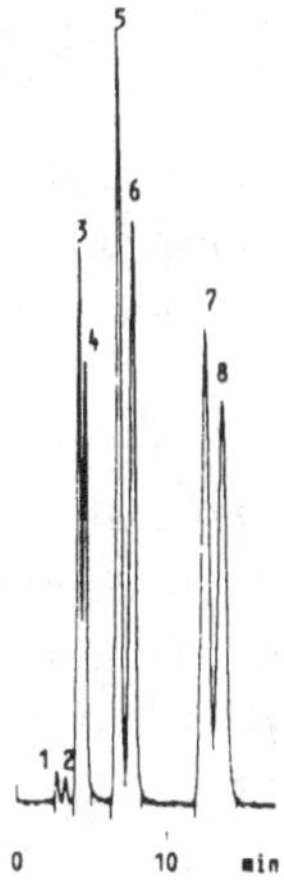

Fig. 1 HPLC of neopterin and monapterin phosphates
1=neopterin, 2=monapterin, 3=NMP, 4=MMP,
5=NDP, 6=MDP, 7=NTP, 8=MTP

Omission of isopropanol in the mobile phase improved the separation of isomers and use of the shorter column enabled the separation in less than 20 min. However, separation of dihydroforms of neopterin and monapterin triphosphate is not possible with this system. Neopterin and monapterin triphosphate obtained after incubation of NH_2TP with NH_2TP 2'-epimerase and subsequent oxidation with iodine were hydrolyzed at room temperature to the corresponding di- , and monophosphates yielding finally neopterin and monapterin as hydrolysis products (Fig. 2).

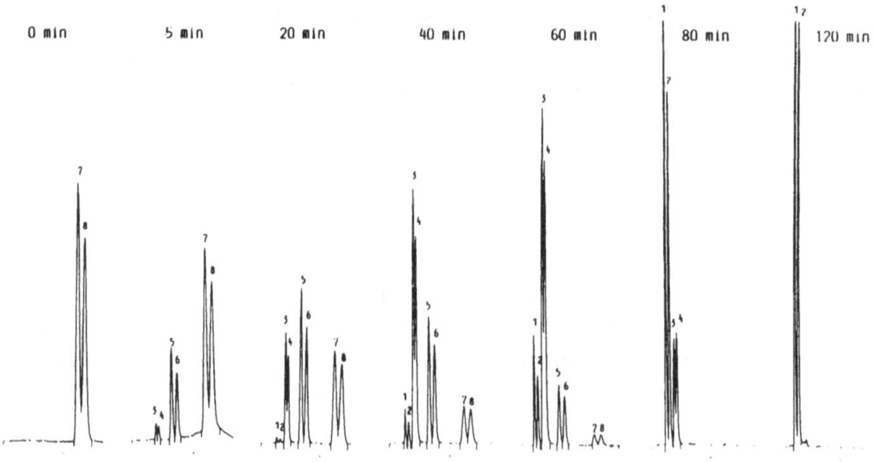

Fig.2 Hydrolysis of neopterin and monapterin triphosphate with alkaline phosphatase. Numbers of the peaks refer to compounds in Fig.1

MH_2TP was detected after incubation of GTP with different preparations of GTP cyclohydrolase I from <u>E. coli</u>. Preparations of NH_2TP containing different amounts of MH_2TP were used in the PPH_4S assay and found to produce variable results. These observations suggested that MH_2TP or one of its metabolites may effect the tetrahydrobiopterin biosynthesis. To find out at which step the inhibition occurs we started a number of experiments by incubating monapterin and its reduced forms with enzymes involved in the biosynthesis of tetrahydrobiopterin. Incubations with enzymes purified from human and salmon liver showed that monapterin but not dihydromonapterin and tetrahydromonapterin inhibit the activity of PPH_4S (Tab. 1). PPH_4S from salmon liver was inhibited to a lesser extent and there was no effect on sepiapterin reductase from human liver. MH_2TP present in some substrate (NH_2TP) preparations shows the same effect but so far we are not able to say if it is an effect of its oxidized form MTP or MH_2TP itself.

TABLE 1

EFFECT OF SOME PTERINS ON ACTIVITY OF 6-PYRUVOYL TETRAHYDROPTERIN SYNTHASE AND SEPIAPTERIN REDUCTASE

Compound	Concentration	Enzyme activity (%)		
		PPH_4S	PPH_4S	SR
	µmol/l	human	salmon	human
none	-	100	100	100
L-Monapterin	24	5	45	100
L-H_2-Monapterin	24	100	-	-
L-H_4-Monapterin	24	100	-	-
D-Monapterin	24	100	-	100
L-MH_2TP	64	79	-	-

PPH_4S-activity : pmol BH_4/(min·mg protein)
SR-activity : µmol S red./(min·mg protein)

Incubations of L-monapterin with mouse myeloma cell lines X63 Ag8 which are not able to produce tetrahydrobiopterin and in which dihydropterin is the main metabolite showed significant decrease in intracellular pterin production (Tab. 2). There was no effect when incubations were performed with D-monapterin. These experiments indicated that in vivo the inhibition of PPH_4S might induce a deficiency of pterins.

TABLE 2

INHIBITION OF PTERIN BIOSYNTHESIS IN MOUSE MYELOMA CELL LINES BY L-MONAPTERIN

Cell line X63 Ag8 incubated with	Number of cells	P	M
		pmol/10 mio cells	
IMDM	17 mio	25	-
IMDM + 1mM D-Monapterin	15 mio	27	299
IMDM + 1mM L-Monapterin	12 mio	10	248

Medium : IMDM + 10% FCS + antibiotics + mercaptoethanol
Incubation : 6 days at 37°C and 5.5% CO_2

To investigate further the effect of monapterin on tetrahydrobiopterin biosynthesis in vivo a group of 5 mice were treated for three days with monapterin. 1.15 mg of monapterin in 0.1 M NaCl, pH 7.3, were injected into a C57 BL/6 male mouse for three days, and 12-h urines were collected. After the monapterin dose the mice were limp and their coat was standing up. One mouse died on the fourth day and its blood, liver and brain were collected. The serum phenylalanine levels remained unchanged upon treatment with monapterin. Urinary biopterin decreased rapidly to about 25% of the initial value, and neopterin which normally could not be detected in mouse was excreted 12 hours after the first injection of monapterin (Fig. 3). This experiment showed for the first time that monapterin can lower urinary excretion of biopterin in mice. Inhibition of PPH_4S in a monapterin treated mouse leads to accumulation of NH_2TP which is degraded and excreted in urine. However, it can not be excluded that monapterin has any effect on the renal function.

By long term treatment with lower doses of monapterin it should be possible to create the animal model of tetrahydrobiopterin deficiency.

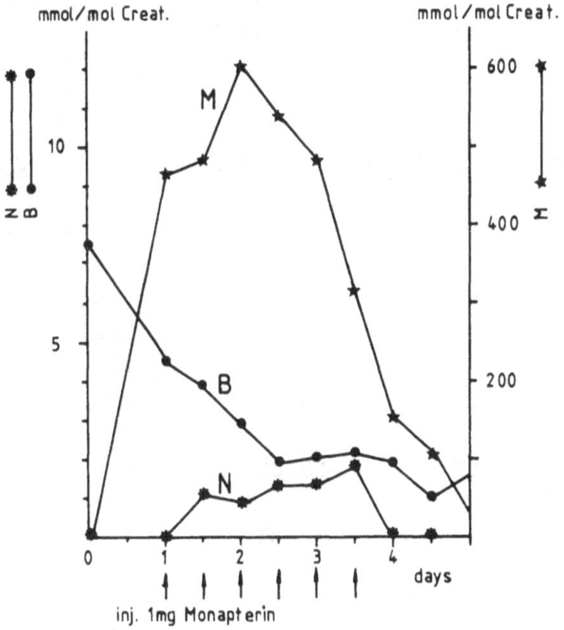

Fig.3 Effect of monapterin on levels of neopterin and biopterin in the urine of mice

NH_2TP 2'-epimerase was purified from 500 g of E. coli. A dialysate of the 50 to 80% ammonium sulfate fraction was purified by DEAE Sephadex A25 and Sephadex G150 column chromatography and by preparative isoelectrofocusing on Sephadex G200 SF gel. Purification steps and enzyme activities are summarized in Table 3. The specific activity of the final enzyme preparation was 126 fold higher compared to the crude extract. Due to the extreme instability of epimerase the overall recovery was less than 1 % and the stabilization of the enzyme will increase the recovery significantly. The Km value for NH_2TP was estimated to be 38.5 μM from a double reciprocal plot. The product formation was linear with time up to 15 min, and the rate of the reaction was directly proportional to the protein concentration when the purified enzyme was used. The molecular weight of the NH_2TP 2'-epimerase was estimated to be 88'000 by gel filtration on Sephadex G150, which is the same as described by Heine and Brown (10). The isoelectric point (pI) of the epimerase was found to be 6.4.

TABLE 3

PURIFICATION SUMMARY OF NH_2TP 2'-EPIMERASE FROM E. COLI

Purification step	Total protein (mg)	Total activity (mU)	Specific activity (mU/mg prot)
Homogenate	33'840	14'184	0.42
$(NH_4)_2SO_4$ 50-80%	26'426	10'389	0.39
DEAE Sephadex A25	150	414	2.75
Sephadex G150	39	188	4.84
Isoelectrofocusing	0.9	49	52.80

mU = µmol MH_2TP / min at 37 C

Experiments are in progress to prove the existence of NH_2TP 2'-epimerase in human tissues. In human liver homogenates there was no measurable activity of epimerase, although it must be stressed that under assay conditions rapid substrate degradation occurs. Efforts to measure the activity in ammoniumsulfate and hydroxyapatite fractions of human liver failed as well. The cause of monapterin occurrence in humans remains still unclear. However, the most plausible explanation will be the enzymatic epimerisation of NH_2TP to MH_2TP with following dephosphorylation and oxidation. It also remains open if there is any biological significance of monapterin in humans.

ACKNOWLEDGEMENT

This work was supported by a grant from the Swiss National Science Foundation, Project No. 3.395-0.86.

REFERENCES

1. H.Rembold, G.Heinz. 1971. Hoppe-Seyler's Z. Physiol. Chem. 352,1271-1272
2. G.Guroff, C.A.Rhoads. 1969. J. Biol. Chem. 244,142-146
3. T.Urushibara, H.S.Forrest, D.S.Hoare, R.N.Patel. 1971. Biochem. J. 125,141-146
4. M.Kobashi, K.Iwai. 1971. Agr. Biol. Chem. 35,47-52
5. M.Viscontini, R.Bühler-Moor. 1968. Helv. Chim. Acta 51,1548-1554
6. K.Iwai, M.Kobashi. 1976. In: Chemistry and Biology of Pteridines (W.Pfleiderer ed.). Elsevier North Holland, p. 341-357
7. J.Cone, G.Guroff. 1971. J. Biol. Chem. 246,979-985
8. J.Dayman, G.Guroff. 1971. Fed. Proc. 30,1141
9. B.Andondonskaja-Renz, H.J.Zeitler. 1986. In: Chemistry and Biology of Pteridines (B.A.Cooper, V.M.Whitehead eds.). Walter de Gruyter, Berlin-New York, p. 431
10. M.C.Heine, G.M.Brown. 1975. Biochim. Biophys. Acta 411,236-249
11. A.Niederwieser, U.Redweik, N.Blau. in preparation.
12. S.Takikawa, H.Ch.Curtius, U.Redweik, S.Ghisla. 1986. Biochem. Biophys. Res. Commun. 134,646-651
13. A.Niederwieser, W.Staudenmann, E.Wetzel. 1984. J. Chromatogr. 290,237-246
14. N.Blau, A.Niederwieser. 1983. Anal.Biochem 128,446-452

BIOSYNTHESIS OF PTERIDINES AND DEAZA FLAVINS IN METHANOGENIC BACTERIA

B. Schwarzkopf*, Q. Le Van*, W. Eisenreich*, P. J. Keller§, H. G. Floss§, and A. Bacher*

*Lehrstuhl für Organische Chemie und Biochemie der Technischen Universität München, Lichtenbergstraße 4, D-8046 Garching, Federal Republic of Germany and §Department of Chemistry, The Ohio State University, Columbus, Ohio 43210

Methanogenic bacteria generate energy by the reduction of carbon dioxide according to the equation

$$CO_2 + 4\ H_2 \longrightarrow CH_4 + 2\ H_2O.$$

They also generate cell mass by reduction of CO_2. For these unusual metabolic activities they utilize a variety of coenzymes whose structures have only recently been elucidated, such as methanopterin (1) (Fig. 1), cofactor F_{420} (2), coenzyme F_{430}, coenzyme M, factor B, and methanofurane (for review see Ref. 3). Little is known about the biosyntheses of methanopterin and coenzyme F_{420}.

We have studied the incorporation of ^{13}C-labeled acetate into methanopterin and the chromophoric moiety (factor F_0) of coenzyme F_{420} by *Methanobacterium thermoautotrophicum*. The products were isolated by chromatographic procedures, and the ^{13}C distribution was determined by quantitative ^{13}C-NMR spectroscopy. ^{13}C-Signals were assigned by ^1H-^1H and ^1H-^{13}C correlation methods (4, 5).

The labeling patterns obtained from a simple precursor such as acetate are not easily interpreted. However, the interpretation of the data is facilitated by comparison with the labeling patterns of other appropriate metabolites from the

Unconjugated Pterins and Related Biogenic Amines

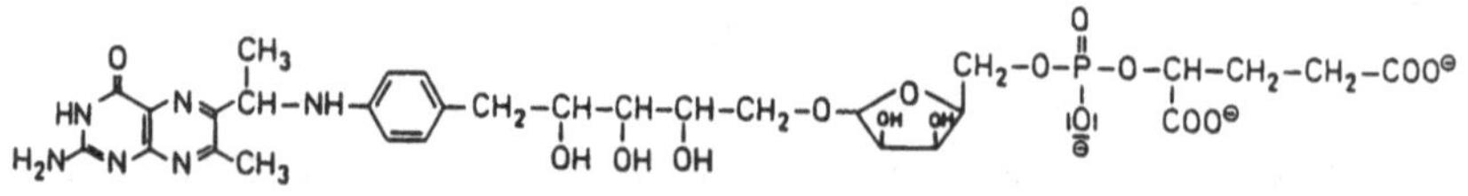

Methanopterin

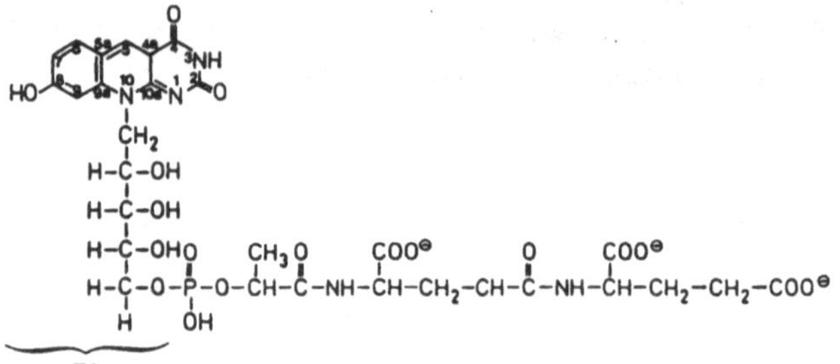

FO

Coenzyme F_{420}

Fig. 1. Structures of methanopterin and coenzyme F_{420}

same fermentation. Such a pattern recognition approach has proved helpful in earlier studies (6). We have therefore isolated and purified guanosine obtained by hydrolysis of cellular RNA and a variety of amino acids obtained by hydrolysis of cell protein. Their ^{13}C enrichments were determined by ^{13}C-NMR spectroscopy. The combined results from separate experiments with [1-^{13}C]acetate (*) and [2-^{13}C]acetate (•) are shown in the subsequent diagrams to facilitate the recognition of labeling patterns. Experimentally observed labeling patterns are shown as structures in boxes. Other labeling patterns have been infered from these experimental data.

The observed labeling pattern of guanosine obtained by hydrolysis of cellular RNA is in line with reported data. Reductive carboxylation of acetate yields pyruvate, which can be converted to glucose phosphate by gluconeogenesis (7). Oxidative decarboxylation yields a pentose phosphate with the labeling pattern observed in the isolated guanosine nucleotide. The observed labeling pattern of the purine nucleus is in line with its formation via serine and glycine (Fig. 2) (8).

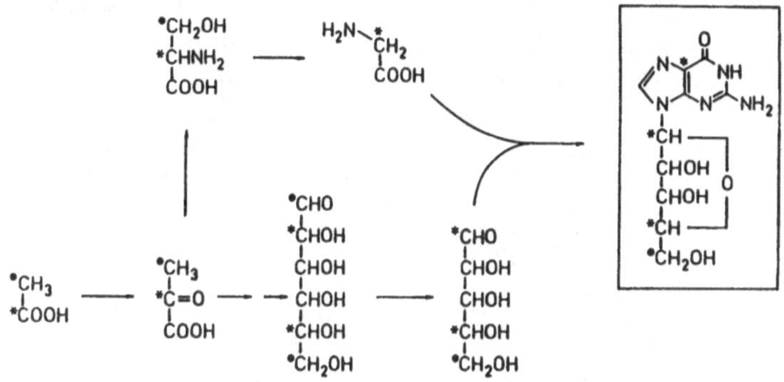

Fig. 2. Biosynthesis of purines in *M. thermoautotrophicum*

Carboxylation of pyruvate yields oxalacetate, succinate and aspartate as shown in Fig. 3. Carboxylation of succinate yields ketoglutarate and glutamate (8). It should be noted that carbon atoms 3 and 4 of ketoglutarate and glutamate receive ^{13}C-label from both [1-^{13}C]acetate and [2-^{13}C]acetate as a consequence of the interposition of symmetric intermediates (fumarate and succinate).

The labeling of the aromatic amino acids is readily explained by the metabolic sequence of the shikimate pathway in line with ^{13}C-incorporation studies by Ekiel et al. (8). 4-Aminobenzoate and 4-hydroxybenzoate could not be observed directly, but their labeling patterns can be infered from the shikimate pathway (Fig. 4).

With this background information at hand, the labeling patterns of the coenzymes became feasible. The pyrimidine ring of methanopterin shows the same labeling pattern as the corresponding ring in guanosine. This suggests that the pyri-

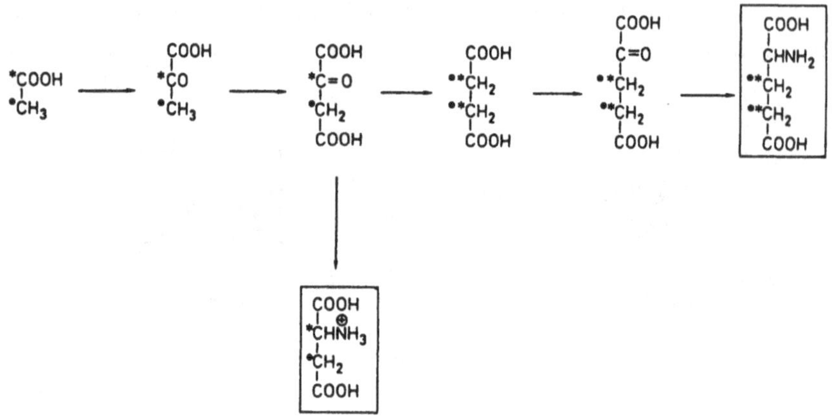

Fig. 3. Biosynthesis of glutamate in *M. thermoautotrophicum*. *, Carbon atom labeled from [1-^{13}C]acetate; ●, carbon atom labeled from [2-^{13}C]acetate.

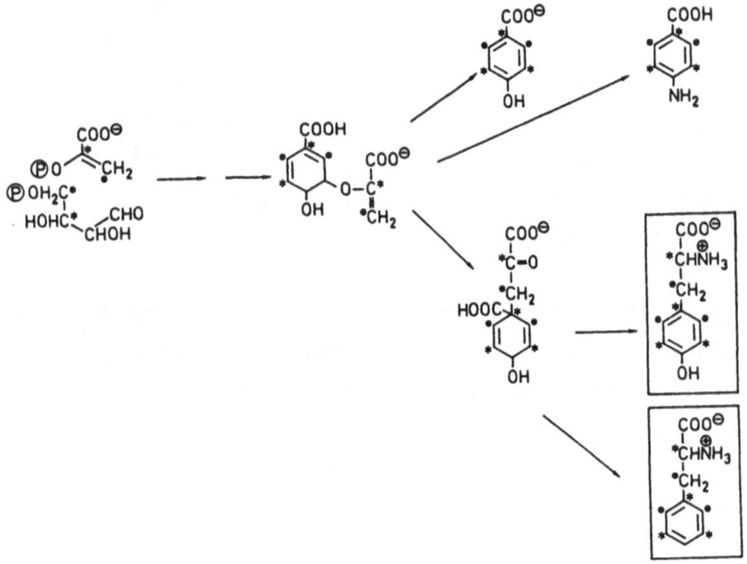

Fig. 4. Biosynthesis of aromatic amino acids in *M. thermoautotrophicum*. *, Carbon atom labeled from [1-^{13}C]acetate; ●, carbon atom labeled from [2-^{13}C]acetate.

midine moiety of methanopterin is biosynthetically derived from GTP in analogy with the biosynthesis of other pteridines (Fig. 5).

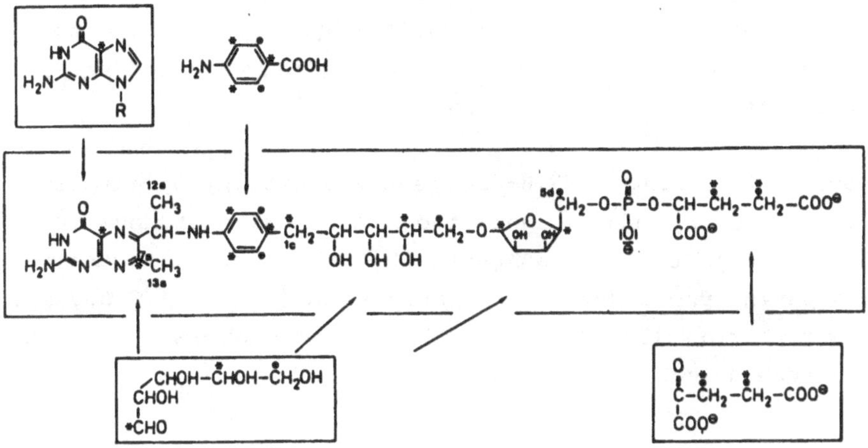

Fig. 5. Biosynthesis of methanopterin in *M. thermoautotrophicum*. *, Carbon atom labeled from [1-^{13}C]acetate; •, carbon atom labeled from [2-^{13}C]acetate.

Position 7a in the pyrazine ring of methanopterin is labeled from [1-^{13}C]acetate. This should be expected if the pyrazine ring is formed by a ring expansion utilizing the ribose side chain of GTP in analogy with the biosyntheses of folate and biopterin (9). In line with this hypothesis, it has been shown that the 13a methyl group of methanopterin is introduced from methionine (10). Our data further suggest that the 12a methyl group is not derived from the pentose moiety, but further evidence is required.

The ribitylaniline moiety of methanopterin shows clearly the expected labeling pattern of a shikimate derivative. Indeed White (11) has obtained evidence for the incorporation of 4-aminobenzoic acid. However, our ^{13}C-data suggest that the 1c position is not derived from the carboxylic group of 4-aminobenzoate which should be unlabeled. Rather, the data are consistent with the hypothesis that all carbon atoms of the ribityl moiety are derived from a pentose precursor.

The glycosidic pentose moiety of methanopterin shows the expected pentose labeling pattern. The hydroxyglutarate moiety shows the same labeling pattern as glutamate and is probably derived from ketoglutarate.

The pyrimidine ring of 7,8-didemethyl-8-hydroxy-5-deazariboflavin (factor F_0) shows again the same label distribution as guanosine (Fig. 6). This suggests that this moiety is derived from a purine precursor in agreement with ^{14}C and ^{13}C studies by other authors (12, 13). The ribityl group shows the expected pentose labeling pattern.

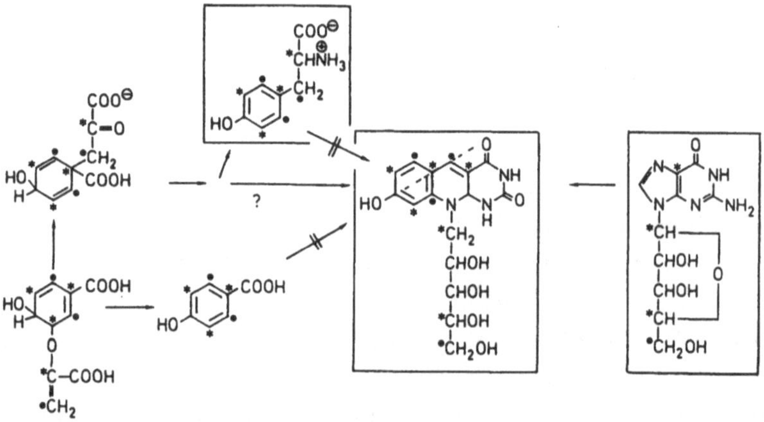

Fig. 6. Biosynthesis of 7,8-didemethyl-8-hydroxy-5-deazaflavin (factor F_0) in *M. thermoautotrophicum*. *, Carbon atom labeled from [1-^{13}C]acetate; •, carbon atom labeled from [2-^{13}C]acetate.

The phenol of the deazaflavin ring duplicates the label distribution of tyrosine and should thus be derived from the shikimate pool. However, 4-hydroxybenzoate can be clearly ruled out as precursor because it should not introduce label into the 5 position of the deazaflavin nucleus. On the other hand, tyrosine has been ruled out as direct precursor by ^{14}C-incorporation experiments (13). We are left with the hypothesis that an intermediate between chorismate and tyrosine could serve as the committed precursor for the biosynthesis of F_0. Specifically, 4-hydroxyphenylpyruvate would be a plausible candidate, but further evidence is required.

Acknowledgement

We thank the Deutsche Forschungsgemeinschaft and the Fonds der Chemischen Industrie for grants to A.B. and NSF for a grant to H.G.F. and P.J.K.

References

1. Van Beelen, P., A. P. M. Stassen, J. W. G. Bosch, G. D. Vogels, W. Guijt, C. A. G. Haasnoot. 1984. Eur. J. Biochem. 138, 563.
2. Eirich, D., G. D. Vogels, R. S. Wolfe. 1978. Biochemistry 17, 4583.
3. Wolfe, R. S. 1985. Trends Biochem. Sci. 10, 396.
4. Keller, P. J., H. G. Floss, Q. Le Van, B. Schwarzkopf, A. Bacher. 1986. J. Amer. Chem. Soc. 108, 344.
5. Le Van, Q., B. Schwarzkopf, A. Bacher, P. J. Keller, S. Lee, H. G. Floss. 1985. J. Amer. Chem. Soc. 107, 8300.
6. Bacher, A., Q. Le Van, P. J. Keller, H. G. Floss. 1983. J. Biol. Chem. 258, 13431
7. Stupperich, E., G. Fuchs. 1984. Arch. Microbiol. 139, 8.
8. Ekiel, I., I. C. P. Smith, G. D. Sprott. 1983. J. Bacteriol. 156, 316.
9. Brown, G. M., M. Williamson. 1982. Adv. Enzymol.53, 345.
10. White, R. H. 1986. J. Bacteriol. 165, 215.
11. White, R. H. 1985. Arch. Microbiol. 143, 1.
12. Scherer, P., V. Höllriegel, C. Krug, M. Bokel, P. Renz. 1984. Arch. Microbiol. 138, 354.
13. Jaenchen, R., P. Schönheit, R. K. Thauer. 1984. Arch. Microbiol. 137, 362.

TETRAHYDROBIOPTERIN BIOSYNTHESIS AND PTERIDINE METABOLISM IN CELLS OF THE BLOOD SYSTEM, LYMPHOID TISSUES AND IN BIOLOGICAL FLUIDS

B. Andondonskaja-Renz, H.-J. Zeitler
Institut für Klinische Hämatologie der Gesellschaft für Strahlen- und Umweltforschung, D-8000 München 2, Germany

Introduction

Reduced forms of pteridines are intermediates in the biosynthesis and catabolism of pterins and serve partly as co-enzymes in the biosynthesis of melanin, neurotransmitters (e.g. serotonin, catecholamines) and prostaglandins.
On the basis of in vitro experiments, the immune response-associated release of dihydroneopterin has been described as an interferon-gamma (IFN-gamma)-mediated interaction between activated T lymphocytes and macrophages (1).
However, our investigations on the in vivo excretion of pteridines into plasma and urine following transplant operations provided evidence of the involvement of at least five pteridines (neopterin, monapterin, biopterin, isoxanthopterin, pterin; mainly in their reduced forms) in the processes during immunological activation episodes (2). It still remains open to question in which cells this "switch-on" of "immune response-mediated" biopterin biosynthesis takes place.
On comparing the intracellular pteridine pattern of some immature monocytic cell lines before and after in vitro incubation with interferon-gamma, the observed IFN-gamma-mediated increase in pteridines indicated the ability of tetrahydrobiopterin biosynthesis to occur in the cell lines investigated. In addition, the increase in pterin metabolism was confirmed by analysing the excretion of cellular pteridines into culture medium (2, 3).

Unconjugated Pterins and Related Biogenic Amines

A systematic investigation was pursued on pteridine metabolism in all of the cell types which may be involved in immunological activation episodes. The aim of this investigation was two-fold: first, to determine quantitatively the levels of pteridines present in normal cells (e.g. human blood, bone marrow, lymphoid tissues) or in cell lines, as well as the amounts excreted into body fluids (CSF, plasma), urine or growth medium; secondly, to determine the intracellular extent of pteridine metabolism (cellular answer) during interaction between the cells and stimulating agents (e.g. IFN-gamma).

In this study we are reporting on the pteridine patterns in biological fluids and normal cells of human blood, bone marrow and lymphoid tissues.

Materials and Methods

Blood fractionation: Blood treated with anticoagulant (fresh peripheral blood, bone marrow, blood units) was separated into plasma and cell fractions by various centrifugation techniques, as described in (4).

Preparation of cells from lymphoid tissues (thymus; tonsils): Lymphoid tissues were homogenized and the lymphocyte fraction was separated on density gradients (Ficoll-Paque; Percoll) (5 - 8). The cell fractions were washed twice (Dulbecco buffer), lysed and lyophilized.

Cerebrospinal fluid (CSF), urine: the samples were deep-frozen immediately after collection and analysed as described in (9).

Oxidation of reduced forms of pteridines and deproteinization of biological materials: The reduced forms of pteridines were oxidized with iodine at pH=1. In all cases, acid extraction of pteridines from biological materials was performed by deproteinization with trifluoroacetic acid (pH=1); for details see (4).

Rp-HPLC analysis of pteridines: All rp-HPLC analyses were performed with fluorescence detection as described previously (4).

The estimation of the portion of reduced pteridines was made on the basis of the difference between the analyses before (oxidized portion of pteridines) and after the oxidation procedure (total amount of pteridines. Tetrahydrofolates are decomposed in part to pterin during acid iodine oxidation).

Results and Discussion

The blood contains many types of cells with very different functions, ranging from the transport of oxygen to the production of antibodies. The main functions of blood cells are outlined in Table 1.

TABLE 1. FUNCTIONS of BLOOD CELLS.

TYPE of CELL	MAIN FUNCTIONS
RED BLOOD CELLS (ERYTHROCYTES)	TRANSPORT O_2 and CO_2
WHITE BLOOD CELLS (LEUCOCYTES) :	
GRANULOCYTES	
NEUTROPHILS	PHAGOCYTOSIS of ANTIBODY-COATED BACTERIA.
EOSINOPHILS	DESTROY LARGER PARASITES and MODULATE ALLERGIC INFLAMMATORY REACTIONS.
BASOPHILS	RELEASE HISTAMINE and SEROTONIN in CERTAIN IMMUNE REACTIONS.
LYMPHOCYTES	MAKE IMMUNE RESPONSES.
MONOCYTES	BECOME MACROPHAGES in the TISSUES.
KILLER CELLS	KILL VIRALLY INFECTED CELLS and SOME TUMOR CELLS.
MEGAKARYOCYTES, GIVING RISE to PLATELETS	INITIATE BLOOD CLOTTING

In Table 2 we see the concentration and distribution of pteridines in human cells of the blood system and lymphoid tissues. RED CELLS from bone marrow differ from peripheral erythrocytes in that they reveal a higher level of intracellular neopterin

TABLE 2. PTERIDINES in HUMAN CELLS.
Rp-HPLC analysis of iodine-oxidized (pH=1) and deproteinized samples; (fluorescence detection, exc/em = 360/460 nm).
Abbreviation: (-), not detectable; n.qu., not quantified.

TYPE of CELLS		XANTHO-PTERIN	NEO-PTERIN	MONA-PTERIN	ISOXANTHO-PTERIN	LUMAZIN	BIO-PTERIN	6-HYDROXY-METHYLPTERIN	PTERIN	NEO/BIO	NEO/MON
RED CELLS, bone marrow (ng/ml red cells)		n.qu.	4.6±1.8	1.1±0.5	n.qu.	n.qu.	1.8±0.5	-	n.qu.	2	4
ERYTHROCYTES, peripheral blood (ng/ml erythrocytes)		n.qu.	2.4±1.1	0.4±0.2	0.3±0.1	n.qu.	1.5±0.7	-	0.7±0.4	1.2	6
BUFFY COAT, bone marrow	pg/ mg lyophylized cells	n.qu.	900±150	250±150	n.qu.	n.qu.	50±10	-	23±4	20	4
BUFFY COAT, peripheral blood		5	350±150	150±50	5	n.qu.	25±10	-	15±5	20	3-6
GRANULOCYTES		n.qu.	315±112	92±31	n.qu.	n.qu.	16±12	-	74±68	20	3-4
THROMBOCYTES		21±12	37±26	9±4	14±12	5±2	9±2	-	35±10	2-4	2-4
LYMPHOCYTES, peripheral blood 1)		16	502	51	24	n.qu.	11	-	30	48	10
2)		43	1228	399	25	n.qu.	20	-	35	62	3
3)		59	2144	363	19	n.qu.	35	-	68	62	6
LYMPHOCYTES, from THYMUS (THYMOCYTES) 90% T-lymphocytes		-	716	32	-	n.qu.	90	-	66	8	22
LYMPHOCYTES from TONSILS (after tonsil-ectomy). 60% B-lymphocytes; 40% T-lymphocytes.		-	906	54	-	n.qu.	140	-	309	7	17

and monapterin. However, the ratio of neopterin to monapterin (N/M) was found to be about 4 to 6 in both compartments. The ratio of neopterin to biopterin (N/B) was comparable (bone marrow, N/B ≃ 2; peripheral erythrocytes, N/B ≃ 1.2).
In the WHITE CELL fractions neopterin was the dominant pterin; both peripheral blood and bone marrow presented a ratio of neopterin to biopterin of about 20, whereas the ratio of neopterin to monapterin was like that in red cells (N/M ≃ 4). As reported for the red cell fractions, the intracellular levels of neopterin, monapterin and biopterin were always higher in bone marrow derived white cells. These findings point to intensive pterin metabolism in cells of the marrow compartment.
High biopterin concentrations in bone marrow cells have also been described by Fukushima et al. in rats (10).
Neutrophils, eosinophils and basophils are jointly classified as GRANULOCYTES. The main functions of granulocytes are, for example, the destruction of invading bacteria. In granulocytes, the pattern of the pteridines investigated was like that of the total white cell fraction ('buffy coat') of peripheral blood, except that there was a higher intracellular pterin.
When purified LYMPHOCYTES from peripheral blood (healthy blood donors) were compared (see Table 2), xanthopterin, isoxanthopterin, biopterin and pterin levels remained relatively constant while striking alterations were observed for neopterin and monapterin.
Because it is a site of lymphocyte production, the thymus is referred to as a central lymphoid tissue. Although many lymphocytes die soon after they have developed in a central lymphoid tissue, others migrate via the blood to the peripheral lymphoid tissues (e.g. lymph nodes, spleen, tonsils). Lymphocytes isolated from LYMPHOID tissues (thymus, 90% T lymphocytes; tonsils, 60% B lymphocytes, 40% T lymphocytes) demonstrated similar pteridine patterns. However, the ratio of neopterin to biopterin (N/B) was much lower than in all the white cells of peripheral blood due to the markedly higher intracellular biopterin level in the lymphoid tissues.

TABLE 3. PTERIDINES IN HUMAN BODY FLUIDS and URINE.
Rp-HPLC analysis of iodine-oxidized (pH=1) and deproteinized samples (fluorescence detection, exc/em 360/460 nm).

PTERIDINES	PLASMA (peripheral blood) (ng/ml plasma)	CEREBROSPINAL FLUID (pg/mg lyophilized CSF).*	URINE (nmol/mmol creatinine)
XANTHOPTERIN	3.0 ± 1.5	-	30 ± 28
NEOPTERIN	2.8 ± 1.3	543	310 ± 120
MONAPTERIN	0.3 ± 0.1	93	41 ± 10
ISOXANTHOPTERIN	0.6 ± 0.2	16	40 ± 20
LUMAZIN	n.qu.	n.qu.	n.qu.
BIOPTERIN	2.2 ± 0.7	393	576 ± 153
6-HYDROXYMETHYL-PTERIN	-	53	5,1 ± 3.6
PTERIN	1.2 ± 0.2	71	60 ± 10
RATIO NEO/BIO	1.2 ± 0.2	1.4	0.5± 0.1
RATIO NEO/MON	7	6	7 - 10

*)1mg lyophilized CSF ≃ 100 mg cerebrospinal fluid. Pteridine values, mean of three analysis of one donor.
Abbreviation: (-), not detectable; n.qu.,not quantified.

The THROMBOCYTES initiate blood clotting and help to repair breaches in the walls of blood vessels. In contrast to white cells, these cells (separated from platelet concentrates) demonstrated relatively low concentrations of intracellular neopterin and monapterin.

The pteridine patterns of human plasma, cerebrospinal fluid (CSF) and urine were also determined fluorometrically; they are outlined in Table 3.

The reduced portions of the three main pterins (neopterin, monapterin, biopterin) in the cellular fractions and in biological fluids were calculated from analyses before and after acid iodine oxidation. They are summarized in Table 4.
Interestingly, the lymphocyte fraction of tonsil tissue pre-

TABLE 4. REDUCED FORMS of PTERIDINES, (%).
The values are the differences between the rp-HPLC analysis of acid iodine oxidized samples (total amount of pteridines) and of non-oxidized samples (oxidized portion of pteridins).

HUMAN	NEOPTERIN	MONAPTERIN	BIOPTERIN
LYMPHOCYTES, peripheral blood	54	54	85
THYMUS LYMPHOCYTES, (thymocytes)	49	47	83
TONSIL LYMPHOCYTES (60% B-lymphocytes, 40% T-lymphocytes)	86	84	98
PLASMA (periph.blood)	54	54	54
CEREBROSPINAL FLUID	88	86	94
URINE	57	65	95

sented the majority of these pterins in their reduced forms ($>$ 84% and 98%, respectively).

As evident from Table 4, more than 50% of the neopterin, monapterin and biopterin present intracellularly and in biological fluids exists in the dihydro- or tetrahydro-form. Using HPLC methods with fluorescence detection, the reduced portion is on the one hand not detectable in non-oxidized samples and therefore decreases the analytical detection markedly, in part even by a factor of about 10. This concerns in particular the CSF and tonsil tissue. On the other hand, most of the biopterin moiety present in CSF, urine or intracellularly (e.g. lymphoblastic and lymphocytic biopterins) is known to exist in the tetrahydro-form. These reduced pterins are very sensitive to oxidation and chemical conversion reactions, even under less acidic conditions and at increased temperatures.

The frequently used (11,12,13,14,15,16,17,18) "extraction" of biopterin and of its reduced forms from unoxidized biological samples by autoclaving (approx., 10 min at 120°C) prior to bioassay with Crithidia fasciculata leads to some conversion of reduced pterins. (Note that some pterins other than bio-

pterin/reduced biopterins are Crithidia-active: L-erythro-neopterin, sepiapterin, 6-hydroxymethylpterin, monapterin, D-erythro-neopterin (34)). The mechanism of autoxidation reactions of tetrahydrobiopterin was elucidated by Pfleiderer and described in (19), likewise Fukushima et al. (20) and Zeitler et al. (4) have demonstrated some schemes on the conversion reactions (neopterin/reduced forms; biopterin/reduced forms) in different buffer systems and at increased temperatures. As reported in detail by Milstien (21), 55% of the biopterin moiety was destroyed after autoclaving (11 - 18) at pH 4.5, and only 18% (!) of original amount of <u>tetrahydro</u>biopterin was converted to oxidized, stable forms.
This means that even the levels of lymphoblastic or lymphocytic tetrahydrobiopterin (after autoclaving of unoxidized white cells from less than 5-6 ml whole blood) would be at or somewhat above the detection limit. In this connection intracellular alterations of less than 10 - 20% can hardly be demonstrated as significant.

<u>Occurence of monapterin (threo-neopterin)</u>.

We have already reported on the occurence of monapterin (threo-neopterin) and its levels in plasma and cells of the peripheral blood and bone marrow in healthy subjects and patients suffering from various diseases (2-4, 9, 22-26) .
We know nothing about the biosynthesis and biochemical role of monapterin in human and animal cells. Iwai et al. reported the reduced monapterin to be active as a cofactor of phenylalanine hydroxylase in rat liver (27). Monapterin was first described in human urine by Fukushima et al. . As early as in 1970, Fukushima proposed that the intermediary metabolites of the biosynthetic pathway from GTP to tetrahydrobiopterin are dihydroneopterin, threo-dihydroneopterin (dihydromonapterin), and sepiapterin (28). Twelve years later, increased urinary monapterin and neopterin concentrations were reported for some cases of tetrahydrobiopterin deficiency in newborns (29).

In our study on normal human blood cells, body fluids and urine (see Tables 2-4), the simultaneous occurrence of neopterin and monapterin as well as their nearly constant ratio in the compartments investigated is evident. Neopterin was always found in higher concentrations than monapterin.
Analogous findings were obtained in patients suffering from malignant diseases (tumor, leukemia): alterations in neopterin were accompanied in most cases by analogous alterations in the monapterin levels before and during treatment with cytostatics and/or antibiotics (22-25). Proglicem, which affects the beta-cells in the human pancreas and induces hyperglykemia, caused a significant decrease in both pterins in the peripheral leucocyte fraction, the plasma and the urine (26).
The fact that the reduced portion of both neopterin and its threo-isomer monapterin was nearly identical in all the cell systems and body fluids/urine investigated is impressive (see Table 4). Therefore, it was quite obvious for us -as reported some years ago (22, 24)- to suggest an enzymic reaction of the EPIMERASE type in monapterin-forming animal cell systems which results in the formation of a reduced and phosphorylated moiety of monapterin (L-) from an appropriate neopterin derivative (probably D-dihydroneopterin triphosphate). The epimerization reaction would explain the formation of L-monapterin (which is closely related to the naturally occuring euglenapterin) from D-neopterin rather than by reduction steps at the level of intermediates in the biosynthesis cascade between dihydro neopterin triphosphate and L-tetrahydrobiopterin:

a) It is assumed now that the phosphate-eliminating reaction (substrate, dihydroneopterin triphosphate) occurs before any reduction reaction (30);
b) The phosphate-eliminating reaction results in the formation of a methyl group at C-3'. However, monapterin still has a hydroxymethyl group ($-CH_2OH$) at C-3'.

The occurrence and effect of dihydromonapterin triphosphate on tetrahydrobiopterin biosynthesis in vitro and in vivo is discussed in this volume elsewhere (35).

There have been many reports on the occurrence of monapterin (or its reduced forms) in several microorganisms. This indicates the potential importance of this pterin, although metabolic roles for monapterin have been assumed in only two of the microorganisms investigated. in Comamonas sp., a reduced form of monapterin was decribed to function as a co-factor in the enzymic hydroxylation of phenylalanine to tyrosine (31), and in Methylococcus capsulatus the cyclic monophosphoester of L-dihydromonapterin is thought to function as a co-factor for alcohol dehydrogenase (32). In Serratia indica D-erythro-neopterin, L-monapterin, L-monapterin 2',3'-cyclic phosphate and 6-hydroxymethylpterin were identified as products from GTP (27). Although Brown et al. (33) reported an enzyme-dependent epimerization of D-dihydroneopterin triphosphate to dihydromonapterin triphosphate in E. coli ('D-erythro-dihydroneopterin triphosphate epimerase'), the biochemical function of monapterin in E. coli remained unknown.

Conclusions

(1) Human bone marrow presented higher intracellular pteridine concentrations than peripheral blood cells.
(2) In peripheral blood cells, intracellular biopterin was found to be approximately within the same range.
(3) Not only biopterin, but also neopterin and monapterin were found to be present m a i n l y in their reduced forms (e.g., CSF, tonsil lymphocytes).
(4) In each of the compartments investigated, the erythro-threo-isomers NEOPTERIN and MONAPTERIN presented
 i) nearly identical portions of their reduced forms,
 ii) nearly constant ratios (neopterin/monapterin; N/M).
 As an interpretation of these findings we suggest an enzymic reaction of the EPIMERASE type between D-dihydroneopterin triphosphate and L-dihydromonapterin triphosphate.

References

1. Huber, Ch., J.R. Batchelor, D. Fuchs, A. Hausen, A. Lang, D. Niederwieser, G. Reibnegger, P. Swetly, J. Troppmair. H. Wachter. 1984. J. Exp. Med. 160, 310.
2. Andondonskaja-Renz, B., H.-J. Zeitler. 1986. In: Chemistry and Biology of Pteridines (B.A. Cooper, V.M. Whitehead, eds.). De Gruyter, Berlin-New York, p. 431.
3. Zeitler, H.-J., B. Andondonskaja-Renz. 1987. In: Biochemical and Clinical Aspects of Pteridines (W. Pfleiderer, J.A. Blair, H. Wachter, eds.) De Gruyter, Berlin-New York. Vol 5 (in press).
4. Zeitler, H.-J., B. Andondonskaja-Renz. 1986. In: Methods in Enzymology (S.P. Colowick, N.O. Kaplan, F. Chytil, D. McCormick, eds.). Academic Press, Orlando, USA. Vol. 122, p. 273.
5. Pertoft, H., A. Johnsson, Wärmegård, B., R. Seljelid. 1980. J. Immunol. Methods 33, 221.
6. Gutierrez, C., R.R. Bernabe, J. Vega, M. Kreisler. 1979. J. Immunol. Methods 29, 57.
7. Ulmer, A.J., H.D. Flad. 1979. J. Immunol. Methods 30, 1.
8. Gmelig-Meyling, F., T.A. Waldmann. 1980. J.Immunol. Methods 33, 1.
9. Zeitler, H.-J., B. Andondonskaja-Renz, G. Küther, A. Struppler. 1986. In: Chemistry and Biology of Pteridines (B.A. Cooper, V.M. Whitehead, eds.). De Gruyter, Berlin-New York, p. 251.
10. Fukushima, T., J.C. Nixon. 1980. Anal. Biochem. 102, 176.
11. Ziegler, I. 1985. In: Biochemical and Clinical Aspects of Pteridines (H. Wachter, H.Ch. Curtius, W. Pfleiderer, eds.). De Gruyter, Berlin-New York, Vol. 4, p. 347.
12. Fink, M., U. Jehn, W. Wilmanns, I. Ziegler. 1982. In: Biochemical and Clinical Aspects of Pteridines (H. Wachter, H.Ch. Curtius, W. Pfleiderer, eds.). De Gruyter, Berlin-New York, Vol. 1, p. 241.
13. Ziegler, I., M. Fink, W. Wilmanns. 1982. Blut 44, 231.
14. Ziegler, I., H.J. Kolb, U. Bodenberger, W. Wilmanns. 1982. Blut 44, 261.
15. Fink, M., U. Jehn, W. Wilmanns, H. Rokos, I. Ziegler. 1983. In: Biochemical and Clinical Aspects of Pteridines (H.Ch. Curtius, W. Pfleiderer, H. Wachter, eds.). De Gruyter, Berlin-New York, Vol. 2, p. 223.
16. Fink, M., I. Ziegler, H. Rokos. 1985. In: Biochemical and Clinical Aspects of Pteridines (H. Wachter, H.Ch. Curtius, W. Pfleiderer, eds.). De Gruyter, Berlin-New York, Vol. 4, p. 387.
17. Fink, M., I. Ziegler, H.J. Kolb, K. Demtröder, V. Nüssler. 1987. 1th Workshop on Unconjugated Pterins and Related Biogenic Amins. February 28 - March 7, Flims, Switzerland.
18. Fink, M., V. Nüssler, K. Demtröder, E. Höniges, I. Ziegler, M. Goldberg. 1987. 6th Winter Workshop on Biochemical and Clinical Aspects of Pteridines. February 14-21, St.Christoph/Arlberg, Austria. (See also this volume).

19. Pfleiderer, W.. 1975. In: Chemistry and Biology of Pteridines (W. Pfleiderer, ed.). De Gruyter, Berlin-New York, p. 941.

20. Fukushima, T., J.C. Nixon. 1979. In: Chemistry and Biology of Pteridines (R.L. Kisliuk, G.M. Brown, eds.). Elsevier, North Holland, p. 31.

21. Milstien, Sh.. 1983. In: Biochemical and Clinical Aspects of Pteridines (H.Ch. Curtius, W. Pfleiderer, H. Wachter, eds.). De Gruyter, Berlin-New York, Vol. 2, p. 65.

22. Andondonskaja-Renz, B., H.-J. Zeitler. 1984. In: Biochemical and Clinical Aspects of Pteridines (W. Pfleiderer, H. Wachter, H.Ch. Curtius, eds.). De Gruyter, Berlin-New York, p. 295, Vol. 3.

23. Andondonskaja-Renz, B., H.-J. Zeitler. 1985. In: Biochemical and Clinical Aspects of Pteridines (H. Wachter, H.Ch. Curtius, W. Pfleiderer, eds.). De Gruyter, Berlin-New York, Vol. 4, p. 559.

24. Zeitler, H.-J., B. Andondonskaja-Renz. 1984. In: Biochemical and Clinical Aspects of Pteridines (W. Pfleiderer, H. Wachter, H.Ch. Curtius, eds.). De Gruyter, Berlin-New York, Vol. 3, p. 313.

25. Zeitler, H.-J., B. Andondonskaja-Renz. 1986. In: Cancer Detection and Prevention (The International Society for Preventive Oncology) (H.E. Nieburgs, ed.). A.R. Liss, Inc., New York, (in press).

26. Zeitler, H.-J., B. Andondonskaja-Renz. 1985. (Unpublished data).

27. Iwai, K., M. Kobashi. 1975. In: Chemistry and Biology of Pteridines (W. Pfleiderer, ed.). De Gruyter, Berlin-New York, p. 341.

28. Fukushima, T.. 1970. Arch. Biochem. Biophys. 139, 361.

29. Niederwieser, A., A. Matasovic, W. Staudenmann, M. Wang, H.Ch. Curtius. 1982. In: Biochemical and Clinical Aspects of Pteridines (H. Wachter, H.Ch. Curtius, W. Pfleiderer, eds.). De Gruyter, Berlin-New York, Vol. 1, p. 293.

30. S. Ghisla. 1987. 1th Workshop on Unconjugated Pterins and Related Biogenic Amines. February 28-March 7, Flims, Switzerland. (See this volume).

31. Guroff, G., C.A. Rhoads. 1969. J. Biol. Chem. 244, 142.

32. Urushibara, T., H.S. Forrest, D.S. Hoare, R.N. Petel. 1971. Biochem. J. 125, 141.

33. Brown, G.M., J. Yim, Y. Suzuki, M.C. Heine, F. Foor. 1975. In: Chemistry and Biology of Pteridines (W. Pfleiderer, ed.). De Gruyter, Berlin-New York, p. 219.

34. Nixon, J.C.. 1985. In: Folates and Pterins (R.L. Blakley, St.J. Benkovic, eds.). J. Wiley&Sons, Inc., New York, Vol. 2 (Chemistry and Biochemistry of Pterins), p. 1.

35. Blau, N., P. Steinerstauch, G. Schoedon, R. Redweik, A. Niederwieser. 1987. 1th Workshop on Unconjugated Pterins and Related Biogenic Amines. February 28 - March 7, Flims, Switzerland. (See this volume).

MODIFIED PTERIDINE SYNTHESIS IN TRANSFORMED CELL LINES AND DURING HEAT SHOCK

J. Seidl[1], M. Przybylski[2], Ch. Meinschad[1*] and I. Ziegler[1]

[1]Institut für Experimentelle Hämatologie der Gesellschaft für Strahlen und Umweltforschung, Landwehrstraße 61, D-8000 München 2, FRG

[2]Institut für Organische Chemie der Universität Mainz, Becherweg 18-20, D-6500 Mainz, FRG

Introduction

A pteridine binding α_1-acid glycoprotein (AGP_M) was found to accumulate in the blood of patients with malignant diseases (1). This AGP_M was characterized by a chromophoric group which could be detached with 6 M guanidine hydrochlorid . Spectral analysis indicated that the chromophoric group is a pteridine. In the following experiments we show that this pteridine is synthesized by transformed cell lines.
It had earlier been claimed that pteridines in transformed cell lines (2,3,4) and normal cells (5) originate from folate cleavage. In this study we have reinvestigated this postulated folate cleavage. Moreover, we have begun to chemically characterize the pteridine which is typically found in malignant cells and to elucidate its biogenesis. Additionally, we have studied its induction by heat shock, since response of cells to heat shock is known to induce a set of proteins which normally become active during certain stages of differentiation (for review see 6).

*New address: Boehringer Mannheim GmbH, Werk Penzberg

Unconjugated Pterins and Related Biogenic Amines

Material and methods

HeLa, HT-29 and 239 cells were cultured in MEM with 10% NBS and 10% FCS, respectively. Reh, HL 60 and T-47D were grown in RPMI with 10% FCS.
To study folate metabolism, medium folate was replaced by ^{14}C-folate (Amersham; final concentration 1,5μM, specific activity 1μCi/18 nmol).
Heat shock was performed by cultivating the cells and bacteria at 43° C for 4 hours.
Reverse phase HPLC determination of pteridines after acidic iodine oxidation is described in (7). The amount of pterin X was calculated by assuming a fluorescence intensity of biopterin.
Fast atom bombardment (FAB) mass spectra were obtained with a modified Finnigan MAT-312 spectrometer, equipped with a highfield laminated magnet system and extended analyzer. A thermal Cesium primary ion source (AMD Intertra, Beekeln/FRG) was employed and operated at ca. 1400 mA and 4.5 kV (relative to source potential). Alternate acquisition of positive and negative ion spectra was carried out using a post-acceleration / ion conversion dynode system (AMP) placed in-line before the 17-stage electron multiplier, employing a dynode voltage of +8 kV for negative conversion.
Lyophilized samples of the chromatographically isolated pterin X of approx. 10 μg were dissolved in ca. 5 μl methanol : 1M HCl (5:1), and 1-μl aliquots were admixed to 0.5 μl of glycerol or tetraethylene glycol (TEG) as matrix substance. After brief treatment in an ultrasonic bath, the resulting, nearly homogenous solution was applied on a 2 mm² stainless steel FAB target at the tip of a self-constructed, thermostated direct insertion probe. Samples were introduced in the ion source and spectra acquired at 17 sec/decade by the Finnigan SS-200 data system; a resolution of ca. 2.000 (10%) was employed.

Results and discussion

Origin of AGP_H chromophore

Fig.1 demonstrates that the chromophore of AGP_H is

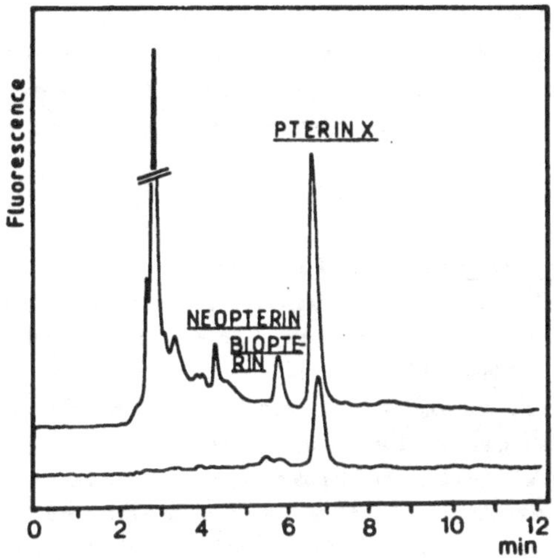

Fig.1 HPLC separation (1% acetonitrile, 7% methanol, 3mM phosphoric acid) of the chromophoric group of AGP_M (lower lane) and of HeLa pteridines (upper lane)

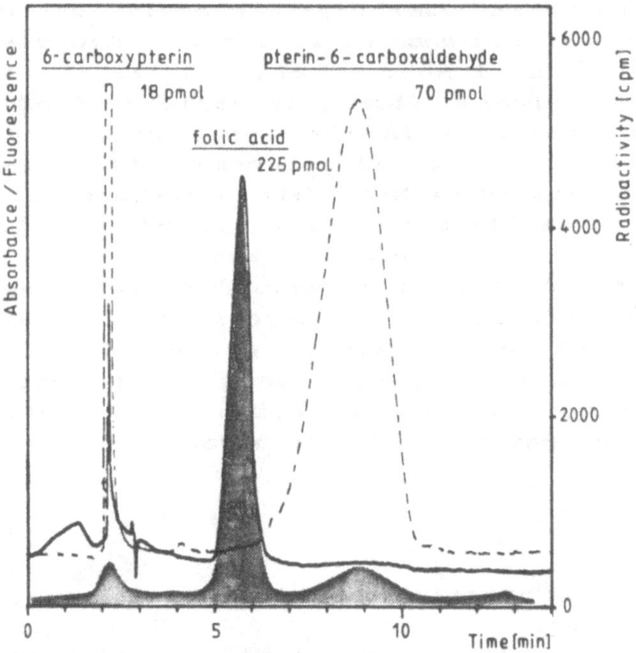

Fig.2 HPLC separation (10% methanol) of commercially obtained ^{14}C-folate
____ absorption at 280nm, - - - fluorescence (exc. 350nm, em. 450nm), dotted area: radioactivity

identical with a pteridine produced, e.g., by HeLa cells. In all of the 6 cell lines investigated (see methods) this pteridine (pterin X) predominated by far. Neopterin and biopterin were present only in minimal amounts.

<u>Folate cleavage in HeLa cell culture</u>

To examine ^{14}C-folate cleavage in HeLa cells, maximal synthesis of pterin X was induced by heat shock (see section below). The isolated pteridines showed about one third of the specific activity expected, if they had originated from folate cleavage.

This gave us the notion that unconjugated pteridines may already be present in the commercially obtainable ^{14}C-folate preparation. Indeed, HPLC separation of freshly manufactured ^{14}C-folate demonstrated that one third of the radioactivity was already present as 6-carboxypterin and pterin-6-carboxaldehyde (fig.2). To eliminate these impurities the sample was purified by preparative HPLC and subsequently incubated for 24 hours at 37° C with HeLa cells (3.6×10^{4} cells/ml) or with medium alone. After this period 40% of the activity was recovered in the fraction of unconjugated pteridines, irrespective of whether the purified sample was incubated with or without HeLa cells.

Earlier investigators (2,3,4,5) had used unpurified ^{14}C-folate preparations and extended the incubation over a period of 3 to 8 days. Therefore, we conclude that the labelling of the unconjugated pteridines was not due to folate cleavage by the cells, but rather to both folate impurities and folate decomposition. The serum portion of the medium may possibly contribute to enzymatic folate degradation (for review see 8).

<u>Induction of pterin X by heat shock and by limited oxygen supply</u>

HeLa cells which had been subjected to heat shock or to limited oxygen supply accumulated at least 3-fold amounts of pterin X as compared to controls (fig.3, tab.1). Neopterin and biopterin levels were not rai-

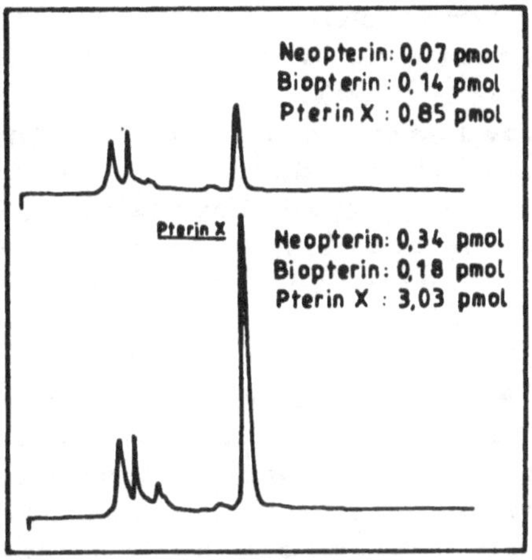

Fig.3 HPLC separation of pteridines from HeLa (10^6 cells) upper lane: control, lower lane: reduced oxygen condition (2 hours)

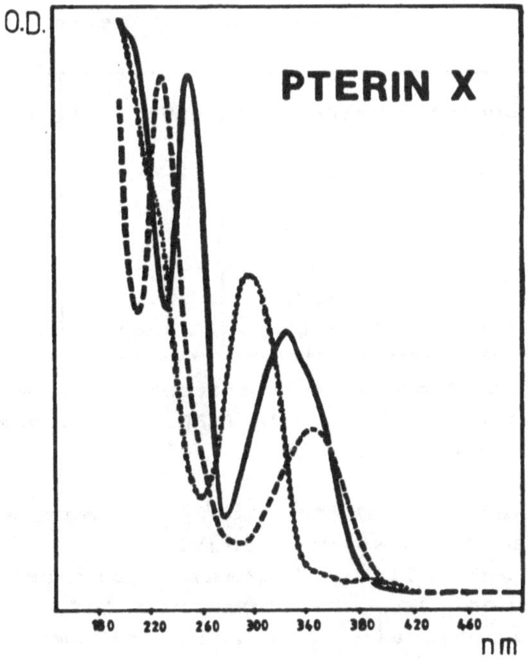

Fig.4 UV-spectrum of pterin X
-·-·- pH1, ——— pH 7, - - - pH 13

sed by stress conditions. This is in contrast to phorbol ester treatment which causes an accumulation of both these pterins (9).
Heat shock-induced synthesis of pterin X is not restricted to eucaryotes. Bacteria, such as E.COLI, also accumulate at least 3-fold levels of pterin X (maximally 60 pmol/mg protein). Levels of monapterin, the only reported pteridine in E.COLI (10), were only minimally raised.
Heat shock-induced synthesis of pterin X can be mimicked by incubation of the HeLa cells at 37°C with sepiapterin (SEP). Transfer of the cells to 43°C together with sepiapterin further increases the amount of pterin X. Maximal levels are reached by this combination (tab.1). Both heat-shock induced and sepiapterin-mediated accumulation of pterin X is inhibited by methotrexate (MTX). This indicates that dihydrofolate reductase (DHFR) activity is essential.

Temp.		Pteridines pmol/10^6cells			
		controls	+SEP(1μM)	+SEP(1μM) +MTX(1μM)	+MTX(1μM)
37°C	Biopterin	0.45	5.43	1.18	n.d.
	Pterin X	1.94	3.07	0.74	n.d.
43°C	Biopterin	0.20	6.92	2.13	0.12
	Pterin X	4.77	9.09	4.63	1.32

Tab.1 Accumulation of pterin X by incubation of HeLa (10^6 cells) with sepiapterin (SEP) and by heat shock; n.d. = not determined

It is known that tetrahydrobiopterin may arise from sepiapterin via successive reduction by sepiapterin reductase and DHFR (11). Increased biopterin levels which were found in the presence of sepiapterin (tab.1) seem to originate from this "salvage pathway". However, the data suggest that a further DHFR-mediated reduction of sepiapterin occurs. Iodine oxida-

tion (at pH 1) of this product yields pterin X.
The product, which results from sepiapterin after incubation with DHFR and NADPH, was identified as dihydrosepiapterin (12). Using the conditions described in (12), a linear progression of dihydrosepiapterin formation was obtained during a period of 20 minutes. Iodine oxidation (at pH 1) of the incubation product, dihydrosepiapterin, under acidic conditions yields pterin X. The K_M for sepiapterin in DHFR-mediated reduction (bovine liver) is 18 µM and the K_I for methotrexate is 2 nM (fig.4). This indicates that the affinity of sepiapterin as a substrate is comparable to that of dihydrofolate (13).
From the results described above we conclude that in transformed cell lines and under heat shock conditions a DHFR-dependent reduction of sepiapterin occurs. It remains to be determined whether resulting dihydrosepiapterin or its oxidation product, pterin x, is present in the living cell. The heat shock-induced formation of dihydrosepiapterin appears to branch off the de novo pathway, which results in tetrahydrobiopterin and can be triggered in transformed cell lines by phorbol ester treatment. In contrary to heat shock induced synthesis of pterin X, phorbolester treatment causes a transient accumulation of neopterin and biopterin (9) which cannot be inhibited by methotrexate. The key compound in the heat shock-induced pathway and the mechanisms of the divergence from the de novo pathway remain to be determined.

Chemical characterization of pterin X

Pterin X is hardly to be distinguished by HPLC from 2-amino-4-oxo pteridine (pterin). Only with phosphate buffer as an eluent (15 mM, pH 6,4) a shoulder indicates a slightly different retention time.
The spectral data are shown in (fig.5). The product of alkaline $KMnO_4$ oxidation comigrates with 6-carboxypterin, thus indicating a carbon side chain on position C6 (14).
Initial molecular weight and structural data for pterin X were sought from FAB mass spectra. In these experiments, the preparation of suitable matrix systems

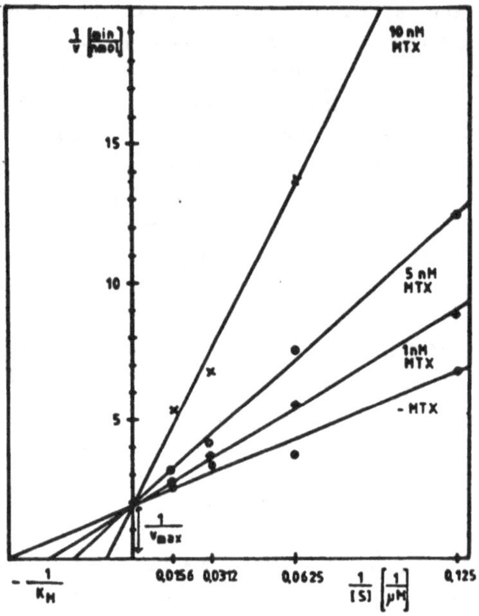

Fig.4 Lineweaver-Burk plot of sepiapterin reduction by DHFR (bovine liver) and NADPH (50 mM) and of MTX inhibition

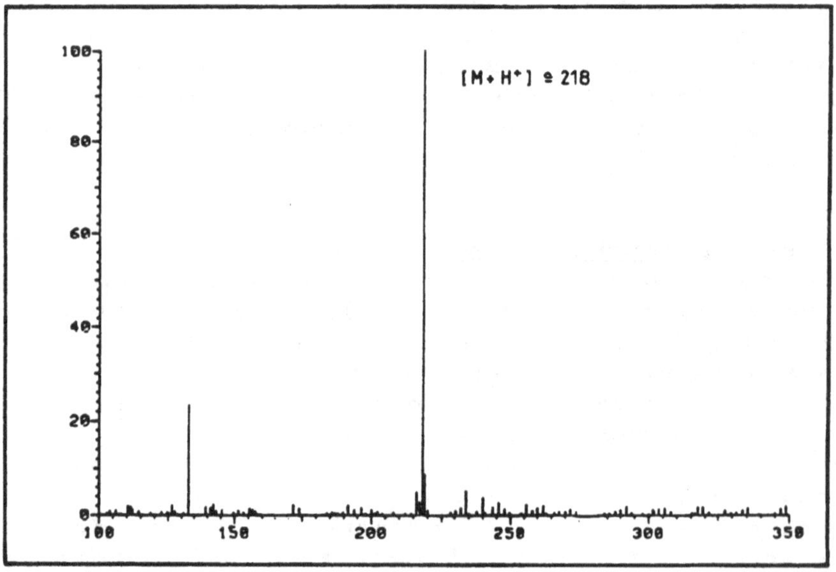

Fig.5 FAB mass spectrum of pterin X

was found to be of crucial importance, and nearly (though not entirely) homogenous solutions were obtained after enforced protonation in methanol/HCl. At these conditions, positive ion spectra with either glycerol or TEG as matrix consistently yielded a most abundant protonated molecular ion MH^+ at m/z 218, despite the presence of considerable amounts of sodium salt in the isolated sample; the negative ion spectra showed, in addition to matrix and background ions a prominent ion at m/z 252/254 (1Cl) attributable to $(M+Cl)^-$, indicating a molecular weight of 217 for pterin X. The lack of any more abundant fragment ions in the positive and negative ion spectra suggests the absence of labile, functional side chain from the sepiapterin structure, but precluded the direct derivation of more detailed structural data, from these spectra.

References

1. Ziegler, I., K. Maier, M. Fink. 1982. Cancer Research 42, 1567.
2. Fukushima, T., and T. Shiota. 1974. J.Biol.Chem. 249, 4445.
3. Halpern, R., B.C. Halpern, B. Stea, A. Dunlap, K. Conklin, B. Clark, H. Ashe, L. Sperling, J.A. Halpern, D. Hardy, A. Smith. 1977. Proc.Natl. Acad.Sci.USA 74, 587.
4. Stea, B., P.S. Backlund, P.B. Berkey, A.K. Cho, B.C., Halpern, R.M. Halpern, R.A. Smith. 1978. Cancer Research 38, 2378.
5. Clynes, M.M., and C. O`Neill. 1980. Cancer Letters 10, 133.
6. Lanks, K.W. 1986. Exp. Cell Res. 165, 1.
7. Ziegler, I. 1985. J.Cell.Biochem. 28, 197.
8. Scott, J.M. 1984. In: Folates and Pteridines (R.L. Blakley and S.J. Benkovic, ed.) John Wiley & Sons. p. 307.
9. Seidl, J., M. Borchert, and I. Ziegler. 1986. Biochem.Biophys.Res.Com. 141, 494.
10. Rembold, H. and G. Heinz. 1971. Hoppe-Seyler`s Z. Physiol. Chem. 352, 1271.

11. Nichol, C.A., C.L. Lee, M.P. Edelstein, J.Y. Chao., D.S. Duch. 1983. Proc.Natl.Acad.Sci.USA 80, 1546.
12. Smith, G.K., and C.A. Nichol. 1986. J.Biol.Chem. 261, 2725.
13. Blakley, R.L. 1984. In: Folates and Pteridines (R.C. Blakley and S.J. Benkovic, ed.), John Wiley & Sons. p. 191.
14. Forrest, H.S., H.K. Mitchell. 1955. J.Amer.Chem.Soc., 77, 4865.

Acknowledgement

We thank Prof. Dr. A. Bacher for most helpful discussions and suggestions.
This research was supported by the Deutsche Forschungsgemeinschaft and by the Bundesministerium für Forschung und Technologie.

SECTION C: PTERINS IN IMMUNOLOGY

DISEASE ASSOCIATED ALTERATIONS OF PTERIN BIOSYNTHESIS: ENHANCEMENT IN INFLAMMATORY DISEASE, HEMOPOIETIC REGENERATION AND ENDOTOXINEMIA

Ch.Huber, M.Herold, J.Troppmair.
Div.Clinical Immunobioloy, Dept.Internal Medicine, University of Innsbruck, A-6020 Innsbruck, Austria

H.Rokos
Henning Berlin GesmbH
Berlin, D-1000 German, Federal Republic

Introduction

Disease associated abnormalities of pterin biosynthesis were first observed in children with inborn metabolic error. Atypical phenylketonuria (PK) is characterized by deficiency to generate appropriate levels of tetrahydrobiopterin (BH4) and clinically presents with CNS dysfunction (1). Subsequently, increased levels of pterins in body fluids were observed in context with disease states not affecting brain function. Elevated levels of neopterin were detected in various inflammatory conditions (2), during hemopoietic regeneration (3) and in shock patients (4). In this article, we summarize previous and recent data elucidating the pathomechanism(s) underlying increased pterin synthesis in the latter states.

Unconjugated Pterins and Related Biogenic Amines

1. Enhancement of Pterin Biosynthesis During Inflammatory Disease

Evidence for enhanced pterin biosynthesis during inflammatory states is mainly derived from the observation of increased neopterin levels in conditions such as viral and bacterial infection, autoimmune disease or allograft rejection. This topic has been recently reviewed (2). In vitro experiments represented the first clue to unravel the mechanisms underlying this phenomenon. We first observed that antigenic stimulation of a mixture of T-lymphocytes and monocytes/macrophages leads to release of massive amounts of neopterin into culture fluids (5). Further studies using highly purified suspensions of T-cells and monocytes/macrophages revealed that the macrophage produces neopterin when challenged with T-cell supernatants (6). Neutralization experiments and stimulation with various recombinant lymphokines finally enabled to characterize interferon-gamma (IFN-gamma) as the active principle contained in these supernatants (6). Subsequently, we and others have in detail studied the capacity of various cytokines released from inflammatory mononuclear cells to enhance pterin biosynthesis (6,12). Results are demonstrated in Figure 1. As shown, T-lymphocytes can not be induced by any of the cytokines tested. Macrophages on the other hand are only stimulated with IFN-alpha and IFN-gamma. Peripheral blood mononuclear cells, which are predominantly composed of T-cells and monocytes, are induced to release neopterin when exposed to IFN-alpha, IFN-gamma and IL-2. Neutralization experiments with anti-IFN antibodies indicated that the latter stimulation is indirect and mediated via induction of interferon-gamma by IL-2.

Figure 1: Cytokine Specificity and Cellular Cooperations Required for Enhanced Pterin Biosynthesis in Mononuclear Blood Cells

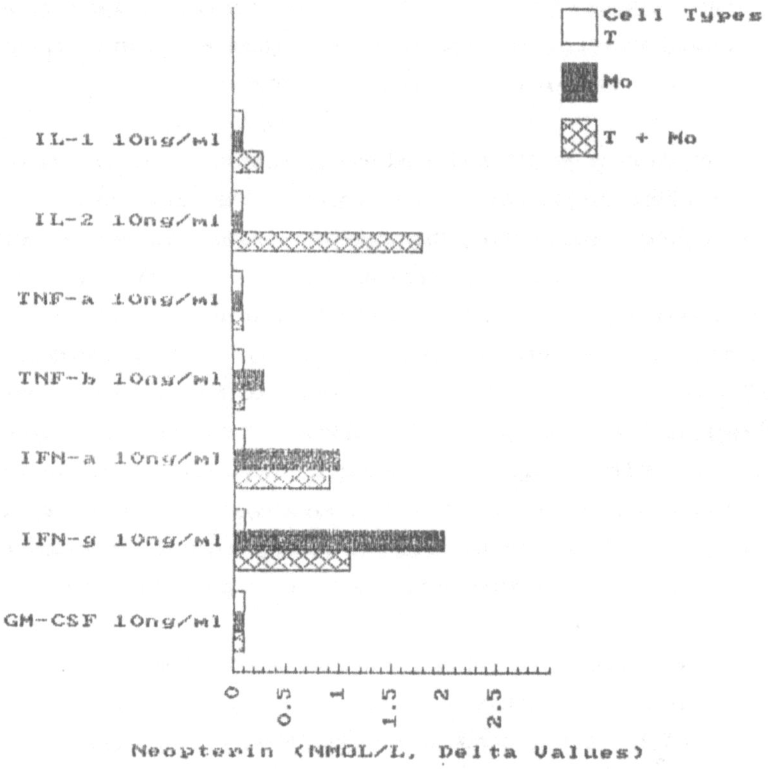

Further experiments, previously published and in detail presented by G.Schoedon et al. in this issue aimed to elucidate the biochemical basis of the phenomenon of IFN-induced enhancement of pterin biosynthesis (7). A systematic analysis of IFN-induced changes of guanosintriphosphate (GTP), GTP-cyclohydrolase I (GTP-CH I), neopterin, biopterin and pterin levels in T-lymphocytes or monocytes/macrophages gained the following results: IFNs increase substrate concentrations and activate the key-enzyme GTP- CH I in both T-cells and monocytes/macrophages. Macrophages, in contrast to T-cells are not capable to further convert dehydroneopterintriphosphate to BH4. This then leads to excretion of large amounts of dehydroneopterin and neopterin into cultur supernatants (7). In conclusion, these in vitro data indicated that both, immune and non-immune IFNs by selectively activating early steps of pterin biosynthesis in cells of the monocyte/macrophage series lead to increased excretion of neopterin and dehydroneopterin which is not associated by enhanced biosynthesis of biopterin.

Further evidence that IFN mediated enhancement of pterin biosynthesis in certain hemopoietic cells might represent the main cause of elevated neopterin levels seen during inflammatory disease states are derived from in vivo studies. First, diseases known to be accompanied by elevated neopterin levels also have detectable levels of circulating IFN (for review see ref.2). Secondly, the view of an intimate relationship between production of endogenous IFNs and induction of pterin biosynthesis is further supported by individual follow up studies. Examples for concordant changes of endogenous IFN-levels and neopterin serum concentrations in a patient with cytomegalovirus disease and a patient rejecting his allograft are shown in Figure 2 a and b. Details of such clinical correlation studies were previously published (8,9).

Third, in vivo application of both, recombinant IFN-alpha and recombinant IFN-gamma induce significant elevation of neopterin levels (10,11). In vivo induction of IFN-mediated pterin biosynthesis, as measured by assessment of neopterin levels is clearly dose dependent (Fig.3).

In conclusion, these results prove that activated cells of the monocyte/macrophage series exhibite enhanced pterin biosynthesis. Pterin metabolism in these activated cells is unique because it selectively involves early steps of this pathway and leads to increased production of neopterin without generation of biopterin. IFNs play a central role by activating the key enzyme GTP-CH I. These in vitro data and the in vivo studies with recombinant interferons indicate that elevated neopterin levels during inflammatory disease reflect the extend of IFN-induced activation of the monocyte/macrophage system.

Figure 2: Correlation of Serum Neopterin With IFN-Gamma Levels in a Patient With CMV-Disease (A) and a Patient With Acute Rejection (AR) of His Renal Allograft (B)

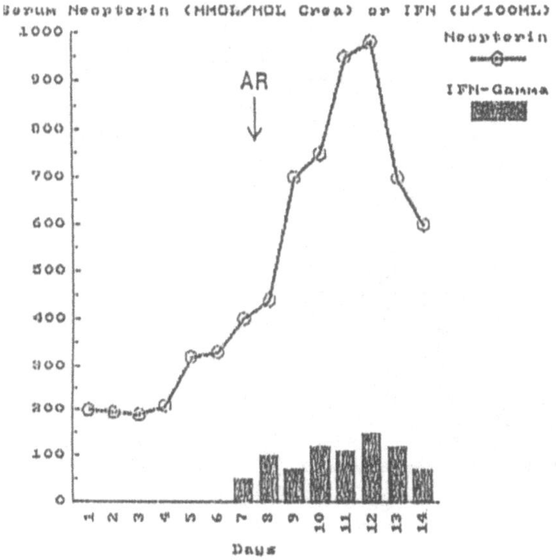

AR- Acute Rejection

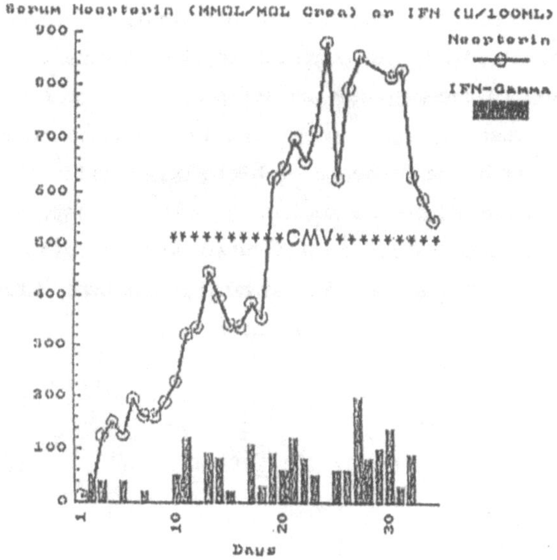

Fig. 3. Dose Dependent Enhancement of IFN-mediated Pterin Biosynthesis

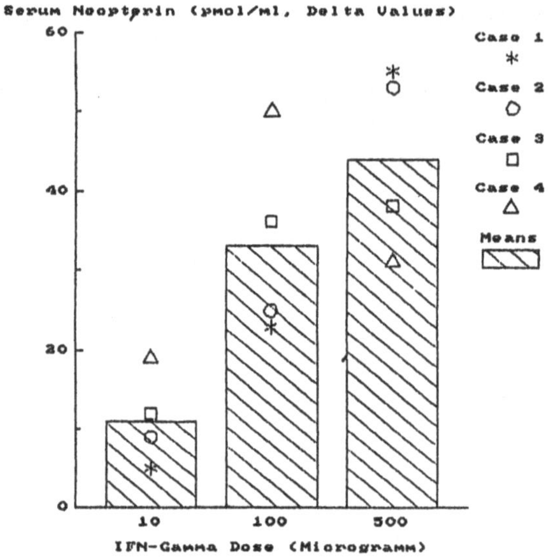

From: W Aulitzky et al (11)

2. Enhanced Pterin Biosynthesis During Hemopoietic Regeneration

Elevated levels of biopterin and/or of neopterin were observed at the time of hemopoietic regeneration subsequent to autologous or allogeneic bone marrow transplantation (3,13). In contrast to the process of allograft rejection discussed in the previous paragraph this enhancement of pterin biosynthesis, at least as far as neopterin levels were concerned, occured independent from the presence of histocompatibility barriers between the bone marrow donor and recipient (13). In fact, similar patterns were seen in tumor patients, who subsequent to letal whole-body irradiation were reconstituted with autologous bone marrow (13). A typical pattern of such a patient is shown in Fig.4.

Figure 4: Serum Neopterin and Interferon Levels During Hemopoietic Regeneration Subsequent to Bone Marrow Transplantation

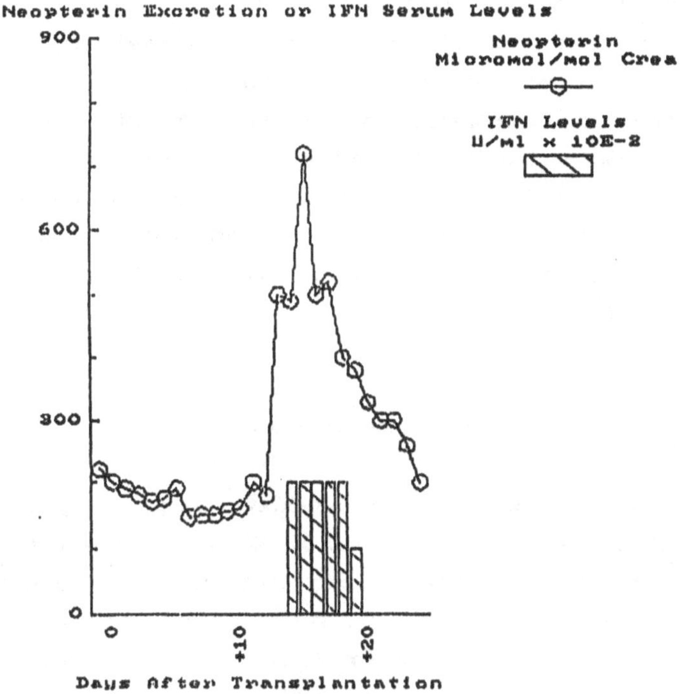

We asked the question, whether similar mechanisms involving IFNs or other cytokines such as colony stimulating factors are involved. To test this we have followed both neopterin and IFN levels in several autologous and allogeneic bone marrow transplant recipients. An example is given in Fig.4. Details were also previously reported (13,14). In the majority of patients, detectable levels of circulating IFNs were demonstrated at the time of hemopoietic regeneration and this coincided with the peak neopterin values. This observations supported the view that endogenous IFNs might be at least partly involved to enhanced pterin biosynthesis during hemopoietic regeneration. In contrast no evidence was obtained for the involvement of colony stimulating factors which also peak shortly before hematological reconstitution.

The negative results obtained by in vitro stimulation of human macrophages with GM-CSF are shown in Figure 1. Although involvement of endogenous IFNs, which presumably are induced by confrontation of newly formed lymphocytes with ubiquitous enviromental pathogens is very likely, this does not explain the elevation of biopterin levels observed around this time (3). One thus would rather believe that enhanced pterin biosynthesis during hemopoietic regeneration is determined by two independent factors: an IFN dependent enhancement of early pterin biosynthesis which involves activated T-cells and monocyte/ macrophages as well as increased biopterin content of immature hemopoietic cells, cells, which appear in the circulation in large numbers at the time of hemopoietic reconstitution (3).

3.Enhanced Pterin Biosynthesis During Endotoxinemia

Enhanced pterin biosynthesis during infectious disease is primarily a feature of systemic virus disease (for review see 2). Elevated levels were however also observed in a variety of bacterial infections. By far the highest levels were detected in patients suffering from virus disease complicated by bacterial infection and in septic shock patients (4,9). A composition of these data is shown in Figure 5.

The latter observation prompted the question whether lipopolysaccharide (LPS) the active principle of endotoxin, which is known to represent a most potent macrophage stimulator, is capable to induce pterin biosynthesis. Previous data by C.F.Nathan indicated that LPS also induces neopterin release from human macrophages (12). We subsequently showed, that LPS induced pterin biosynthesis can not be inhibited by anti-IFN gamma antibodies (results not shown). This finding indicated a direct action of LPS on human macrophages. We further investigated, whether IFNs might sensitize macrophages to respond to endotoxin challenge with increased release of neopterin or of other activation specific cytokines such as tumor necrosis factor (TNF).

Figure 5: Neopterin Excretion in Various Infectious Disease States:

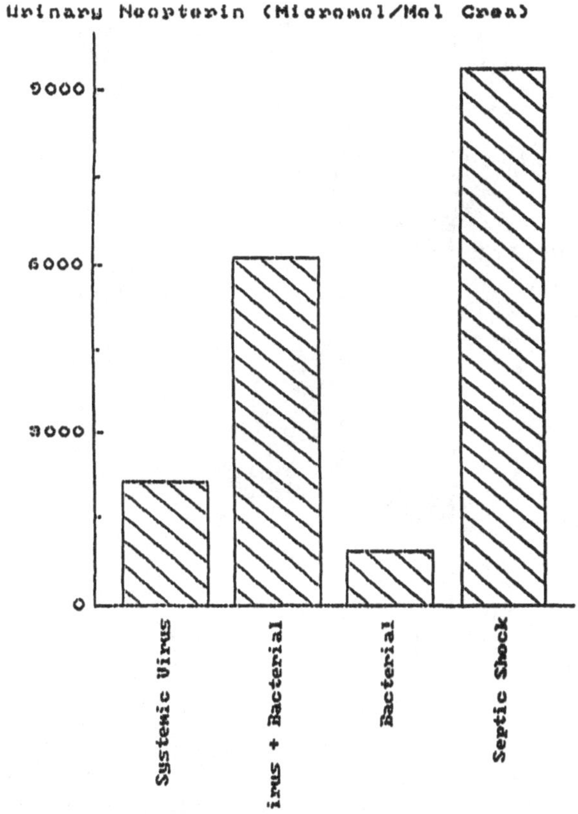

Results shown in Figure 6 clearly demonstrate such a costimulation. In conclusion it appears, that both IFNs and LPS can potentiate enhancement of pterin biosynthesis in macrophages and that such a mechanism might underly the extraordinarily high neopterin levels observed in context with septic shock and bacterial superinfection of systemic virus disease.

Figure 6: IFN and LPS Costimulate Macrophages to Release Neopterin and TNF

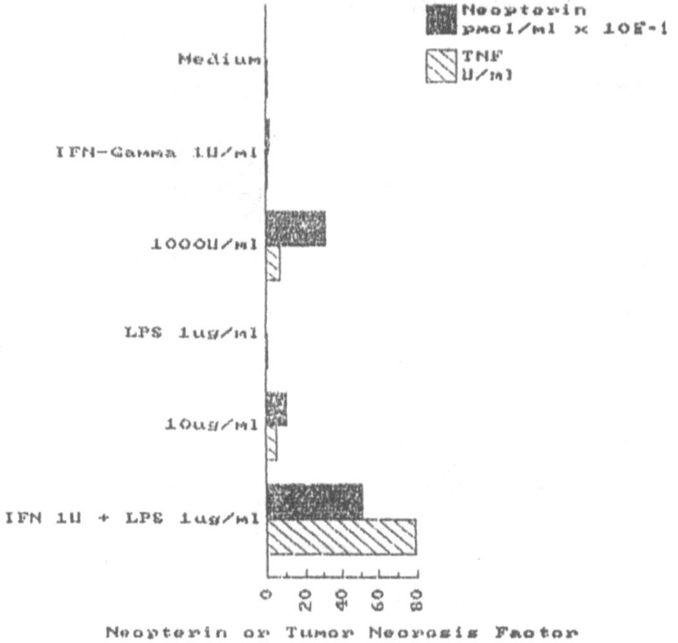

Conclusions

During the last years evidence accumulates, that biosynthesis of pterins is enhanced in inflammatory disease, during hemopoietic regeneration and endotoxinemia. Enhancement in these conditions is primarily restricted to the early steps of this pathway and does not lead to synthesis of BH4. Macrophages stimulated with interferons and/or with endotoxin seem to play a central role in this phenomenon.

This work was financially supported by the Austrian Research Funds "Zur Förderung der wissenschaftlichen Forschung", project no. 5635. We are indebted to Mrs.Maria Price-Placzek for secretarial help during the preparation of this manuscript.

References

1. Niederwieser, D., H.Ch.Curtius, O.Bettoni, J.Bieri, B.Schircks M.Viscontini, J.Schaub. 1979. Lancet i, 131.

2. Huber, Ch., J.Troppmair, H.Rokos, H.Ch.Curtius. 1987. Dtsch.med. Wschr. 112, 107

3. Fink, M., U.Jehn, W.Wilmanns. 1983. In :Biochemical and Clinical .Aspects of Pteridins. (H.Ch.Curtius, W.Pfleiderer, H.Wachter, eds) Walter de Gruyter & Co.Berlin, New York,

4. Strohmayer W., H.Redl, G.Schlag, D.Inthorn. 1987 In: Progress of Clincal and Biological Research, Subseries: Vienna, Shock-Forum, (Schlag G, Redl H eds.) Allan R.Liss Inc., New York (in press)

5. Huber, Ch. D.Fuchs, A.Hausen, R.Margreiter, G.Reibnegger, M. Spielberger, H.Wachter. 1983. J.Immunol.130, 1047.

6. Huber, Ch., J.R.Batchelor, D.Fuchs, A.Hausen, A.Lang, D.Niederwieser, G.Reibnegger, P.Swetly, J.Troppmair, H.Wachter. 1984. J.Exp.Med. 160, 310

7. Schoedon, G., J.Troppmair, G.Adolf, Ch.Huber, A.Niederwieser. 1986. J.Interferon.Res., 6, 697

8. Woloszczuk, W., J.Troppmair, E.Leiter, R.Flener, M.Schwarz, J.Kovarik, E.Pohanka, R.Margreiter, Ch.Huber. 1986. Transplantation, 41, 716.

9. Tilg, H., M.Scriba, R.Margreiter, Ch.Marth, D.Niederwieser, W.Aulitzky, M.Spielberger, H.Wachter, Ch.Huber. 1987. Clinical Transplantation, 1, 37.

10. Lang, A., D.Niederwieser, Ch.Huber, P.Swetly, D.Fuchs, A.Hausen, G.Reibnegger, H.Wachter. 1984. In: Biochemical and Clinical Aspects of Pteridines, Vol.3, (H.Ch.Curtius, W.Pfleiderer, H.Wachter, eds.). Walter de Gruyter & Co, Berlin, New York, 251-254.

11. Wo.Aulitzky, G.Gastl, W.E.Aulitzky, K.Nachbauer, B.Lanske, G.Kemmler, R.Flener, J.Frick, C.Huber Gamma-Interferon for the Treatment of Metastatic Renal Cancer: Dose-Dependent Stimulation and Downregulation of Beta-2-Microglobulin and Neopterin Responses. Immunobiology (subm.)

12. C.FM. Nathan. 1986. Peroxyd and pteridin: A hypothesis on the regulation of macrophage antimicrobial activity by interferongamma. In: IFN-7. (Ion Gresser, eds.), Academic Press, Boston, Toronto Sidney, Tokio.

13. Niederwieser, D., CH.Huber, A.Gratwohl, P.Bannert, D.Fuchs, A. Hausen, G.Reibnegger, B.Speck, H.Wachter (1984). Transplantation 38, 497

14. Niederwieser D., J.Troppmair, G.Adolph, R.Margreiter, Ch.Huber, 1985. J.Exp.Hematol, 13:131-132,

METABOLISM OF PTERINS IN THE CELLULAR IMMUNE SYSTEM OF MAN AND MOUSE

Gabriele Schoedon, Alois Niederwieser+ , Hans-Christoph Curtius

Division of Clinical Chemistry, Department of Pediatrics, University of Zurich, Zurich, Switzerland

Adriano Fontana

Clinical Immunology, Department of Internal Medicine, University of Zurich, Zurich, Switzerland

Jacob Troppmair, Christoph Huber

Clinical Immunobiology, Department of Internal Medicine, University of Innsbruck, Austria

Introduction

An understanding of the difference in tetrahydrobiopterin (BH_4) biosynthesis between primates and lower mammals is of fundamental importance for understanding the role of pteridines in the immune system. One has to be careful in interpreting results concerning any essential role of a pteridine measured in a certain cell type, since this pteridine may not be found in the same cell type of another species (1). Also within a species, expression and regulation of BH_4 biosynthesis varies from tissue to tissue and from cell type to cell type (2).

Generally, the biosynthesis proceeds from GTP via the three enzymes GTP cyclohydrolase I (GTPCH), 6-pyruvoyl tetrahydropterin synthase (PPH_4S), and sepiapterin reductase(SR) to BH_4 (Figure 1). In primates, selective accumulation of dihydroneopterin triphosphate (NH_2TP) occurs, which leads to the appearence of neopterin.

Since neopterin is not normally found in non-primates (1), this indicates that the BH_4 biosynthesis is regulated differently in primates due to different rate-limiting steps of the biosynthetic enzymes.

In this study, the regulation of BH_4 biosynthesis in human immune cells is compared with that in mouse immune cells.

Unconjugated Pterins and Related Biogenic Amines

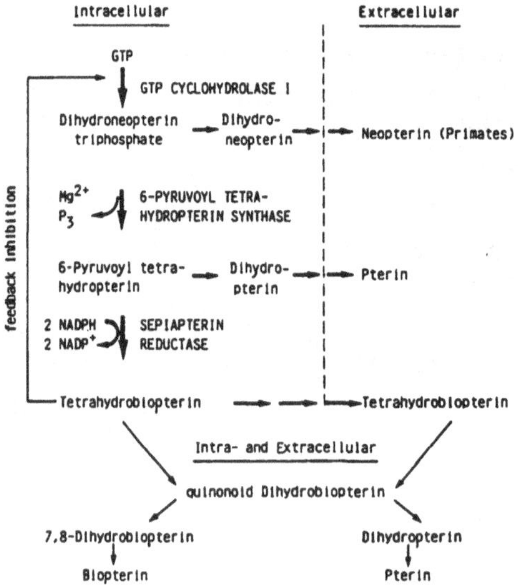

Fig. 1: Biosynthesis and metabolism of pterins in peripheral blood mononuclear cells (enzymatic and non-enzymatic steps).

Materials and Methods

Isolation of human T cells and monocytes from peripheral blood and culture and stimulation techniques of these cells are described in (3). The mouse immune cells used in the experiments are EL-4, a leukmia cell line; an ovalbumin specific T cell lineage (OVA-T), whose growth is strictly dependent from IL-2, and peritoneal macrophages from B_{10}-BR mice. Culture and stimulation techniques of mouse immune cells are discribed in (4). Pterins (5) and GTP (6) were measured by means of HPLC. The assays for GTPCH, PPH_4S, and SR were performed as described (7).

Results and Discussion

In the human immune system, neopterin, biopterin, and pterin are found as metabolites, but there is a different pattern in T cells and macrophages (Figure 2).

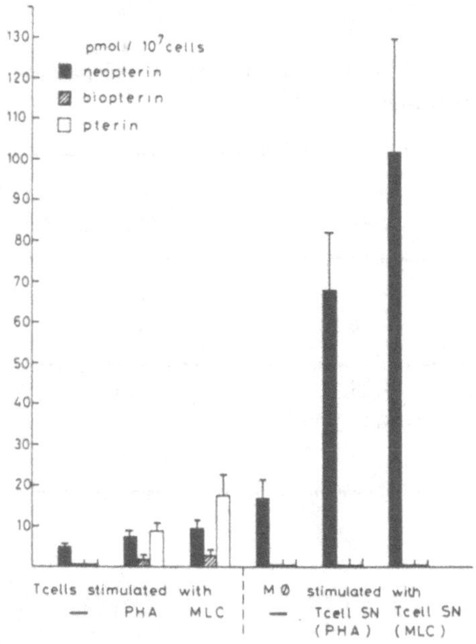

Fig. 2: Intracellular pterins in human T cells and macrophages (MØ). Highly purified T cells were stimulated either with 1 ug/ml phytohemagglutinine (PHA) or with $1x10^6$ allogeneic T cells (MCC). Purified MØ were stimulated by addition of 25% (v/v) culture supernatant (SN) from the activated T cells. Stimulation for 72 h, means ± SEM.

Low concentrations of all pterins are found in the T cells, in macrophages only neopterin is detected, and its concentration rises drastically after stimulation. We suggested there must be a difference in activities of the BH_4 biosynthetic enzymes, thus we measured simultaneously the activities of the three enzymes GTPCH, PPH_4S and SR as well as GTP levels in human T cells and macrophages.

As shown in Table 1, in T cells stimulated with lectin, GTPCH activity is significantly increased. There is a low, but significant PPH_4S activity and a high activity of SR, but there is no change in activites of these enzymes upon stimulation.

In macrophages, GTPCH activity is 10 times higher than in T cells, but there is no PPH_4S activity found either before or after stimulation. Inhibitors of PPH_4S in macrophags were ruled out by recovery of known PPH_4S activity from human liver added to the cell lysate.

Table 1:
Simultaneous determination of the enzymes of BH_4 biosynthesis as well as starting substrate, GTP, in human peripheral blood mononuclear cells

Cell type	stimulation	GTPCH μU/mg protein	PPH_4S μU/mg protein	SR mU/mg protein	GTP nmol/10^7 cells
T cells	no	$\angle$ 0.03	0.3	0.15	0.9
	PHA	0.7	0.3	0.18	4.0
MØ	no	0.4	n.d.	0.19	3.0
	T cell SN PHA	9.5	n.d.	0.15	7.0
	IFN γ	4.2	n.d.	0.15	6.0

1 μU of GTPCH = 1 pmol NH_2TP/min at 37°C
1 μU of PPH_4S = 1 pmol BH_4/min at 37°C
1 mU of SR = 1 nmol S_{red}/min at 37°C

results are means, n = 3

SR activity is the same as in T cells and also didn't change after stimulation. Both in T cells and macrophages, the rise of intracellular GTP always correlates with the rise of GTPCH activity.

PPH_4S is not expressed in human macrophages, this leads to the accumulation of high levels of NH_2TP, which in turn is a consequence of the high GTPCH activity.

In the mouse immune system, BH_4 biosynthesis shows two fundamental differences (Table 2):

First, no intermediates of BH_4 biosynthesis can be detected. High levels of biopterin, which was found to be all BH_4 by electrochemical detection, are measured after stimulation.

Second, mouse macrophages do not lack PPH_4S activity, since BH_4 is found in high concentration in these cells.

Table 2: Intracellular concentration of pterins and GTP in mouse mononuclear cells before and after stimulation.

Cell type	stimulation	neopterin	biopterin	pterin	GTP
		(pmol/mg protein)			(nmol/10^7 cells)
OVA-T	no	nd	32	16	1.5
	IL-2 (2.5% v/v)	nd	124	16	4.2
EL-4	no	nd	10	26	5.6
	IL-2 (2.5% v/v)	nd	4	3	-
B_{10}-BRMØ	no	nd	93	nd	-
	LPS (10 ug/ml)	nd	214	nd	-

nd: not detected. -: not measured. Detection limits of neopterin and biopterin, 0.08 pmol/107 cells; pterin 0.18 pmol/107 cells. Results are given as means (n 3), or from single experiments.

Since no neopterin is found in any of the stimulated mouse immune cells, we suggest that GTPCH is the rate limiting enzyme. From a comparant study it is known that in mice or rats PPH_4S activity is up to 50 times higher than in primates (8). Therefore NH_2TP is at once metabolized further without any accumulation.

In stimulated mouse OVA-T cells the rise of biopterin levels correlates with the rise of GTP (Figure 3), and this can be compared with the correlation of neopterin and GTP in the human immune cells (7).

In both, man and mouse immune cells, GTPCH is responsible for the rising pteridine levels.

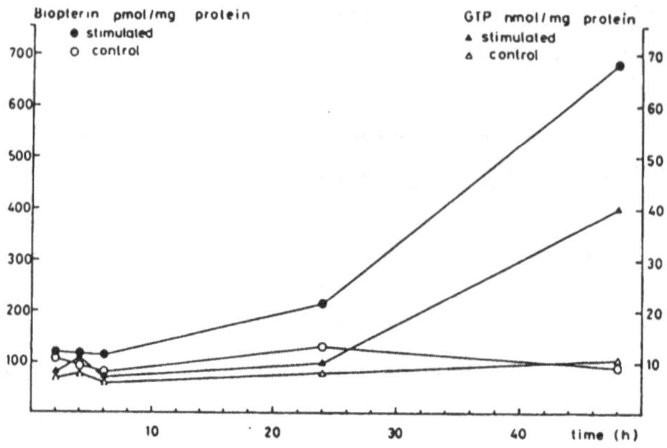

Fig. 3: Kinetics of biopterin biosynthesis in mouse ovalbumin specific T-cells (OVA-T).
Change of biopterin concentrations and GTP in OVA-T cells stimulated with 2.5% (v/v) IL-2 and in unstimuated controls during 48 hours.

We studied the regulation of GTPCH activity in stimulated mouse OVA-T cells by adding the protein biosynthesis inhibitor cycloheximid to the culture medium.

Table 3: Effect of the inhibition of protein biosynthesis by cycloheximide in IL-2 stimulated OVA-T cells on GTP, GTPCH activity and intracellular concentration of biopterin after 48 hours of incubation.

OVA-T cells incubated with	GTP nmol/mg protein	GTPCH uU/mg protein	biopterin pmol/mg protein
-	10	nd	100
IL-2 (2.5% v/v)	40	0.2	680
IL-2 (2.5% v/v) + cycloheximide (10^{-6}M)	66	0.2	630

nd: not detected. Results from single experiments. Detection limit for biopterin, 0.08 pmol/107 cells; GTPCH activity, 0.03 uU/mg protein 1 uU GTPCH = 1 pmol NH_2TP/mg at 37°C.

As shown in Table 3, GTP concentration after stimulation plus cycloheximid is two times as high as without cycloheximid, while GTPCH activity and BH_4 biosynthesis remain unchanged. We conclude that in immune cells GTPCH is not induced, but is activated by the high GTP levels, which are a consequence of cell stimulation.

We can conclude that the occurence of NH_2TP and neopterin in the immune system of man is first caused by the rate limiting enzyme beeing PPH_4S, and second, by the lack of this enzyme's activity in human macrophages. A physiological role of NH_2TP of NH_2 in the immune system can probably be ruled out.

The only enzyme sensible to immune stimulation in man and mouse is GTPCH, and since GTP rises simultaneously, we conclude that the increase of GTPCH activity in the immune cells is unspecific and a consequence of very high substrate concentrations.

The role of BH_4 in the immune system is also not yet clear. There is some evidence for an interaction of BH_4 with IL-2 in the T-lymphocyte activation (9). But general statements like "BH_4 is needed for cell proliferation" or "for cell differentiation" can not be made since there are highly proliferating cell lines which lack BH_4 biosynthesis due to GTPCH deficiency, and mature human macrophaghes fulfil all their special functions without any BH_4.

Acknowledgements

This work was supported by the Swiss National Science Foundation project no. 3.601-0.84. Dihydroneopterin triphosphate was kindly provided by Mr U. Redweik. We are greatful to Mrs L. Kierat for skillful technical assistance.

References

1. Duch, D.S., S.W. Bowers, J.H. Woolf, C.A. Nichol. 1984. Life Sci. 35, 1895.

2. Nichol, C.A., C.-C. Lee, M.P. Edelstein, J.Y. Chao, D.S. Duch. 1983. Proc. Natl. Acad. Sci. 80, 1546.

3. Schoedon, G., J. Troppmair, Ch. Huber, A. Niederwieser. 1986. J. Interferon Research 6, 697.

4. Fontana, A., H. Hengartner, N. de Tribolet, E. Weber. 1984. J. Immunol. 132, 1837.

5. Niederwieser, A., W. Staudenmann, E. Wetzel. 1984. J. Chromatogr. 290, 237.

6. Duch, D.S., C.L. Lee, M.P. Edelstein, C.A. Nichol. 1983. Molec. Pharm. 24, 103.

7. Schoedon, G., J. Troppmair, A. Fontana, Ch. Huber, H.-Ch. Curtius, A. Niederwieser: Eur. J. Biochem. (in press)

8. Hasler, Th., A. Niederwieser. 1986. In: Chemistry and Biology of Pteridines 1986 (B.A. Cooper and V.M. Whitehead, eds.). de Gruyter Berlin, pp. 319-322.

9. Ziegler, I., U. Schwulera, I. Ellwart. 1986. Exp. Cell. Res. 167, 531.

TETRAHYDROBIOPTERIN SYNTHESIS IS TRIGGERED BY INTERLEUKIN 2 BINDING AND MODULATES RECEPTOR AFFINITY

Irmgard Ziegler

Gesellschaft für Strahlen- und Umweltforschung, Institut für Experimentelle Hämatologie, Landwehrstr. 61, D-8000 München 2, FRG

Introduction

Interleukin 2 (IL-2) promotes the growth of activated lymphocytes (for review see ref. 1). Internalization of the cellular growth hormone (2) and transduction of the mitogenic signal (see ref. 1) depends on its binding to high affinity receptors which show apparent K_d values of 2-10 pM. They comprise about 10% of total receptor number. The low affinity receptor chain was defined as a 55 kD glycoprotein (p55) with K_d values 1000 fold higher (3). Ligand binding to p55 appears to cause a conformational change in the receptor, increasing its affinity to the converter (4). Recently, evidence has been presented that the high affinity IL-2 receptor is a heterodimer which consists of a 75 kD chain (p75), non-covalently linked to the earlier characterized, low affinity p55. (5, 6). In analogy to earlier work on enzyme regulation (see ref. 7) an allosteric interaction is discussed (4, 6).

Lymphocytic tetrahydrobiopterin (BH_4) was found to modulate IL-2 activity (8). The enhancement of IL-2 activity results in a co-mitogenic effect during lectin stimulation of lymphocytes (9, 10). Further analysis indicated that the cooperation between the lymphokine and the pteridine occurs at the receptor level (11).

Unconjugated Pterins and Related Biogenic Amines

In this study we have begun to examine the regulatory role of pteridines during formation of the heterodimeric high affinity receptor complex and for the subsequent internalization of the signal polypeptide.

Methods

Human T cells were isolated and cultivated as described in (11). The quantitative binding studies (recombinant ^{125}I-IL-2; NEN, Boston) were performed at 0° and were based on the method described by Robb (3). A rapid separation of bound from unbound label was achieved by the use of 96-well millititer GV plates (Millipore). Binding data from triplicate sets of experiments could thus be obtained at 10 min intervals. The parameters were estimated by the method of iterative weighted least squares (14). The non-linear regression procedure NLIN of the statistical computer software SAS (15) was used. In the following, a typical experiment is described. Statistical analysis of the Hill plots from 12 independent experimental sets will be published elsewhere.

The progress of internalization was followed at 37° after the radiolabelled IL-2 (at concentrations of 95-105 pM) had been bound by the extracellular receptor domain at 0° (20-40 min). Released ^{125}I-tyrosine was determined in the medium supernatant. Internalized ^{125}I-IL-2 was distinguished from surface-remaining growth hormone by its resistance to a 30 sec pH 4.0 treatment. Cell-bound label was separated by silicon oil density gradient centrifugation.

Results and Discussion

Pteridine synthesis by lymphocytes. Peripheral blood mononuclear cells and T lymphocytes undergo 3 episodes of pteridine accumulation which are reviewed in Table 1.

Replacing IL-2 with phorbolester, pteridines peak at maximally 20 fold levels after 10-20 min in IL-2 receptor^{+} T cells, and return to initial values during the following 30-40 min. Resting human peripheral blood mononuclear cells do not react within this short period (13). The capability of primed T cells for a rapid but transient pteridine synthesis substantiates a regulatory effect of tetrahydrobiopterin on IL-2 binding, described below.

Table 1.
Episodes of pteridine synthesis in human peripheral blood mononuclear cells and in T cells

Cell type	stimulant	time of maximum level	period of decline	reference
resting pbmc	lectin	3 days	1 day	(12)
primed T cells	IL-2	8-11h	4-6 h	(11)
primed T cells	phorbol-ester	10-20 min	10-20 min	(13)

Effect of BH_4 on IL-2 binding data. In order to analyze the effect of exogenously added pteridine on binding kinetics, internalization and endogenous pteridine synthesis had to be minimized and protracted. This was achieved by keeping the cells at 0° which does not affect binding to the extracellular receptor domain.

Fig. 1a demonstrates that saturable binding of control cells represented a normal saturation curve. The Scatchard plot

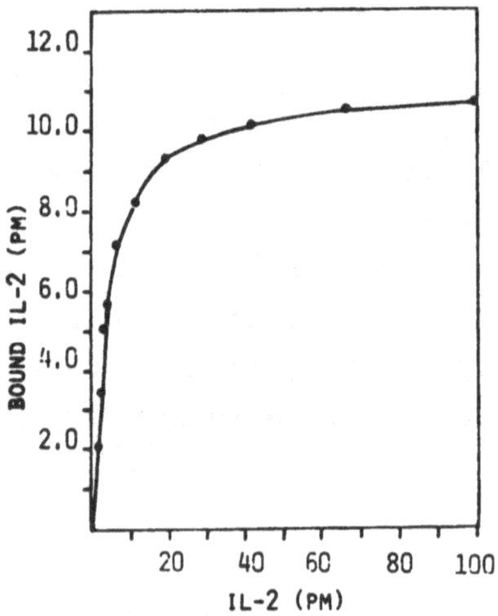

.Fig.1 Saturable binding (40 min; 0°) of $r^{125}I$-IL-2 in control cells (left hand) and from incubation together with (6R)-tetrahydrobiopterin (8×10^{-7}M) (right hand)

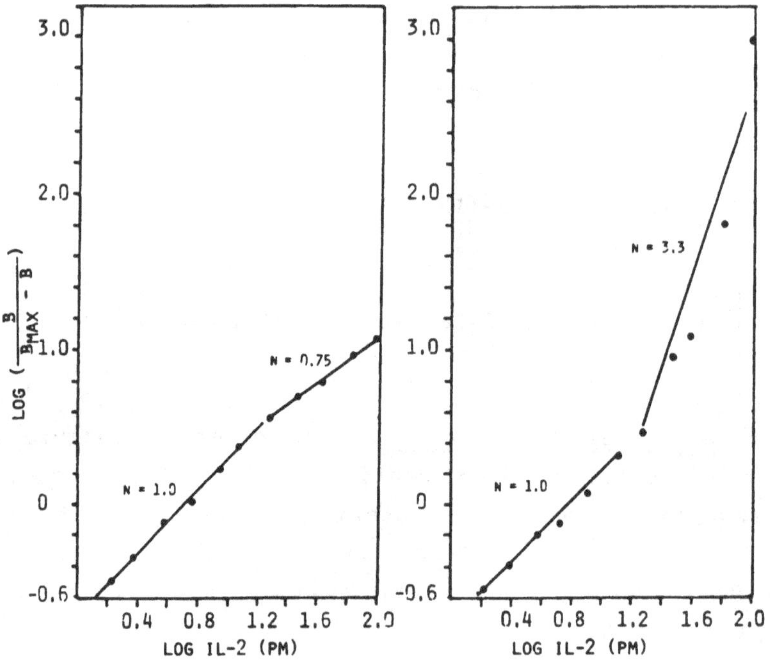

Fig.2 Hill plot of IL-2 binding data from Fig.1. Left hand: control cells; right hand: incubation together with (6R)-tetrahydrobiopterin

(not shown) yielded a linear regression line, indicating 3,600 maximal binding sites and a K_d of 9.1 pM. The Hill plot of the data (Fig. 2a) showed that initial binding occurs by homogenous, non-interacting sites. At concentrations $>$ 15 pM, a slight negative cooperation was indicated. Its actual extent by statistical evaluation will be presented elsewhere.

Addition of BH_4 together with the cytokine did not markedly change the initial slope. Beyond half maximum IL-2 concentrations, however, ligand binding did not level off as in con-

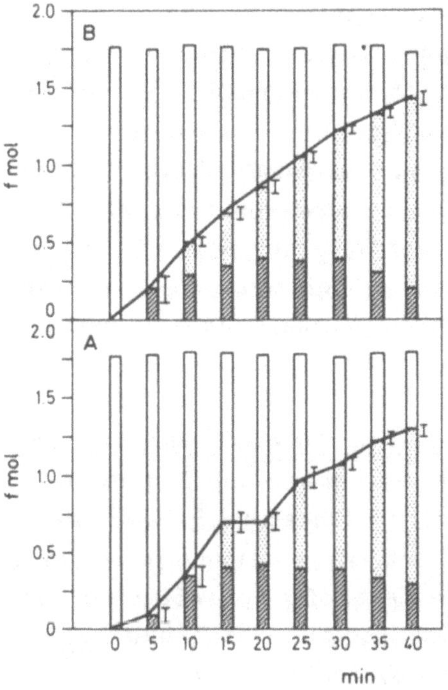

Fig.3 Internalization and processing of $r^{125}I$-IL-2.
Total column height: amount of IL-2 bound at the external membrane domain during preincubation at 0°. Subdivisions: distribution of radioactivity during further processing at 37°. ▯ IL-2 at the external receptor domain; ▨ intracellular IL-2; ▩ released ^{125}I-tyrosine.
A = IL-2 (105 pM); B = IL-2 (105 pM) + ((6R)-tetrahydro-biopterin (8 x 10^{-7}M)

trols but continued to increase markedly (Fig. 1b). Plotting of the data according to Scatchard showed that best fit regression was no more a straight line but divided into 2 sections (not shown). Like in controls the Hill plot (Fig. 2b) indicated stoechiometric, independent binding for the initial slope. Further binding (IL-concentrations > 15 pM) however, yielded a Hill number of 3.3 which indicates a marked cooperativity. The axiom of noninteracting, homogenous binding sites to be present at concentrations up to 200-300 pM IL-2 (3, 5) cannot be held if BH_4 is added together with the lymphokine. Under these conditions, IL-2 occupancy beyond half-saturation rather increased the affinity of the remaining sites. The data agree with the concept that the ligand first saturates pre-existing high affinity receptors which comprise only part of those which are estimated by Scatchard analysis. Subsequently, IL-2 encounters single p55 and p75 receptor subunits and may cause their allosterically regulated interaction (for review see ref. 7). This may result in additional high affinity binding, if BH_4 is simultaneously present. Its mode of operation and the explanation of the molecular basis have to await experiments with the isolated receptor proteins.

Effect of BH_4 on IL-2 internalization. Internalization of the ligand depends on both metabolic events and on previous high affinity binding. Therefore, the increasing portions of intracellular IL-2 at 37° together with released ^{125}I-tyrosine reflect the extent of high affinity binding at the extracellular receptor domain, which had been achieved by previous incubation of the cells at 0° and at 105 pM radiolabelled IL-2 concentrations. Fig. 3 shows that internalization proceeds with biphasic kinetics in control cells. It becomes linear when BH_4 is added together with the cytokine.

The data are consistent with the quantitative binding data presented above. It appears that pteridines enhanced association of receptor subunits contributes essentially to a homogeneous upward slope of ligand internalization, only occurring after a period of 15-20 min in controls. The heterogenous receptor distribution of the control cells is accounted for the gradual entry into the S phase (17). It becomes linear if BH_4 is added together with the cytokine (11). As final result, a co-mitogenic effect of reduced pteridines is noticed after 72 h (9, 10) which had provided initial evidence for the participation of pteridines in the regulation of T cell proliferation.

References

1 Greene, W.C., and Leonard, W.J. 1986. Ann. Rev. Immunol. 4, 69.

2 Weissman A.M., Harford, J.B., Svetlik, P.B., Leonard, W.L., Depper, J.M., Waldmann, Th.A., Greene, W., and Klausner, R.D. 1986. Proc. Natl. Acad. Sci. (USA) 83, 1463.

3 Robb, R.J., Warner, C., Greene, W.C., and Rusk, C.M. 1984. J. Exp.Med. 160, 1126.

4 Kondo, Sh., Shimizu, A., Saito, Y., Kinoshita, M., and Honjo, T. 1986. Proc. Natl. Acad. Sci. (USA) 83, 9026.

5 Smith K.A. 1987. Immunology today 8, 11.

6 Sharon, M., Klausner, R.D., Cullen, B.R., Chizzonite, R., and Leonard, W.J. 1986. Science 234, 859.

7 Frieden C. 1967. In: Regulation of enzyme activity and allosteric interactions. (J.P. Changeaux, E. Kramme, and A. Pihl eds.) Academic Press London and New York, pp 69-71.

8 Ziegler, I., Schwuléra, U., Sonneborn, H.H., and Müller, W.J. 1985. Naturwiss. 72, 330.

9 Ziegler, I., Hamm, U., and Berndt, H.J. 1983. Cancer Res. 43, 5356.

10 Webber, S., and Yaye, D.L. 1986. In "Chemistry and Biology of Pteridines" (B.A. Cooper and V.M. Whitehead, eds.) W. de Gruyter, Berlin, pp. 235-238.

11 Ziegler, I., Schwuléra, U., Ellwart, J. 1986. Exp. Cell Res. 167, 531.

12 Ziegler, I. 1985. Biochem. biophys. Res. Comm. 132, 404.

13 Seidl, J., Borchert, M., and Ziegler, I. 1986. Biochem. biophys. Res. Comm. 141, 494.

14 Hartley, H.D. 1961. Technometrics 3, 269.

15 SAS user's guide: Statistics, version 5 ed., SAS institute, N.C. (1985).

16 Greene, W.C., and Leonard, W.J. 1986. Ann. Rev. Immunol. 4, 69.

17 Cantrell, D.A. and Smith, K.A. 1984. Science 224, 1312.

Acknowledgements. The experiments were skilfully assisted by Mrs. C. Altmann and G. Dachs. We thank Dipl.Math. M. Herbold, MEDIS-Institut der GSF, for advice in computational statistics. The work was supported by the Deutsche Forschungsgemeinschaft and the Bundesministerium für Forschung und Technologie.

KINETICS OF PRODUCTION AND RELEASE OF NEOPTERIN. COMPARISON OF HUMAN PBMC WITH THE PERMANENT MONOCYTIC CELL LINE U937 AND ITS SUBCLONES.

K. Rokos, R.O.F. Kunze, M.A. Koch

Robert Koch-Institut, Department of Virology, Nordufer 20, D-1000 Berlin 65, Germany

H. Rokos

Research Laboratory, Henning Berlin GmbH
Komturstraße 58-62, D-1000 Berlin 42, Germany

K. Nilsson

University of Uppsala, Department of Pathology, University Hospital, S-75185 Uppsala, Sweden

Human monocytes/macrophages can be induced by interferon-γ (IFNγ) to produce and release neopterin (N) by activation of the key enzyme of pterin biosynthesis guanosinetriphosphate-cyclohydrolase (GTP-CH). The functions of dihydroneopterintriphosphate (NH_2P_3), the immediate product of GTP-CH, or its degradation product N are not known in macrophages, where biosynthesis of pterins seems not to proceed to tetrahydrobiopterin, for which NH_2P_3 is an intermediate in other cells. (For reviews see 1,2). N release from macrophages has been postulated as a mechanism for overflow regulation of the GTP-pool (3). Investigations concerning this hypothesis and/or other functions of N or its derivatives are scarce, possibly due to difficulties in obtaining peripheral monocytes/macrophages in sufficient numbers and uniform quality. Induction of GTP-CH activity by IFNγ and consequently N production is subject to individual variation (4). The use of a permanent cell line would facilitate <u>in vitro</u> experiments to a considerable extent. So far no permanent cell lines capable of N production

Unconjugated Pterins and Related Biogenic Amines

to any significant extent - in contrast to biopterin production - have been found. The capacity of the parent cell line U937, a human monocytic cell line growing in suspension (5), and its subclones for production of N upon activation with IFNγ was examined.

Materials and Methods

Peripheral blood mononuclear cells (PBMC) were isolated from heparinized blood with a Ficoll-Paque gradient (Pharmacia). PBMC were cultured in Linbro-plates in RPMI 1640, supplemented with 2 mM glutamine, 20 % autologous plasma and antibiotics at 37 °C, 4 % CO_2 and at a density of 10^6/ml. U937 and subclones were cultured in RPMI 1640 supplemented with 10 % FCS, 2 mM glutamine and antibiotics at a cell density of $2 \cdot 10^5$/ml if not stated otherwise. Concentrations used were: Recombinant IFNγ (Biogen Research Cambridge, Mass.) 1000 U, 2000 U or 5000 U/ml, phytohemagglutinin (PHA) (Wellcome) 10 µg/ml and recombinant interleukin 2 (IL-2) (NEN) 1000 U/ml.
Determination of N in cells according to (4), and in supernatants was performed after 72 h (if not specified otherwise) in duplicate by radioimmunoassay (Henning Berlin).
Flowcytometric analysis was performed with an Epics C (Coulter). Cells were labelled with fluorescence-conjugated monoclonal antibodies (mAb) T4, HLA-DR and Leu15 (Coulter Clone and Becton Dickinson).

Results and Discussion

N release of subclones of U937 is shown in Table 1. N is released into supernatants by some subclones in different amounts; the parent cell line and subclones 1 and 5 do not release N. N values found in supernatants depend on cell density

	cells/ml	medium	1000 U IFNγ	5000 U IFNγ
Clone 2	$2\cdot10^5$	< 1	5.9	5.1
	$5\cdot10^5$	< 1	6.6	7.2
Clone 3	$2\cdot10^5$	3.3	5.7	8.6
Clone 4	$2\cdot10^5$	2.6	9.0	9.2
	$5\cdot10^5$	3.6	9.2	11.0
Clone 10	$2\cdot10^5$	1.2	1.9	2.3
	$5\cdot10^5$	1.4	2.9	3.1
Clone 16	$2\cdot10^5$	< 1	3.6	6.0
	$5\cdot10^5$	1.0	5.4	6.7
Clone CTB	$2\cdot10^5$	1.8	3.4	3.4
	$5\cdot10^5$	2.4	2.3	5.3

Table 1. N release by subclones of U937 at different cell densities induced by various concentrations of IFNγ.

and concentration of IFNγ as with monocytes/macrophages. The kinetics of N accumulation in culture medium were investigated for clones 4 and 16 (Fig. 1), of intracellular N for clone 16 (Fig. 2). N release from cells starts between 20 and 48 hours, whereas intracellular N is already elevated at 20 h, staying at a high level for further 48 h. In clone 16 slightly elevated intracellular N and N release can be induced by transfer to fresh medium. N in supernatant increases highly at day 2 and 3, explaining slight variability of N found in different experiments.

For comparison kinetics of N release and intracellular N of human PBMC is shown in Fig. 3 and 4. N in PBMC increases highly between 4 and 20 hours, whereas N accumulation in the supernatant is appreciably only at 48 h. Not only IFNγ and PHA, but exposure to culture medium alone increases intracellular N (4) and N release.

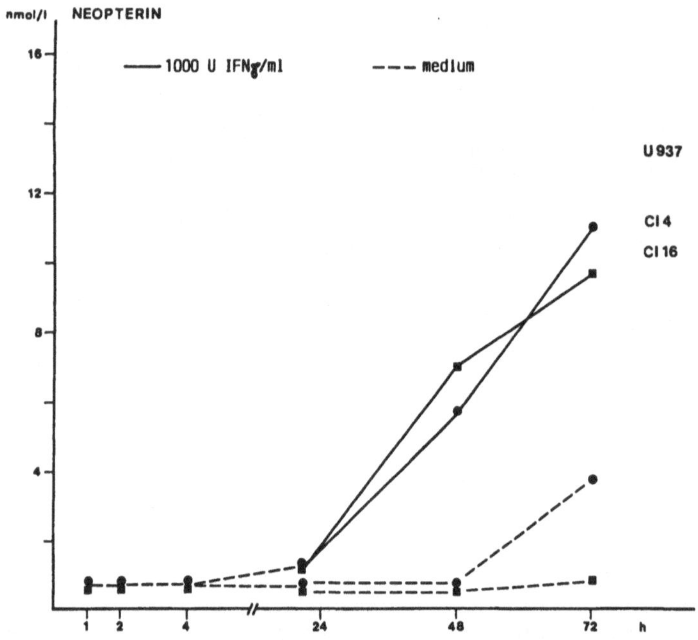

Fig.1. Kinetics of N release by clone 4 and clone 16 induced by IFNγ and transfer to fresh culture medium.

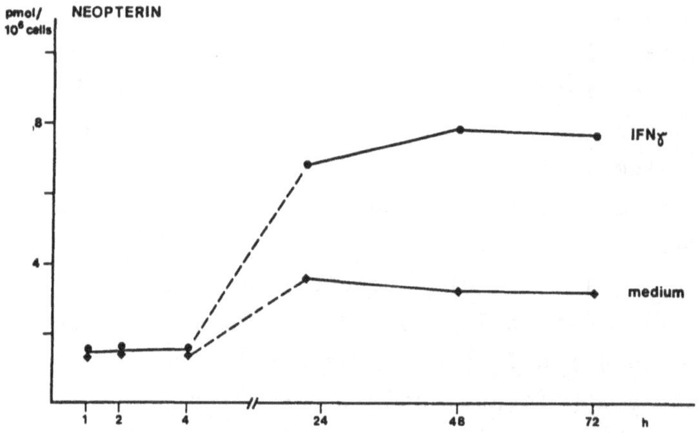

Fig.2. Kinetics of intracellular N accumulation in clone 16 induced by IFNγ and transfer to fresh culture medium.

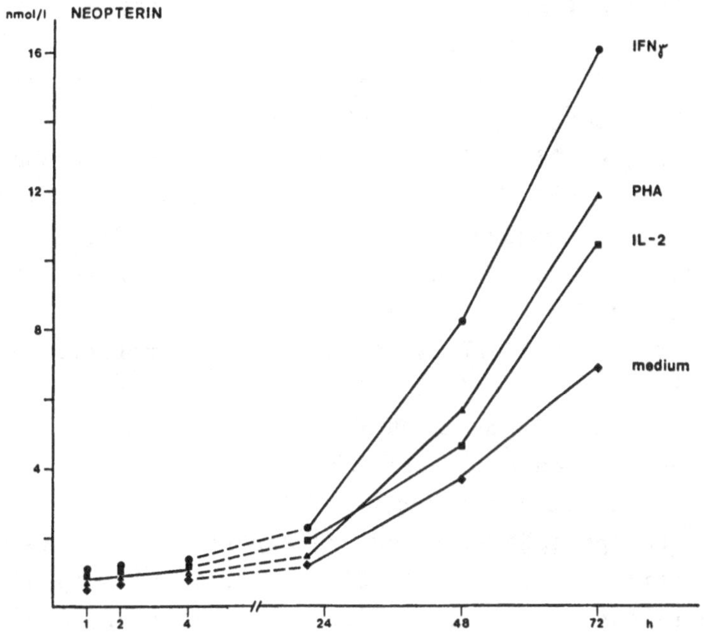

Fig.3. Kinetics of N release by human PBMC induced by IFNγ (●), PHA (▲), IL-2 (■), and culture medium (◆).

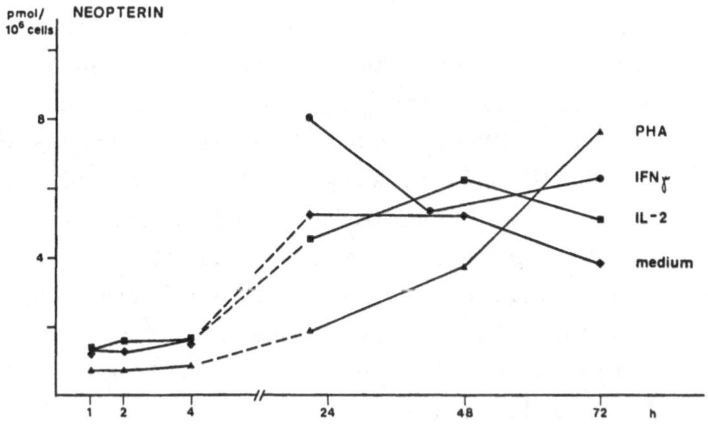

Fig.4. Kinetics of intracellular N accumulation in human PBMC induced by IFNγ (●), PHA (▲), IL-2 (■), and culture medium (◆).

N release from cells starts well after elevation of intracellular N, which takes several hours to build up. Intracellular N and release of N can be reduced by washing and centrifugation of PBMC after a 2 h preincubation, PBMC were then cultured for further 18 h (A), compared to PBMC cultured undisturbed for 20 h (B). (Medium alone: A 2.8 pmol/10^6 cells, 1.3 nmol/l; B 8.2 pmol/10^6 cells, 2.7 nmol/l.) These findings suggest that isolation procedures affect the ability of cells to synthesize N for some hours.

With respect to intracellular N, amount of N released upon stimulation with IFNγ and kinetics of N production clones 4 and 16 are similar to human PBMC and adherent cells (4). Clones 3 and 4 release as well small but measurable amounts of N after transfer to fresh medium (Table 1).

Monocytes from peripheral blood are reported to be capable of IFNγ production in response to LPS; control cultures of monocytes may release IFNγ without further activation, however in significantly smaller quantities (6). These observations suggest, that N synthesis of PBMC, isolated adherent cells and some subclones of U937 upon exposure to culture medium may be induced by IFNγ produced endogenously.

Since U937 is proliferating in contrast to fully differentiated monocytes/macrophages, the effect of IFNγ on increase of cell numbers was determined. Cell density increased from initial $2 \cdot 10^5$/ml to $6 \cdot 10^5$/ml (clone 4 and 16), whereas with 1000 (2000) U IFNγ/ml cell density reached only $3 \cdot 10^5$ ($2.5 \cdot 10^5$)/ml with clone 4 and $4.5 \cdot 10^5$ ($3.5 \cdot 10^5$)/ml with clone 16, showing a dose dependent inhibitory effect of IFNγ on cell proliferation as reported for the parent cell line U937 (7).

Intracellular N and N release increased with dosis of IFNγ. Clone 4: 10.0 (13.4) pmol/10^6 cells; 6.6 (7.1) nmol/l. Clone 16: 7.6 (13.2) pmol/10^6 cells; 5.4 (6.7) nmol/l.

IFNγ is reported to induce increased expression of Ia antigens on monocytes/macrophages (8). To investigate whether subclones of U937 resemble peripheral monocytes in this respect, expres-

	$CD4^+$	$HLA\text{-}DR^+$	$CD11^+$	Neopterin
Clone 1	↓	(↑)	↑	–
Clone 2	–	(↑)	↑↑	↑↑
Clone 4	–	↑↑	↑↑	↑↑
Clone 5	–	↑↑↑	↑↑↑	–
Clone 10	–	–	(↑)	↑
Clone 16	–	↑	↓↓	↑↑
Clone CTB	↓	↑↑	↑	↑

– no alteration
↑, ↑↑, ↑↑↑ increase } of percent positive cells and/or mean of fluo-
↓, ↓↓ decrease } rescence or neopterin release respectively.

Table 2. Alterations of expression of surface antigens on subclones of U937 induced by IFNγ together with changes of N release.

sion of HLA-DR antigens together with binding of the mAb T4, recognizing $CD4^+$ cells, and Leu15, recognizing the antigen $CD11^+$, the membrane complement receptor type 3 (CR_3) expressed on various cell types, e.g. monocytes and subsets of natural killer cells, were determined by flowcytometry in cells cultured with and without IFNγ (Table 2). Cell marker profiles of subclones of U937 are given in (9). In all clones investigated there is little change in the binding of mAb T4, but considerable alterations in binding of mAb HLA-DR and mAb Leu15 are found in several clones. N release in response to IFNγ is given simultaneously. The alterations of the four parameters together are distinct for each clone. No correlation is seen between capacity of clones to produce N or change of expression of surface antigens upon activation by IFNγ.
Some subclones of the leukemic cell line U937 are able to synthesize and release N upon stimulation with IFNγ like monocytes/macrophages. Since amount of N produced and kinetics of N synthesis are comparable in PBMC and clone 4 and 16, these subclones seem to be suitable for more detailed investigations of influences of lymphokines and other mediators on pterin synthesis in monocytes.

Acknowledgement

Thanks are due to Mrs. H. Luczak for competent secretarial help.

References

1. Ziegler, I., H. Rokos. 1986. EOS J. Immunol. Immunopharmacol. 6, 168-177.
2. Huber, C., J. Troppmair, H. Rokos, H.-C. Curtius. 1987. DMW 112, 107-113.
3. Troppmair, J., A. Lang, C. Huber et al. 1985. In: Biochemical and Clinical Aspects of Pteridines (H. Wachter, H.-C. Curtius, W. Pfleiderer, eds.) Walter de Gruyter, p. 363-377.
4. Rokos, K., R. Kunze, W. Lange, M.A. Koch. 1986. In: Chemistry and Biology of Pteridines 1986. (B.A. Cooper, V.M. Whitehead, eds.) Walter de Gruyter, p. 415-418.
5. Sundström, C., K. Nilsson. 1976. Int. J. Cancer 17, 565-577.
6. Nugent, K.M., J. Glazier, M.M. Monick, G.W. Hunninghake. 1985. Am. Rev. Respir. Dis. 131, 714-718.
7. Hattori, T., M. Pack, P. Bougnoux et al. 1983. J. Clin. Invest. 72, 237-244.
8. Stiehm, E.R., L.H. Kronenberg, H.M. Rosenblatt et al. 1982. Ann. Intern. Med. 96, 80-93.
9. Åsjö, B., I. Ivhed, M. Gidlund et al. 1987. J. Virol. (in press).

Section D: Pterins in Different Diseases

VARIATIONS OF NEOPTERIN, DIHYDRONEOPTERIN AND IMMUNOLOGICAL PARAMETERS DURING PHYSICAL EXERCISE (JOGGING)

H. Rokos

Research Laboratory, Henning Berlin GmbH
Komturstraße 58-62, D-1000 Berlin 42, Germany

K. Rokos, R.O.F.Kunze

Robert Koch-Institut, Department of Virology, Nordufer 20, D-1000 Berlin 65, Germany

Introduction

In lymphocyte subsets from peripheral blood diurnal variations as well as variations induced by physical and psychological stress are found correlating with the well known alterations of serum cortisol levels (1-3). First results showed that diurnal as well as stress related variations of serum neopterin paralleled cortisol levels (4). This prompted further studies on the influence of physical exercise (jogging) on neopterin (N), dihydroneopterin (NH_2), cortisol, and other immunological parameters.

Materials and Methods

All volunteers gave informed consent. Blood was obtained by venipuncture or, for serial measurements, with an indwelling cannula; serum or plasma was separated within 1 hour and was used immediately or kept frozen at -25 °C.
Serum N and NH_2 were determined by RIA (Neopterin-RIAcid, Henning Berlin), the latter after acidic iodine oxidation. Serum cortisol was measured by RIA (IBL, Hamburg), serum creatinine by a modified Jaffé method (Merck, Darmstadt).

Unconjugated Pterins and Related Biogenic Amines

Total lymphocytes were determined by differential blood cell count. Lymphocytes were isolated from heparinized blood by Ficoll-Pacque (Pharmacia, Freiburg) density centrifugation. For flowcytometric analysis (Epics C, Coulter) cells were labelled with monoclonal antibodies (mab) coupled to fluorescein isothiocyanate for the determination of CD2, CD4, CD8, and CD11 antigen carrying cells (Epics, Coulter Electronics). Peripheral blood mononuclear cells (PBMC) were cultured for 72 hours in Linbro plates (37 °C, 4 % CO_2 in RPMI 1640 supplemented with 2 mM glutamine, 10 % FCS and antibiotics).

Results

Serum levels of N, cortisol, and creatinine before, during and after a run of ca. 20 km are shown in Fig.1. Values in one volunteer and diurnal variations are given in Fig.2. Maximal N

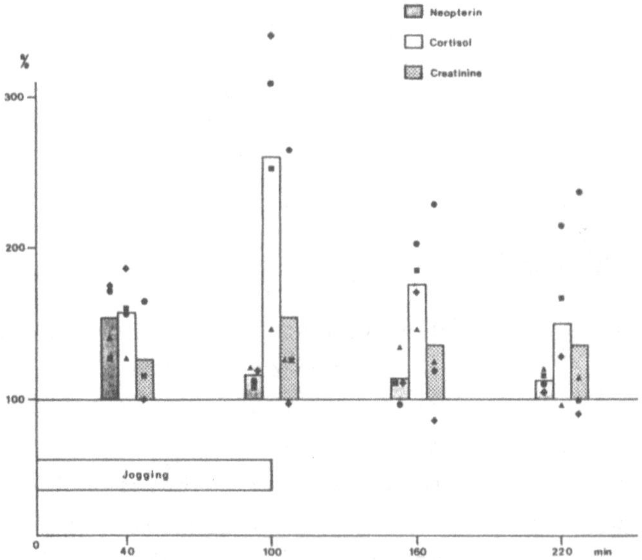

Fig.1. Variation of serum neopterin, cortisol, and creatinine of 4 healthy male volunteers during jogging in percent of initial values. Individual values are given, bars show the mean.

elevation is already seen in the first sample drawn during the run, whereas cortisol increased further during jogging; large individual variations are seen, particularly in creatinine levels.

Variations in absolute numbers of leukocytes, lymphocytes, and T-lymphocyte subsets during jogging are shown in Fig.3. The increase in leukocyte numbers, starting with physical stress and continuing in the recovery phase is remarkable. Percentage as well as absolute numbers of lyphocytes start to increase with exercise, but decrease in the later part of the run to 50 % of the initial value and remain low for further 2 hours; granulocytes change inversely to lymphocytes (data not shown). During exercise $CD4^+$ cells decrease and $CD8^+$ cells increase, yielding a strong decrease of the $CD4^+/CD8^+$ ratio. In the recovery phase the percentage of $CD4^+$ cells increase first, resulting in increased $CD4^+/CD8^+$ ratios; absolute numbers remain low because of pronounced lymphopenia.

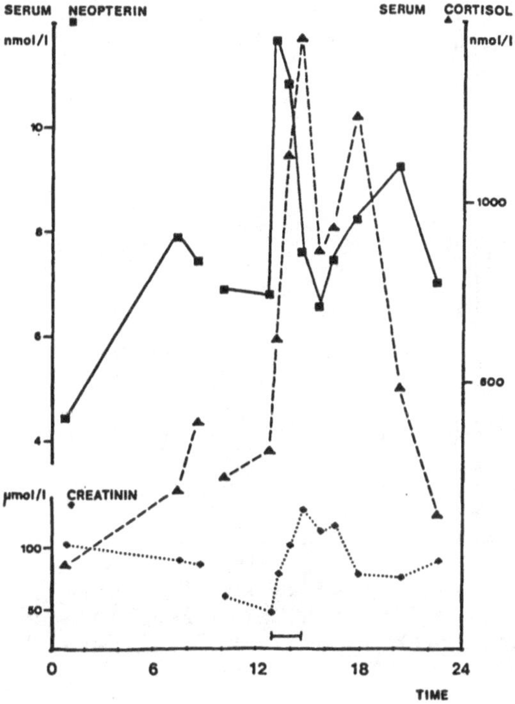

Fig.2.

Serum neopterin (■), cortisol (▲), and creatinine (●) of 1 healthy male volunteer during a 24 hour period which included the 20 km-run (⊢⊣).

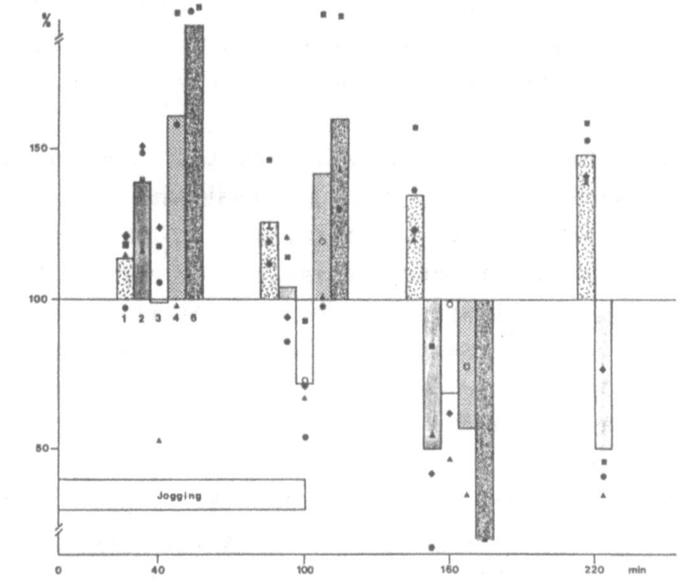

Fig.3. Variation in the numbers of leucocytes (1), lymphocytes (2) and T-cell subsets $CD4^+$ (3), $CD8^+$ (4), $CD11^+$ (6) during jogging in percent of individual initial values. Individual values are given, bars show the mean.

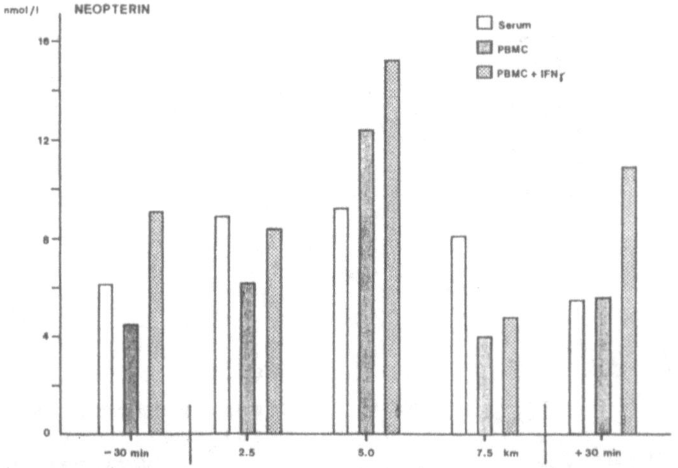

Fig.4. Comparison of neopterin levels in serum and in supernatants of PBMC (10^6 cells/ml) from heparinized blood (obtained during jogging) after incubation for 3 days with or without IFNγ (1000 U/ml).

Subsets of natural killer cells (NK) express CD11, the membrane complement receptor type 3 (CR3); these cells are elevated during exercise, but depressed in the recovery phase. NK labelled with mab Leu7 behave like $CD11^+$ cells (data not shown). Percentage of $CD2^+$ cells - a general marker recognizing most lymphocytes (rosette forming antigen) - practically did not vary (data not shown). Monocytes were found in small numbers (0-7 %) without recognizable trend (data from differential blood counts). B-lymphocytes labelled with the mab B1 vary between 3 and 11 % with no trend being observed during jogging.

culture 1		culture 2	
medium	5.9	medium	2.8
medium	5.9	IFNγ	16.1
IFNγ	16.8	medium	26.3
IFNγ	17.2	IFNγ	37.5

Table 1. Neopterin content (nmol/l) in supernatants of PBMC (10^6 cells/ml) with or without interferon-γ (1000 U/ml).

After 3 days (culture 1), non-adherent cells were washed by centrifugation in PBS, and returned to the adherent cells remaining in the well of the culture plate; culture 2 was then resumed for another 3 days.

For cell culture experiments blood was drawn from one volunteer during a further run (7.5 km, 35 min). N levels in cell culture supernatants and initial sera are shown in Fig.4. For comparison, PBMC obtained without physical stress were preincubated with and without IFNγ, then washed and again incubated as indicated (Tab.1). PBMC either obtained during jogging or preincubated with IFNγ, released more N into supernatants; additionally, less further activation could be achieved by IFNγ compared to PBMC obtained before or after exercise. N produc-

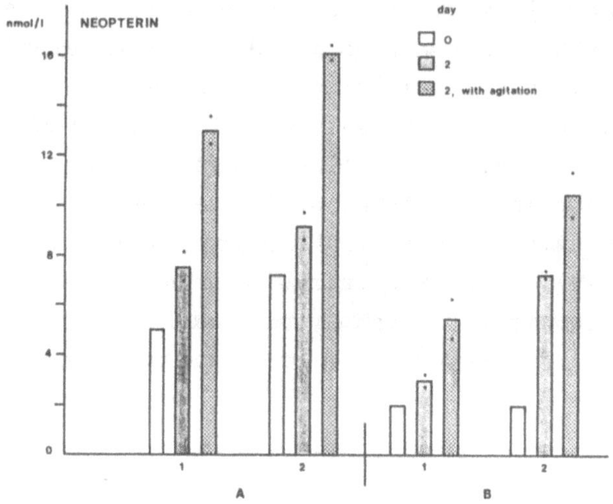

Fig.5. Neopterin in supernatants of cell culture. **A.** heparinized blood; **B.** heparinized blood diluted with medium (1/1) of 2 healthy normals (1 and 2) on day 0, day 2 without or with gentle shaking. Bars represent the mean of 2 experiments, shown as points.

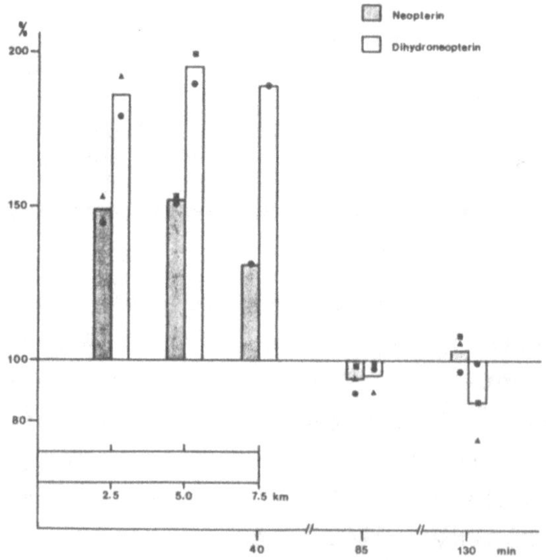

Fig.6. Serum neopterin and dihydroneopterin levels in 3 healthy male subjects during jogging.

tion of blood cells during cell culture increased if incubation was carried out with gentle shaking (Fig.5).
During physical exercise, serum NH_2 levels are more elevated than N levels (Fig.6).

Discussion

Physical exercise can induce changes of the neuroendocrine system resulting in amenorrhoe in women (5) and raised levels of testosteron, prolactin and cortisol in men (6). Increased catecholamine levels and binding of mab T8 and Leu7 to the respective receptors on T lymphocytes were reported during ergometry (2).
Alterations of T-lymphocyte subsets induced by jogging are extensive, leading to T helper/inducer-T suppressor/cytotoxic ratios below 1, as usually found in infectious disease.
Changes of lymphocyte distribution in peripheral blood may be due either to shift of the subset into (increase of $CD8^+$) or out of the circulation (decrease of $CD4^+$), respectively, or to rapid changes of expression of surface antigens in response to mediators. In vitro adenosin decreased numbers of cells which can be labelled with mab OKT4 in 5 minutes and increased numbers of T-lymphocytes labelled with OKT8 (7). Both mechanisms are compatible with the quantitatively different alterations and recovery times of T-lymphocyte subsets.
N is released by monocytes/macrophages upon activation by IFNγ or lipopolysaccharides (8); possible functions of N or NH_2 in the immune system are so far not known (9), but N is an easily measurable parameter of activation of monocytes/macrophages which have a key role in cellular immune response. Dihydroneopterintriphosphat (NH_2P_3) is an intermediate in the biosynthesis of tetrahydrobiopterin (BH_4), the cofactor of aromatic amino acid hydroxylases, which are involved in catecholamine synthesis (10). N was elevated during the first part of the

run to levels usually found in diseases, e.g. viral infections. N levels decreased during later stages of the run, whereas cortisol increased steadily. Both parameters peaked again in the late recovery phase (ca. 3 and 5 hours after the run for cortisol and N, respectively). During the run the NH_2/N ratio in serum was increased, possibly indicating that release of NH_2 from cells was faster than oxidation in blood (Fig.6).

In renal insufficiency, N in serum is elevated because of reduced excretion as well as creatinine (4). During physical exercise serum creatinine may be markedly elevated, but no correlation with N is seen during jogging (Fig.2).

Whether there is a direct relation between elevation of N and alteration of lymphocyte subsets remains to be shown. Corticotropin Releasing Factor (CRF) induced slight elevation of serum N in parallel to the cortisol elevation (4); a causal relation between cortisol or a factor of the pituitary-hypophyseal axis and N elevation remains to be demonstrated. In this respect the putative involvement of BH_4 in the biosynthesis of steroids in the adrenal cortex seems of interest (11).

Cell culture experiments (Fig.4) showed higher N release into supernatants of peripheral blood mononuclear cells (PBMC) from blood obtained during the run compared to PBMC obtained before and after the run. Capability to release N upon stimulation by IFNγ decreased during the run. Results of cell culture experiments could be explained by higher GTP-cyclohydrolase (GTP-CH) activity of PBMC obtained during the run, because of a linear correlation of N production and GTP-CH activity in PBMC (12). Effects of exercise on N release of PBMC could be paralleled by activation of PBMC by IFNγ <u>in vitro</u> (Table 1). Mediators not produced by PBMC can be excluded in both cell culture experiments, thus showing that short time physical exercise can induce PBMC to increase N production <u>in vitro</u>, too. Therefore, the elevated N levels in serum seem to be at least partly derived from peripheral blood cells.

Gentle shaking of cell cultures yielded higher N levels in supernatants (Fig.5). These experiments show the ease of activation of PBMC, the actual agent still has to be identified.

The immune system of healthy individuals responds to physical exercise similar as to infection. This is supported by the observation that physical exercise induced secretion of interleukin-1 by PBMC (13), indicating again activation of monocytes/macrophages. The N elevation as response of monocytes/ macrophages to physical exercise is paralleled by values of other immunological parameters, such as ratio of T-lymphocyte subsets, leukocyte counts and release of polymorphonuclear elastase (14), which are in the range found in infectious or inflammatory diseases.
A possible link between levels of steroids, as sexual hormones and cortisol, and pterins remains to be established. Whether physical exercise, in analogy to the beneficial effects upon the cardiovascular system, exerts a similar training effect on the immune system, needs to be further investigated.

Acknowledgements

We like to thank the jogging volunteers (H.G., D.S., T.W.) as well as Mrs. H. Luczak for secretarial and Mrs. B. Rogalla and K. Dierks for technical assistance.

References

1. Ritchie, A.M.S., J. Oswald, H.S. Micklem et al. 1983. Br.Med.J. 286, 1773-1775.

2. Landmann, R., M. Wesp, and F. Gudat. 1985. Cancer Detection and Prevention 8, 543.

3. Baker, G.H.B., N.A. Byrom, M.S. Irai et al. 1984. Lancet i: 574.

4. Rokos, H., H. Frisius, R. Kunze. 1986. In: Chemistry and Biology of Pteridines 1986 (B.A. Cooper, V.M. Whitehead, eds.). Walter de Gruyter, Berlin, New York, p. 411-414.

5. Jacobs, H.S. 1982. Br.J.Obstet.Gynaecol. 89, 498-500.

6. MacConnie, S.E., A. Barkan, R.M. Lampman et al. 1986. N.Engl.J.Med. 315, 411-417.

7. Birch, R.E. and S.H. Polmar. 1986. Int.J.Immunopharmacol. 8, 329-337.

8. Huber, C., J.R. Bachelor, D. Fuchs et al. 1984. J.Exp.Med. 160, 310-316.

9. Ziegler, I. and H. Rokos. 1986. EOS J.Immunol.Immunopharmacol. 6, 169-177.

10. Nichol, C.A., G.K. Smith, D.S. Duch. 1985. Ann.Rev.Biochem. 54, 729-764.

11. Abou-Donia, M.M., D.S. Duch, C.A. Nichol et al. 1983. Endocrinology 11, 2088-2094.

12. Rokos, K., R. Kunze, W. Lange und M.A. Koch. 1986. Bundesgesundheitsblatt 29, 382-387.

13. Cannon, J.G., M.J. Kluger. 1983. Science 220, 617-619.

14. Schaefer, R.M., K. Kokot, A. Heidland, R. Plass. 1987. N.Engl.J.Med. 316, 223-224.

BIOPTERIN AND NEOPTERIN IN PLASMA AND LYMPHO-MONONUCLEAR BIOPTERIN IN GRAFT VERSUS HOST DISEASE AND INFECTION AFTER BONE MARROW TRANSPLANTATION.

M. Fink, V. Nüssler, K. Demtröder, E. Höniges, H.J. Kolb
III. Med. Klinik der Ludwig Maximilians Universität, Klinikum Großhadern, D-8000 München 70, Germany

I. Ziegler
Gesellschaft für Strahlen- und Umweltforschung, Institut für Experimentelle Hämatologie, D-8000 München 2, Germany

Introduction

Biopterin and neopterin are pterins with established clinical relevance. Biopterin in its tetrahydro-form is a hydroxylation-cofactor (for review see: 1), and recent data also indicate an active participation of tetrahydro-biopterin in immunologic reactions as modulator of interleukin 2 (2,3,16). For neopterin, so far an active biological function is unknown. It is, however, released from macrophages after stimulation with interferon gamma (4) and thus useful for monitoring immune reactions in transplantations and viral infections.
In bone marrow transplantation (BMT), typical complications are infections and graft versus host diseases (GvHD; 10), which lead to a lethal outcome in about 20% of all cases. In a previous publication, we demonstrated not only an elevation of neopterin in plasma, but also an elevation of lympho-mononuclear biopterin in the two weeks before a severe intestinal GvHD with a subsequent fall of the latter 2-3 days before onset of clinical symptoms (9); in vitro studies following this observation resulted in a cooperation of tetrahydro-biopterin with interleukin 2 (2,3,16). On the

Unconjugated Pterins and Related Biogenic Amines

other hand, we also found a continuous elevation of lympho-mononuclear biopterin in a probable cutaneous GvHD (8), so that a different pterin pattern in cutaneous and intestinal GvHD was likely. The further elucidation of these findings was the aim of this study.

Material and methods

Methods employed have already been described in earlier publications (9,11). Venous blood (4 ml) was drawn into sodium citrate (1:10). The first 2 ml of the sample were diluted with PBS (1:1) and then lympho-mononuclear cells were isolated by ficoll-hypaque density gradient centrifugation. The second 2 ml were centrifuged without any admixture and plasma pipetted off for determination of neopterin by RIA (Neopterin-RIAcid Henning Berlin). Only in pat. F.S. (fig. 3), neopterin was determined in the plasma obtained by ficoll-separation and biopterin in the plasma-fraction of the second 2 ml sample.

Total biopterin was determined by means of the crithidia assay without oxidation of the samples (12); thus, a partial loss of the tetrahydro-form of biopterin has to be taken into account (13). The constancy of this loss, however, has been confirmed by an interassay variation of only 6,4% for total biopterin in our laboratory.

Take of the transplanted marrow has been defined as the first day with at least 10^9 leucocytes/l blood. Grading of GvHD has been performed according to (10).

Results

In Pat. H.F., m., 20 yrs., BMT has been performed because of severe aplastic anemia. A mild intestinal and cutaneous GvHD developed on day 9 after BMT and was treated successfully with prednisolone. On days 35-40 after BMT, two different events complicated the clinical course: a wound infection on the right shoulder occurred and ciclosporin A-level dropped because of irregular intake by the patient. This was followed by a simultaneous rise of plasma-neopterin and lympho-mononuclear biopterin with a spontaneous decrease of the latter 2 days before the beginning of a grade III-IV intestinal GvHD (fig.1). A definitive decrease of neopterin, however, has first been achieved 2-4 days after treatment with i.v. prednisolone together with azathioprine. In the following time, diarrhoea slowly receded and the patient could be discharged in good condition.

Patient G.B., f., 31 yrs., has been transplanted in the first remission of an acute myeloblastic leukemia. After transplantation, a grade II cutaneous GvHD occurred and was treated with prednisolone beginning day 16 after BMT; both plasma neopterin and lympho-mononuclear biopterin were

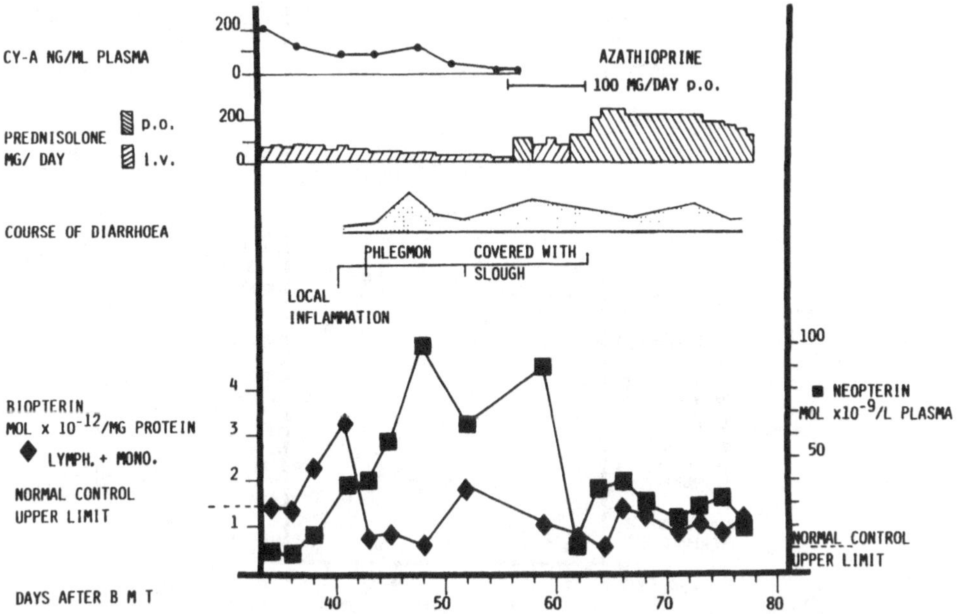

Fig.1. Decrease of lympho-mononuclear biopterin 2-4 days before the beginning of intestinal GvHD with simultaneously occuring wound-infection.

distinctly elevated before treatment and slowly decreased under successful therapy (data not shown). The patient could be discharged on day 30 after BMT, but 5 days later she had to be readmitted because of generalised exanthema and severe diarrhoea as signs of cutaneous and intestinal grade III-IV GvHD. Neopterin and lympho-mononuclear biopterin were highly elevated at this time, but rapidly decreased under i.v. prednisolone and ciclosporine A therapy. Poor absorption of these drugs under oral treatment following day 44 resulted in increasing diarrhoea and a high peak of both pterins on day 49. 2 days later, lympho-mononuclear biopterin decreased to nearly normal values, while plasma-neopterin only dropped after resumption of i.v.-therapy with ciclosporin A and prednisolone (fig.2). Hemolytic-uremic syndrome and aspergillus pneumonia were further complications, which led to the death of the patient on day 86 after transplantation.

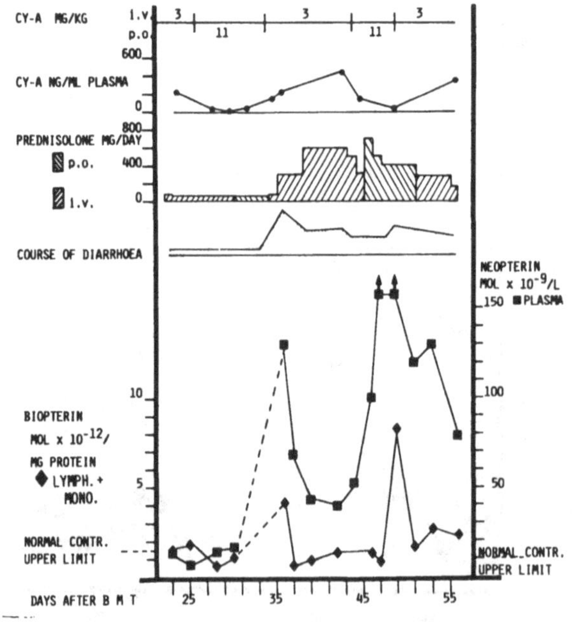

Fig.2. Short peaks of lympho-mononuclear biopterin at the beginning (day 36) and at deterioration (day 49) of a mixed cutaneous and intestinal GvHD.

In patient F. S., m., 17 yrs., BMT has been performed because of a refractory acute lymphoblastic leukemia. In the aplastic phase 11 days after transplantation, the patient developed high temperature and staphylococcus aureus was found in the blood culture. Lympho-mononuclear biopterin exhibited a short rise with a maximum 2 days after the rise of temperature, while plasma biopterin was rising at the same time when lympho-mononuclear biopterin decreased (fig. 3). At the time of defervescence, plasma biopterin decreased one day later than neopterin. In this patient, the take (day 22) was not accompanied by the rise of any pterins.
In pat. M. L., m., 20 yrs., bone marrow was transplanted during the first relapse of an acute lymphoblastic leukemia.

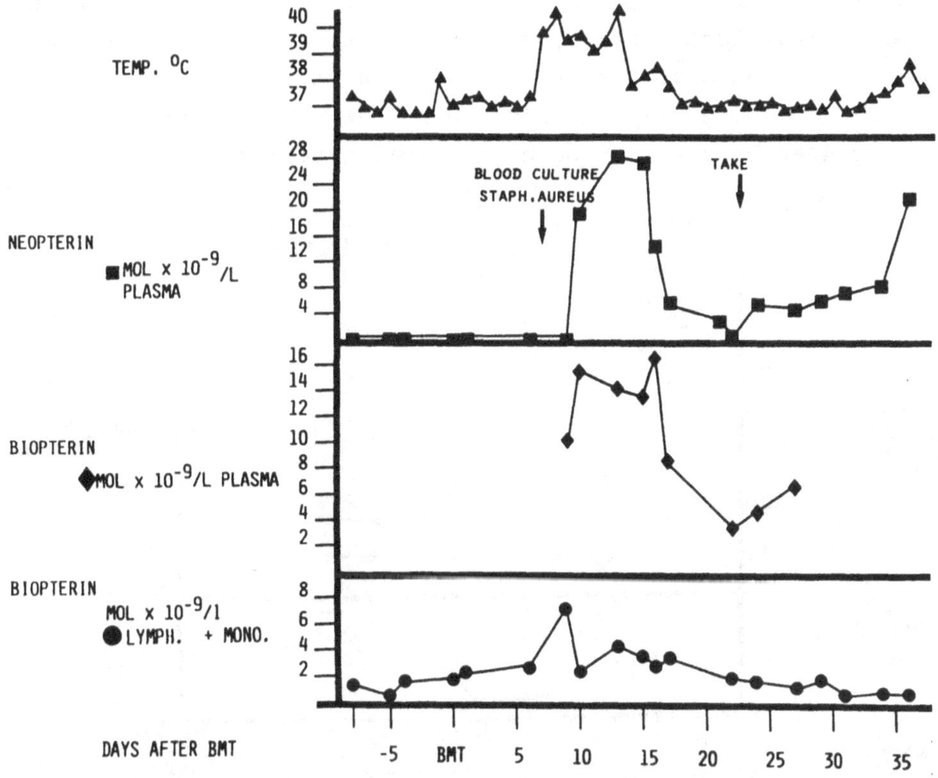

Fig.3. Simultaneous decrease of lympho-mononuclear biopterin and increase of plasma biopterin in the early phase of septicemia after BMT.

On day 7, exanthema occurred, and on day 9, cutaneous GvHD grade II-III has been proven histologically. In this patient methotrexate, which interacts with biopterin metabolism (9), has not been given for prophylaxis of GvHD. Doubling ciclosporin A-dose on day 8 had no therapeutic effect, but treatment with prednisolone was followed by improvement of GvHD within a few days. The course of GvHD was accompanied by a simultaneous rise of plasma neopterin and lympho-mononuclear biopterin and a subsequent decrease under prednisolone treatment (fig.4).
Highly elevated levels of lympho-mononuclear biopterin ($20x10^{-12}$mol/mg protein), relatively low plasma biopterin values ($13x10^{-9}$mol/l) and high neopterin levels ($80x10^{-9}$mol/l) have also been observed in a therapy-resistant cutaneous GvHD. A similar pterin pattern has been found in the late stage of a pneumocystis carinii pneumonia where plasma biopterin decreased and lympho-mononuclear biopterin rose at clinical deterioration.

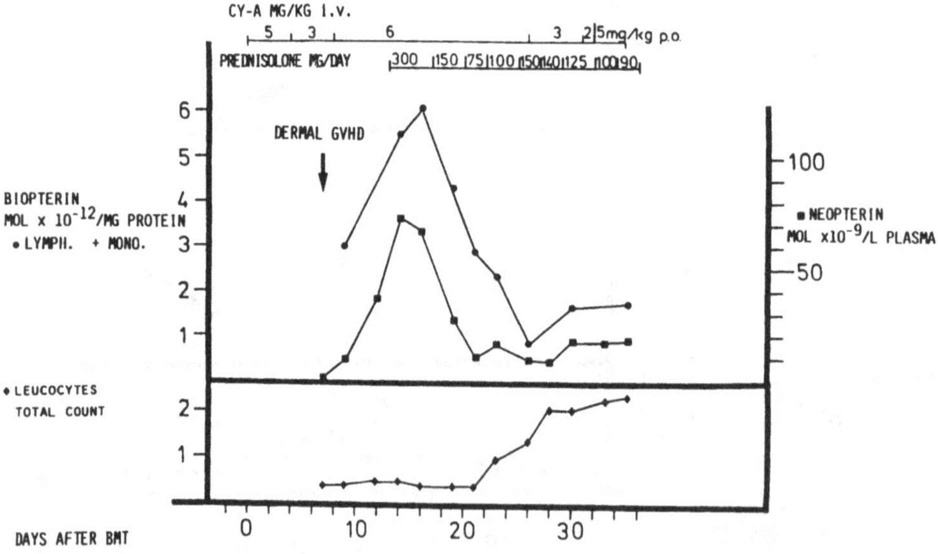

Fig.4. Simultaneous rise of lympho-mononuclear biopterin and plasma neopterin in a dermal GvHD.

Discussion

Tetrahydro-biopterin as hydroxylation-cofactor is the pterin with the best established functional relevance. However, its active participation in immunologic reactions is controversely discussed. On the one hand, tetrahydro-biopterin has been found to participate in lymphocyte stimulation (6,14,15) and to enhance interleukin 2 activity (2,3,16). Furthermore, in 3 of 3 children with atypical phenylketonuria, caused by tetrahydro-biopterin deficiency, a lethal pneumonia has been reported (20). This leads to the question, whether the respiratory infections caused by swallowing difficulties in these children have been enhanced by a defect immune system. On the other hand, in a patient with a deficiency of GTP-cyclohydrolase I, only low levels of biopterin were found in the plasma, while his immunoreactive cells were unable to produce pterins. Interestingly, gross abnormalities in the patient's immune system could not be observed (17).

In bone marrow transplantation, complications involving the immune system such as infections and graft versus host-diseases are frequent (10). Therefore, this procedure is suited for studying biopterin in lympho-mononuclear cells in vivo at the beginning of immunologic reactions; in vitro, this is the period, during which tetrahydro-biopterin modulates interleukin 2 activity (16). Infection and GvHD are interrelated in BMT patients (18). Furthermore, in GvHD a lack of interleukin 2 (18) and interleukin 2 receptor dysfunction (19) has been demonstrated. These special characteristics have to be considered in interpreting our results.

In our preceeding study, an important finding was the spontaneous decrease of the previously elevated lympho-mononuclear biopterin 2-3 days before the beginning of a grade IV intestinal GvHD (9). This pattern has exactly been found again in our patient H.F. (fig.1) and thus may be typical for intestinal GvHD. Plasma neopterin in both cases linearly increased beginning 2 weeks (9) resp. 8 days (fig. 1) before clinical manifestation of GvHD and decreased only after high dose prednisolone treatment (fig.1 after changing from p.o.- to i.v.-treatment) of 2-3 days duration. This neopterin pattern in GvHD is identical with that found in urine by other authors (7).

In a mixed cutaneous and intestinal GvHD, however, only brief periods of elevated lympho-mononuclear biopterin were found at the beginning or at deterioration of clinical symptoms (fig.2). A similar peak occurred in a patient, who suffered from a septicemia in the aplastic phase after BMT (fig.3). In this case, a rise of plasma biopterin was found simultaneously with the decrease of lympho-mononuclear biopterin. This pattern is consistent with a release of biopterin from lympho-mononuclear cells with beginning immune reaction, which has also been demonstrated in vitro (5). Biopterin-levels in plasma, however, were about 10 times lower than those demonstrated to be optimal in vitro (2,16), which may indicate a local activity of the released tetrahydro-biopterin.

In a cutaneous GvHD, which occurred before take of the transplanted marrow, (fig.4) another pterin pattern has been observed: lympho-mononuclear biopterin and plasma neopterin rose and then decreased simultaneously under prednisolone therapy. An additive influence of the beginning take can not be excluded in this case; it is, however, less probable because of the interval of one week between the maximum of pterins and the take. Additionaly, a similar pattern has been found for lympho-mononuclear biopterin in the early cutaneous GvHD of our patient G.B. (data not shown) and in a severe scrotal inflammation, which was probably due to a cutaneous GvHD (8). A single determination in a therapy-resistent cutaneous GvHD also resulted in highly elevated lympho-mononuclear biopterin while plasma biopterin was only mildly elevated.

In a pilot study on pterins after hepatitis vaccination, a decrease of lympho-mononuclear biopterin and a rise of plasma neopterin were found two weeks after vaccination. Fatigue and temperature in one of these patients were correlated with elevated lympho-mononuclear biopterin (9). These data indicate, that two pterin-patterns are possible in immune reactions: the first includes accumulation of lympho-mononuclear biopterin with subsequent release before or at the beginning of clinical symptoms (intestinal GvHD, septicemia), while the second is characterized by a continuous accumulation of lympho-mononuclear biopterin with only mildly elevated plasma biopterin levels (cutaneous GvHD). These two patterns are probably not restricted to BMT-patients.

Acknowledgement

We thank the nursing staff of our transplantation unit F 12 for cooperation. This study was supported by grant no. Fi 324/2-2 from the Deutsche Forschungsgemeinschaft.

References

1. Nichol, C.A., G.K. Smith, D.S. Duch. 1985. Ann.Rev. Biochem. 54, 729.

2. Ziegler, I., U. Schwulera, H.-H. Sonneborn, W.J.P. Müller. 1985. Naturwissenschaften 72, 330.

3. Ziegler, I., H. Rokos. 1986. EOS rivista di immunologia ed immunofarmacologia 6, 169.

4. Huber, Ch., J.R. Batchelor, D. Fuchs, A. Hausen, A. Lang, D. Niederwieser, G. Reibnegger, P. Swetly, J. T. Troppmair, H. Wachter. 1984. J.Exp.Med. 160, 310.

5. Ziegler, I. 1985. Biochem.Biophys.Res.Comm.132, 404.

6. Ziegler, I., U. Hamm, J. Berndt. 1983. Cancer Res. 43, 5356.

7. Niederwieser, D., Ch. Huber, A. Gratwohl, P . Bannert, D. Fuchs, A. Hausen, G. Reibnegger, B. Speck, H. Wachter. 1984. Transplantation 38, 497.

8. Fink, M., U. Jehn, W. Wilmanns, H. Rokos, I. Ziegler. 1983. In: Biochemical and Clinical Aspects of Pteridines, Vol. 2 (H.Ch. Curtius, W. Pfleiderer and H. Wachter, eds.). W. de Gruyter, Berlin, p.223.

9. Fink, M., I. Ziegler, H. Rokos. 1985. In: Biochemical and Clinical Aspects of Pteridines, Vol. 4 (H. Wachter, H.Ch. Curtius and W. Pfleiderer, eds.). W. de Gruyter, Berlin, p. 387.

10. Thomas, E.D., R.A. Clift, A. Fefer, F.L. Johnson, P.E. Neiman, K.G. Lerner, H. Glucksberg, C.D. Buchner. 1975. N.Engl.J.Med. 292, 832.

11. Ziegler, I., H.J. Kolb, U. Bodenberger, W. Wilmanns. 1982. Blut 44, 261.

12. Baker, H., O. Frank, A. Shapiro, S.H. Hutner. 1980. In: Methods in Enzymology, Vol. 66 (D.B. McCormick and L.D. Wright, eds). Academic Press New York, p. 490.

13. Milstien, S. 1983. In: Biochemical and Clinical Aspects of Pteridines, Vol. 2 (H.Ch. Curtius, W. Pfleiderer and H. Wachter, eds.). W. de Gruyter, Berlin, p. 65.

14. Schoedon, G, A. Niederwieser, J. Troppmair, Ch. Huber. 1985. In: Biochemical and Clinical Aspects of Pteridines, Vol. 4 (H. Wachter, H.Ch. Curtius and W. Pfleiderer, eds.). W. de Gruyter, Berlin, p. 369.

15. Webber, S., D.L. Jaye. 1986. In: Chemistry and Biology of Pteridines (B.A. Cooper and V.M. Whitehead, eds.). W. de Gruyter, Berlin, p. 235.

16. Ziegler, I., U. Schwulera, J. Ellwart. 1986. Exp.Cell.Res. 167, 531.

17. Niederwieser, A, N. Blau, M. Wang, P. Joller, M. Atares, J.Cardesa-Garcia. 1984. Eur.J.Pediatr.141, 208.

18. Witherspoon, R.P., L.G. Lum,L. Storb. 1984. Sem.Hematol. 21, 2.

19. Joseph, L.J., T. Iwasaki, T.R. Malek, G.M. Shearer. 1985. J. Immunol.135, 1846.

20. Smith,I., B.E. Clayton, O.H. Wolff. 1975. Lancet, 1108.

EFFECTS OF PIRENZEPINE, ACICLOVIR AND TRIMETHOPRIM ON PTERINS IN BLOOD CELLS AND PLASMA.

M.Fink, V. Nüssler, K. Demtröder, E. Höniges
III. Med. Klinik der Ludwig Maximilians Universität, Klinikum Großhadern, D-8000 München 70, Germany

M. Goldberg
Institut für Tierphysiologie der Ludwig Maximilians Universität, D-8000 München 22, Germany

Introduction

In clinical studies, a rise of total biopterin in serum (1) and blood cells (2) under treatment with folic acid antagonists has been reported. In vitro, methotrexate preferably led to an increase of oxidized cellular biopterin and dihydro-biopterin (3,4). Influence of other drugs in therapeutic concentrations on pterin metabolism have hardly been investigated so far.

In a previous study, we found a rise of lympho-mononuclear biopterin and plasma neopterin in 2 patients after BMT (bone marrow transplantation) after withdrawal of pirenzepine, aciclovir and treatment with folic acid antagonists (2). In one of these cases, the alterations of pterin levels were accompanied by a septic shock (klebsiella pneumoniae). This reaction was unexpected because of an already established take with $3,5x10^9$ leucocytes/l blood before clinical deterioration. Thus, the question rose, if a pharmacological interaction with pterin metabolism had led to promotion of this septic shock. To analyse the influence of these drugs on pterins was the aim of this study.

Unconjugated Pterins and Related Biogenic Amines

Material and Methods

Methods employed have already been described in earlier publications (2,5). Venous blood (4 ml) was drawn into sodium citrate (1:10). The first 2 ml of the sample were diluted with PBS (1:1) and lympho-mononuclear cells were isolated by Ficoll-Hypaque density gradient centrifugation. The second 2 ml were centrifuged without any addition and plasma was used for the determination of neopterin (Neopterin-RIAcid, Henning Berlin) and biopterin.
Total biopterin was determined by means of crithidia assay without oxidation of the samples (6). Thus, a partial loss of the tetrahydro-form of biopterin has to be taken into account (7). The constancy of this loss, however, has been confirmed by an interassay variation of only 6.4% for total biopterin in our laboratory.
Pterins in urine were acidified and oxidized (8). Oxidized pterins were purified and determined by HPLC (9). Urinary creatinine was measured by using the creatinine kit from Sigma.
The drugs were given in the usual oral dose to probands; in one patient, aciclovir has been administered in the usual i.v. dose.

Results

Five probands have daily received 2x50 mg pirenzepine per os for four days (fig.1). One day after the beginning of treatment, lympho-mononuclear biopterin levels dropped down signicifantly, while plasma neopterin showed a tendency to rise. In the days after withdrawal of the drug, the increasing tendency of lympho-mononuclear biopterin was interrupted only on day 2, while plasma biopterin rose significantly on this day after a significant decrease on the day before. In the next 2 days lympho-mononuclear

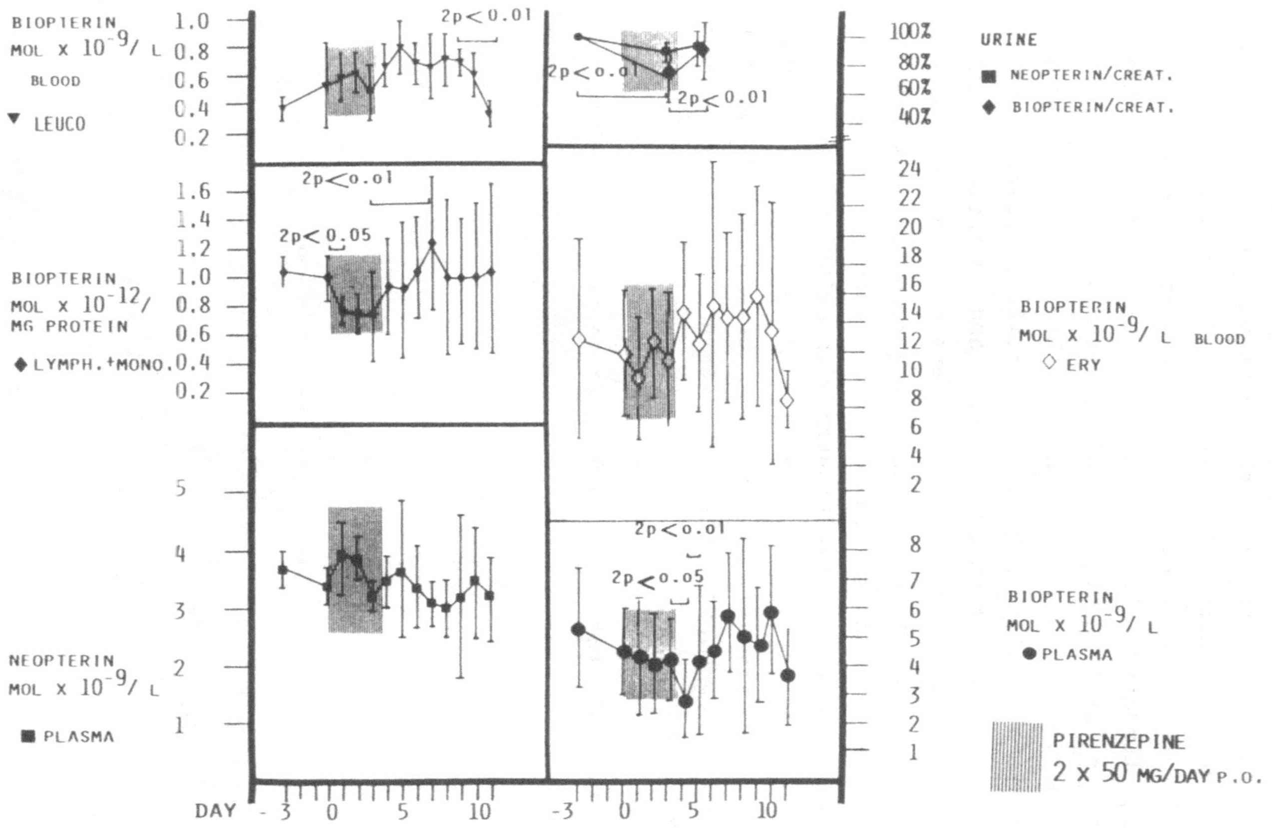

Fig.1. Effect of pirenzepine on biopterin and neopterin in different cells and body fluids in 5 probands (neopterin=4).

biopterin further ascended, the value on the last day of treatment being significantly different to the 4th day after withdrawal. In this period plasma neopterin exhibited a decreasing tendency. On day 6-8 after treatment, biopterin decreased in erythrocytes and significantly in leucocytes. Data concerning the alteration of the lympho-mononuclear biopterin were paralleled by urinary biopterin levels, where a significant decrease under treatment and an increase after treatment were determined. One test person was suffering from grippal infection; his neopterin values rose to a maximum of $37x10^{-9}$mol/l plasma and are therefore excluded in fig.1.
In a patient, who had been treated with 3x250 mg aciclovir i.v./day because of therapy-resistant fever in the aplastic phase of acute myeloblastic leukemia (herpes infection not proved), lympho-mononuclear biopterin decreased under aciclovir therapy from $20x10^{-12}$ to $1x10^{-12}$mol/mg protein within 6 days. For further examination of this effect, aciclovir was given to 5 probands in a daily dose of

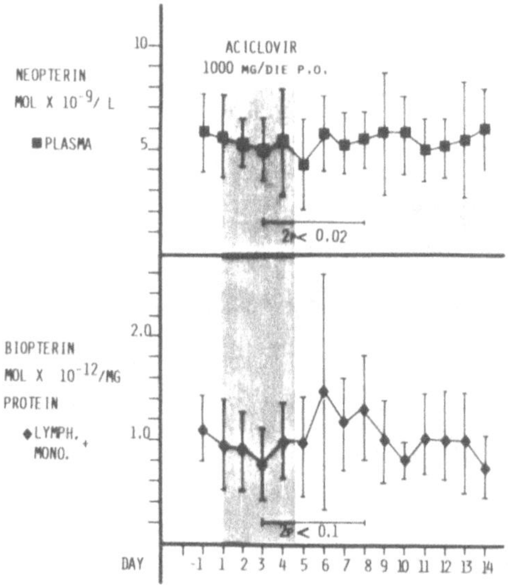

Fig.2. Effect of aciclovir on lympho-mononuclear biopterin and plasma neopterin in 5 healthy test persons.

1000 mg p.o. (fig.2). So far, only lympho-mononuclear biopterin and plasma neopterin have been determined in this experiment. During the observation of the individual follow-ups, some rebound responses seemed to become evident on different days after discontinuation of treatment. In totally, however, no clear changes in plasma neopterin and lympho-mononuclear biopterin were observed.

A combination of aciclovir and pirenzepine was given one healthy test person for 5 days in the same oral dose as mentioned above for the single drugs (fig.3). Initial plasma neopterin values were remarkably low in this proband and exhibited no clear alterations under treatment. Beginning 3 days after withdrawal of the drugs, plasma neopterin exhibited high peaks in the evening values, whereas morning

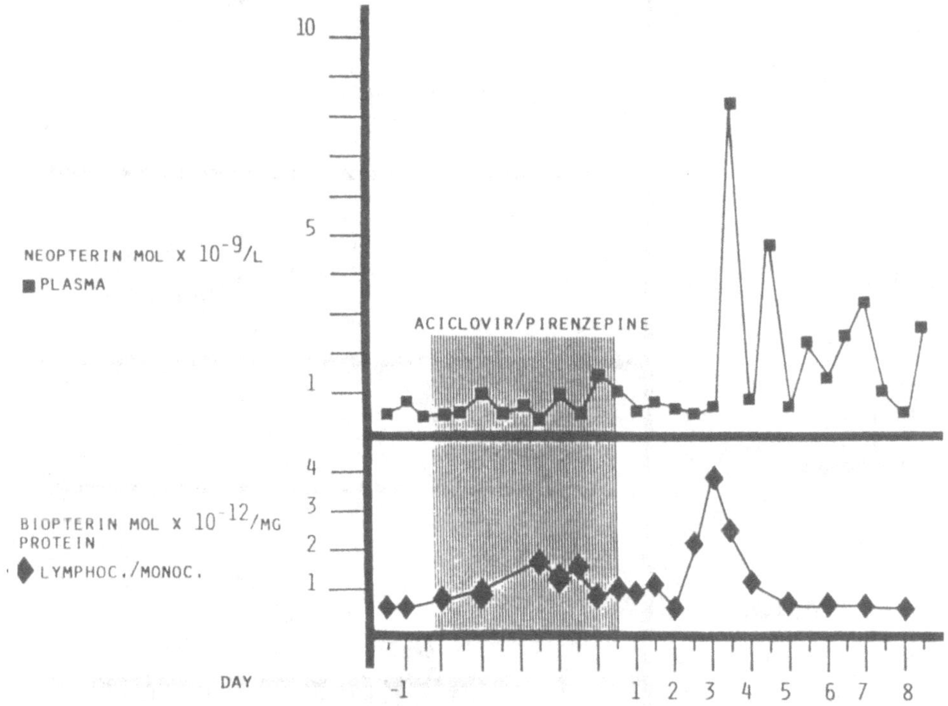

Fig.3. Effect of aciclovir and pirenzepine on pterins in a proband (note low initial plasma neopterin values).

values remained low. In lympho-mononuclear biopterin, a weak rise in the first days was followed by a decrease still under treatment; 2.5 days after discontinuation of the drug, lympho-mononuclear biopterin rose to a marked peak value and reverted to normal 24 hours later. Another proband with higher plasma neopterin values (in normal range) exhibited a similar but less marked response of both pterins after withdrawal of these drugs (data not shown).

Trimethoprim has been applicated for 13 days in the usual dose of 3x100 mg/day p.o. to a healthy test person (fig.4). Biopterin exhibited a marked elevation in all determined cells immediately after the beginning treatment and decreased some days later. The initial elevation lasted in the leuco-

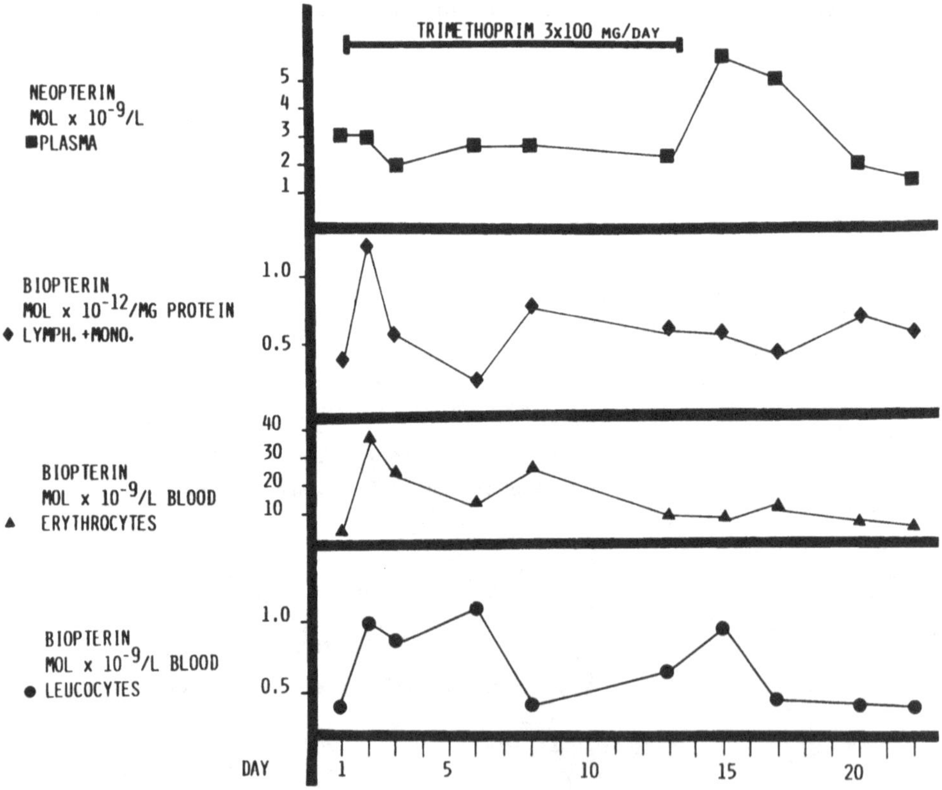

Fig.4. Effect of trimethoprim on pterins in a healthy test person.

cytes for a longer period than in the other fractions measured. After withdrawal of trimethoprim a rise was observed in plasma neopterin and leucocytic biopterin.

Discussion

The reason for this study was the question, whether a septic shock in a patient after BMT (2) was enhanced by pharmacological interaction with the pterin system. This patient was given prednisolone continuously (not shown in: 2), whereas treatment with folic acid antagonists, pirenzepine and aciclovir have been altered in the days before septic shock.

Pirenzepine is a gastric secretion inhibitor (blocking muscarinic receptors) without structural similarity to the pterin ring system. When studying the influence of pirenzepine on pterins, lympho-mononuclear biopterin exhibited a significant decrease under treatment with a rebound response in the days after discontinuation of the drug (fig.1). The biopterin pattern in lympho-mononuclear cells and plasma in the first days after withdrawal of pirenzepine suggests, that biopterin is first of all restored in the cells and excreted to a lower extent into the plasma at this time. Urinary biopterin determined by HPLC also decreased significantly under treatment, indicating a decrease of biopterin production of the whole organism with subsequent rise after withrawal of the drug. Leucocytic biopterin rose in the first 2 days after withdrawal of pirenzepine and significantly decreased 4-6 days later. This may lead to the hypothesis, that biopterin metabolism is different in granulocytes and lymphocytes. Plasma neopterin showed an opposite action to the lympho-mononuclear biopterin in the first days of treatment, which is consistent with a decreased feed back inhibition of GTP-cyclohydrolase by lower tetrahydro-biopterin levels (14). This would, however,

require a feed back among different lympho-mononuclear cell types, because neopterin is produced by macrophages and biopterin by lymphocytes (15).
Aciclovir is an anti-herpetic substance with a partially identical structure as guanosine; additionally, aciclovir is triphosphorylated (10) in herpes-infected cells and may thus interact with guanosinetriphosphat, the precursor of pterin metabolism (11). After finding a marked decrease in lympho-mononuclear biopterin in a patient under i.v. treatment with aciclovir, a study of the influence of orally given aciclovir on pterins has been carried out; it has to be mentioned here, that oral application of aciclovir results in a resorption of only 20% (12). Aciclovir in probands did not lead to a decrease of pterins under treatment but to a mild increase in the days after withdrawal (fig.2). The combination of aciclovir and pirenzepine in one test person, however, was followed by a marked rise of pterins 3 days after the end of treatment (fig.3); this was also the time of septic shock in the BMT-patient (2) and supports the hypothesis of an active participation of pterins in this case.
While in different studies a rise of total biopterin has been demonstrated under influence of folic acid antagonists (1,3,4), we found in a previous study a rise with subsequent decrease of biopterin under treatment with trimethoprim/sulfamethoxazol in human leucocytes (2). In our present investigation of blood pterins under longterm-treatment with trimethoprim alone, we confirmed this finding (fig.4) which is consistent with the induction of dihydrofolate reductase in leucocytes after 4-8 days treatment with folic acid antagonists (13). Additionally, a rise of plasma neopterin and leucocytic biopterin in the days after withdrawing the drug has been observed. The mechanism of this increase is not clear so far, but it is conspicuous, that this increase also appeared a few days after withdrawal of trimethoprim in the same period of time, as the septic shock occured. An inter-relation of pterin metabolism and

septic shock is furthermore likely, because in another study highly elevated neopterin levels have been described before occurrence of this complication (16).
In summary, we are still far from understanding the interrelations of the pterin system and septic complications. The modulation of interleukin 2 activity by tetrahydrobiopterin (17) in connection with the findings of Strohmaier (16) and our data, however, indicate, that such interrelations are probable and should further be investigated.

Acknowledgement

This study was supported by grant Fi 324/2-2 from the Deutsche Forschungsgemeinschaft.

References

1. Leeming, R.J., J.A. Blair, V. Melikan, D.J. O'Gorman. 1976. J.Clin.Path.29,444.

2. Fink, M., I. Ziegler, H. Rokos. 1985. In: Biochemical and Clinical Aspects of Pteridines, Vol. 4 (H. Wachter, H.Ch. Curtius and W. Pfleiderer, eds.). W. de Gruyter, Berlin, p. 387.

3. Culvenor, A.J., L.P. Miller, R.A. Levine, W. Lovenberg. 1984. J.Neurochem. 42, 1707.

4. Duch, D.S., Ch-L. Lee, M.P. Edelstein, Ch.A. Nichol. 1983. Mol.Pharmacol. 24, 103.

5. Ziegler,I., H.J. Kolb, U. Bodenberger, W. Wilmanns. 1982. Blut 44, 261.

6. Baker, H., O. Frank, A. Shapiro, S.H. Hutner, 1980. In: Methods in Enzymologie, Vol. 66 (D.B. McCormick and L.D. Wright, eds.). W. de Gruyter, Berlin, p. 65.

7. Milstien, S. 1983. In: Biochemical and Clinical Aspects of Pteridines, Vol. 2 (H.Ch. Curtius, W. Pfleiderer and H. Wachter, eds.). W. de Gruyter, Berlin, p. 65.

8. Fukushima, T., J.C. Nixon. 1980. Anal.Biochem. 102, 176.

9. Goldberg, M., F. Gassner, M. Merkenschlager. 1987. J.Vet.Med.J, in press.

10. Miller, W.H., R.L. Miller. 1982. Biochem.Pharmacol. 31, 3879.

11. Nichol, C.A., G.K. Smith, D.S. Duch. 1985. Ann. Rev. Biochem. 54, 729.

12. de Miranda, P., M.R.Blum. 1983. J. Antimicrob. Chemother. 12, Suppl. B, 29.

13. Martelli, M.F., F. Aversa, P. Rambotti, A. Veldari. 1978. Enzyme 23, 154.

14. Bellahsene, Z., J.L. Dhont, J.P. Farriaux. 1984. Biochem. J. 217, 59.

15. Schoedon, G., A. Niederwieser, J. Troppmair, Ch. Huber. 1985. In: Biochemical and Clinical Aspects of Pteridines, Vol. 4 (H. Wachter, H.Ch. Curtius and W. Pfleiderer, eds.). W. de Gruyter, Berlin, p. 369.

16. Strohmaier, W., H. Redl, G. Schlag, D. Inthorn. 1987. Crit.Care Med., in press.

17. Ziegler, I., U. Schwulera, J. Ellwart. 1986. Exp.Cell Res. 167, 531.

Section E: Tetrahydrobiopterin Deficiencies

HETEROGENEITY OF TETRAHYDROBIOPTERIN DEFICIENCY

Jean-Louis DHONDT

Laboratoire de Biochimie, Faculté Libre de Médecine,
56 rue du Port, F-59046 LILLE Cédex, France

INTRODUCTION

Even if few problems still exist concerning the physio-pathology of phenylketonuria (PKU) and the follow-up of PKU patients, it is now evident that the neonatal screening of the disease and its dietary treatment are efficient to prevent a severe mental retardation. Thus, about 10 years ago, when evidence of variant forms of PKU due to cofactor deficiency had been brought to light, physicians who were taking care of PKU children became afraid to not be able to recognize such variants which were first described as hyperphenylalaninemic babies who, despite early dietary control of blood phenylalanine concentration, developed a severe and progressive neurologic disease (1,51).

Fortunately, within the next few years, it was established that these atypical PKU cases were quite easy to detect.

Since that time, the list of cases has grown in numbers, and insofar as the precise metabolic pathway of tetrahydrobiopterin synthesis is known, three enzyme defects have been recognised. Biopterin-synthetase (BS) deficient patients were first described by Bartholomé et al (1), Smith et al (51) and Rey et al (45). Kaufman et al (28) and Niederwieser et al (39) published that in such patients abnormally low amounts of biopterin and high amounts of neopterin are excreted. The enzyme defect (phosphate eliminating enzyme, recently named 6-pyruvoyl-tetrahydropterin-synthetase) was then localised by Niederwieser et al (44). Deficiency in Dihydropteridine-reductase (DHPR), an enzyme ensuring the salvage of the active form of the cofactor, was first described by Kaufman et al (27). Evidence that the patients excrete only oxidized forms of biopterin was reported by Milstien et al (38). A third defect on the main route of tetrahydrobiopterin synthesis was described by Niederwieser et al (40) and proved to be due to a lack of GTP-cyclohydrolase (GTP-ch) activity (42). All these defects lead to a lack of the tetrahydrobiopterin-dependent hydroxylase activities (phenylalanine-hydroxylase, tyrosine-hydroxylase and tryptophan-hydroxylase) and therefore to an impaired biosynthesis of the neurotransmitters dopamine, norepinephrine and serotonine, in addition to the hyperphenylalaninemia.

As in other inborn errors of metabolism, the practice of a systematic search for a dysmetabolic situation led to the description of new forms of the disease which introduce difficulties

Unconjugated Pterins and Related Biogenic Amines

in diagnostic and treatment decisions. The aim of the present work is to present data demonstrating the great variability of clinical, biochemical and therapeutical aspects of THB deficiencies.

CLINICAL HETEROGENEITY

With cases already published in the literature and those collected in the "International Register" (15,16) more than 100 patients, born between 1961 and 1986, are presently recorded. The nomenclature is still evolving because of the variability of clinical and biochemical findings. Consequently, the classification presently used is an approximate consensus. Especially, some patients with biochemical evidence of impaired THB metabolism but who never presented clinical symptoms have been reported. These cases will be separately discussed. The term of "typical" form used in this paper only refer to cases with neurological signs, whatever their importance. It is also important to stress that the conclusions of the present analysis have to be tempered by the fact that clinical information which are available from the literature or the register are often sparse, mainly for the older cases which have a long story of mental retardation.

Table I : Classification of tetrahydrobiopterin deficient patients

defect	GTP-ch	DHPR	Biopterin-synthetase (BS)		
			typical	partial	peripheral
nb of patients	4	38	44	7	3
nb of families	4	32	36	6	3
consanguinity *	1/2	12/24	4/19	0/4	-
sex ratio	1F/2M	14F/19M	14F/20M	3F/2M	1F

* ratio per families

Except for older cases, most of the patients have been neonatally screened for hyperphenylalaninemia and then recognised because they developed abnormal neurological signs despite early dietary treatment. However, an increasing number of patients is now detected during the neonatal period since systematic investigation of THB metabolism become usual in all hyperphenylalaninemic babies. In general, patients have been noticed to deviate from normal development at about 4 months of age. Birth was generally uneventful, except a higher incidence of prematurity and lower birth weight in typical BS deficiency (53). Nevertheless, careful examination can reveal some minor neurological signs at birth (hypotonia, feeding difficulties...) and the figures reported in Table II are probably underestimated.

Table II : Clinical status at birth

	birth weight	prematurity	clinical signs
GTP-ch	2600 - 3500 (n=3)	?	1/3
typical BS	2869 ± 599 (n=24)*	5/19	8/20
atypical BS **	3528 ± 257 (n=5)	0/3	0/8
DHPR	3320 ± 471 (n=27)	0/21	0/25

* $p < 0.005$ typical BS vs DHPR
** partial and peripheral forms

No difference in the clinical presentation emerged when comparing patients according to the etiology of the tetrahydrobiopterin deficiency. The most frequent findings (Table III) were mental and motor regression, along with convulsions, which often were myoclonic, and disturbances of tone and posture, including truncal hypotonia, hypertonia of the limbs, spastic quadriplegia. Microcephaly is frequently reported. Drowsiness, irritability, abnormal movements, episodes of hyperthermia, hypersalivation and swallowing difficulties also seemed frequent. Fatal outcome has been observed in 6/36 BS, 8/33 DHPR and 1/3 GTP-ch.

However, despite this general picture of a severe disease, a wide spectrum of symptoms is observed. For example some untreated patients never had convulsions or only 1 or 2 fits whereas some others were badly controlled with a heavy antiepileptic therapy.

Table III: Summary of clinical presentation in typical forms

	BS (n= 25)		DHPR (n= 31)	
mental retardation	25		31	
convulsions, myoclonia	14	56%	17	55%
trunc hypotonia, limb hypertonia	18	72%	18	58%
hyperreflexia, spasticity	14	56%	6	19%
hypersalivation	4		4	
swallowing difficulties	2		9	
microcephaly *	7	28%	7	23%
abnormal movement, ataxia	2		2	

* head circumference below the mean for age minus 2 SD

BIOCHEMICAL HETEROGENEITY

Several methods have been proposed for the biochemical diagnosis and classification of tetrahydrobiopterin deficiency. However, data also show a great variability.

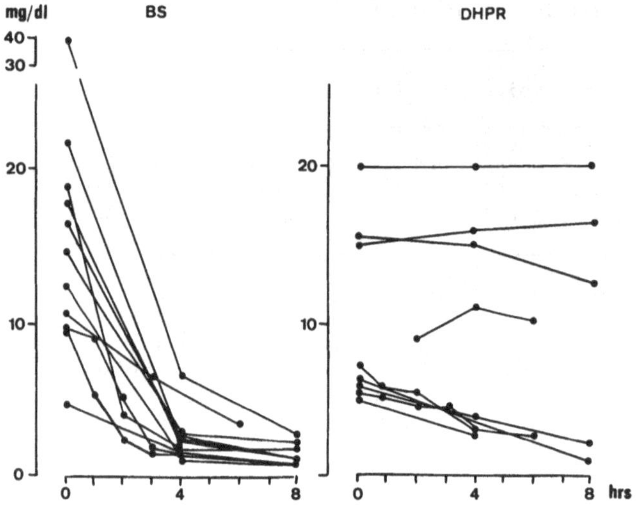

Figure 1 : Response of blood phenylalanine (mg/dl) to oral administration of THB in patients with BS and DHPR deficiency

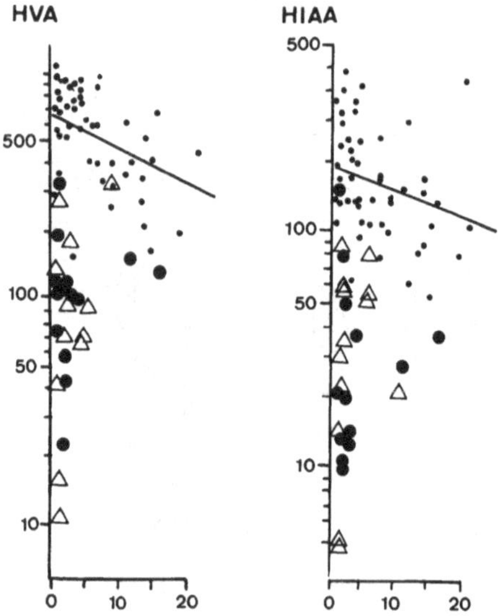

Figure 2 : Amine metabolites concentration (nmol/l) in CSF from △ typical BS and ● DHPR deficient patients (• control values according to Seifert (49))

Neonatal blood phenylalanine levels were usually high; however, 8 out of 44 cases (18%) were near the cut-off value (4 mg/dl). Breast feeding may sometimes explain the low blood phenylalanine, since human milk provides low protein intake and some tetrahydrobiopterin (12). However, when the blood phenylalanine concentration in the screening sample was below 10 mg/dl, the value in the control sample was found again below 10 mg/dl only in 2 cases (1/9 BS and 1/7 DHPR), but above 20 mg/dl in 12 out of the 16. In one case (4), the phenylalanine level remained at 4 mg/dl or below during the first 6 months of life, then the introduction of a higher protein intake led to a rapid increase of phenylalaninemia and to the appearance of neurological signs.

The tetrahydrobiopterin loading test should be ideal for routine detection of THB deficiency (10,11). It theoretically normalises blood phenylalanine levels within 4 to 8 hours by supplying the missing cofactor for phenylalanine hydroxy-lation. The graphs of the available data (figure 1) show marked differences between the slopes obtained in the two groups BS and DHPR patients. However, the value of the phenylalanine before the load has to be taken in account. Blood phenylalanine needs to be elevated for this test to be interpretable; when blood phenylalanine is low even on a normal diet a combined THB-phenylalanine loading test can be used (14,23). The second parameter which can explain the variable response is the THB dose. Initially, loads have been performed with 2 mg THB/kg, but recently a dose of 7.5 mg/kg has been recommended. Nevertheless the load has been shown definitely inefficient in some DHPR deficient patients. For one of them, 3 loads have been performed without significant fall of blood phenylalanine (32). These results when compared to the brisk fall in phenylalanine to normal in BS deficient patients can be explained by the need of a 1:1 molar ratio of THB vs phenylalanine for the hydroxylation process in absence of DHPR.

The measurement of neurotransmitter metabolites in CSF (figure 2) confirms the impaired monoamine neurotransmission. However, variations in the intensity of the neurotransmitter deficiency is observed; in some cases the values are just at the lower limit of the normal range. In regard to the different methods which have been used, the important inter-individual variations, and the age-dependent changes in controls, it is rather difficult to determine which neurotransmission (serotoninergic or dopaminergic) is the most affected in patients.

HETEROGENEITY OF DHPR DEFICIENCY

DHPR deficiency appears the most homogenous group on a clinical point of view. The significance of low DHPR activities in patients with varying degrees of neurological and mental deficiency remains to be clarified (47). Few DHPR patients have been reported to have residual DHPR activity which was linked to less severe clinical symptoms or a better response to the specific treatment, no definite evidence of such assumption is available. On the contrary other studies have shown different molecular defects in patients with similar clinical presentation (19).

Cotton et al (9) have recently reported that THB-unresponsive DHPR deficient patients have in their cells residual mutant DHPR molecules (CRM positive, 3 cases), whereas THB-responsive patients are CRM negative (4 cases).

HETEROGENEITY OF BIOPTERIN-SYNTHETASE DEFICIENCY

Even in "typical" BS deficiency, variation in the intensity of the block may exist, for example *in vivo* phenylalanine-hydroxylation activity of 0.6% and 2.3% have been reported (15). But the most important information came from the description of atypical forms without clinical symptoms and which do not require a specific treatment. However, their distinction is not easy.

Partial and peripheral forms:

The partial form is well documented (table IV). Five of the cases have been discovered by a reinvestigation of their mild hyperphenylalaninemia, and two by neonatal screening for THB deficiency. One patient (DC) was mentally retarded, but its hyperphenylalaninemia has not been controlled during infancy, the sibling (DE) was clinically normal (46). The patient HVJ (23) presented focal spike activity on the EEG trace. Patients TH and TS (13,14) were considered as presenting all the biochemical criteria of a typical BS deficiency: low biopterin level in urine, positive THB loading test. The two patients have been treated for 2 years with neurotransmitter precursors; then the finding of a high dietary phenylalanine tolerance, the normal levels of CSF neurotransmitter metabolites after the therapy has been stopped for 2 weeks, and the normal mental and motor development without treatment since 3 years led to classify these patient as partial forms.

The term of peripheral form has been used in two cases (table V) with abnormal pteridine profile in urine whereas CSF cofactor and neurotransmitter metabolites levels were within the normal limits. A positive THB test supported the hypothesis of impaired THB metabolism, at least at the liver level. The first case has been only published as an abstract (26), the second is more documented (patient JS (25)) and presently aged 4 years has a normal development. The difference between peripheral and partial forms is not evident since CSF data at the neonatal period were not available in the cases with "partial" form. Nevertheless, it is interesting to note that, at 2 years of age, the patient JS still have a low biopterin excretion (about 150 nmol/mmol of creatinine) and that the CSF profile of pterins was not normal (N/B= 4.7) even with a normal biopterin concentration. Its phenylalanine tolerance was large but limited since the use of aspartame resulted in increased of blood phenylalanine levels.

Table IV: Partial form of biopterin synthetase deficiency

patients	AL (46)	DE	DC	TYL (46)	HVJ (23)	TH (13)	TS (14)
		(46)					
date of birth	1973				1973	1980	1981
sex	M	M			F	F	F
circumstances of diagnosis	< reinvestigation of mild HPA >					<neonatal > <screening>	
at age	5 y	10 y	-	2 y	9 y	2 m	2 m
PKU screening (mg/dl)	10	>20		8	4	20	7
clinical signs	no *	no *	yes	no EEG **	no	no	no
THB loading test					pos ***	pos	pos ***
urinary pterins ****							
neopterin	-	1100	2200	16000	3170	11050	16840
biopterin	-	300	300	450	280	350	410

* sleepy after a phenylalanine load, ** focal spike activity
*** combined phenylalanine-THB load, **** nmol/mmol creatinine

Table V: Peripheral form of BS deficiency

reference	Hreidarsson (26)	Hoganson (25)	ML (50)
neonatal Phe	positive	5.8 mg/dl	5.9 mg/dl
Phe at 2w	22 mg/dl	22.4 mg/dl	15 mg/dl
N/B ratio	85 at 14d	144 at 15d	86 at 15d
	id at 5m	29.1 at 15m	22 at 10m
BH4 load	positive	positive	positive
clinical signs	none	none	yes at 6 m
CSF (nmol/l)			
neopterin	94.4	125.7	115.9
biopterin	59.9	40.1	18.4
N/B	1.58	3.13	6.30
at age		4 m :	6 m :
neopterin	-	127.2	137.3
biopterin	-	38.0	5.2
N/B	-	3.35	26.4

Lower urinary excretion of biopterin are noted in "typical" forms (151 ± 130 nmol/mmol creatinine, 18 patients) than in partial forms (308 ± 92, 6 patients), but the overlap of values does not authorised a clear distinction between the two groups on a case to case basis. The measurement in the liver or erythrocytes of 6-pyruvoyl-tetrahydropterin synthetase is now possible (43), but no data comparing activities in the different forms of biopterin-synthetase deficiency has been yet published.

The risk of a "rapid" classification of a new patient is illustrated by the recent report by Scriver et al (50). The patient ML (table V) was believed to have a peripheral form of the disease since CSF data were normal at the neonatal period. Unfortunately, at 6 months of age neurological signs have appeared. Its 6-pyruvoyl-tetrahydropterin synthetase activity was 4% of control values.

Transient forms:

Transient forms of BS deficiency have been also reported (Table VI) but are often too shortly documented. These children are asymptomatic and have been described as presenting a neonatal hyperphenylalaninemia and a high N/B ratio in their urine. Only 2 of them responded to a tetrahydrobiopterin loading test. For the 4 other cases, it is likely that the transient hyperphenylalaninemia resulted from a reduction of phenylalanine hydroxylase activity and the high N/B from a delayed maturation of tetrahydrobiopterin synthesis, as it is usual in PKU babies (17).

HETEROGENEITY OF THE THERAPEUTIC EFFECTIVENESS

The dramatic effect of neurotransmitter precursors on these patients was first described by Bartholomé (2) in a 13 month old child with biopterin-synthetase deficiency (KM). The goals of the treatment are:

1- the control of the blood phenylalanine levels by using a PKU diet or low doses of tetrahydrobiopterin (1-5 mg/kg/d) which can restore a normal phenylalanine hydroxylase activity. In most of the cases, a high tolerance to dietary phenylalanine was observed (14/17 in BS, 6/6 in partial BS and 19/21 in DHPR);

2- the restoration of a normal monoamine neurotransmission can be approached by

* giving neurotransmitter precursors (Dopa and 5-hydroxy-tryptophan) in combination with an inhibitor of peripheral aromatic aminoacid decarboxylase (carbidopa), to bypass the impaired activity of tyrosine- and tryptophan-hydroxylase. However, such treatment is difficult to manage, adverse effects of the drugs have been often reported and a correct dosage can be precisely evaluated only by CSF neurotransmitter metabolites analysis.
* replacement of the missing cofactor. Indeed, a tetrahydro-biopterin supplementation has appeared the most rational therapy in BS deficiency, since DHPR is normal and can regenerate the exogenous cofactor. After a period of

Table VI : Transient form of BS deficiency

reference	Matalon (35)		Clemens (7)	Sawada (48)	Takahashi (54)	Rey *
neonatal Phe (mg/dl)	34.4	pos.	42.6	4	>35	pos.
N/B ratio (age)	8.2 (19d)	16	8.5 (7w)	76.9 (25d)	50.7 (19d)	10.1 (1m)
	1.3 (40d)	1.1 (9m)	0.7 (5m)	1.2 (18m)	?	0.7 (8m)
BH4 load	neg.	pos.	neg.	not done	neg.	pos.
clinical signs	none	tremor at 3m	none	none	none	none
treatment	diet	neurotr. (3 to 9m)	diet	none	diet stop at 6m	diet stop at 8m

diet: PKU diet
* unpublished

controversial reports on the blood brain barrier permeability to THB, it has been demonstrated the definite efficiency of THB on clinical and biochemical signs. However, large doses are necessary to obtain a significant passage through the brain blood barrier and variable long term efficiency of tetrahydrobiopterin monotherapy has been noticed. Only two patients clearly responded to a THB monotherapy (YO, YZ (41)). In other cases, tetrahydrobiopterin administration participates to the clinical improvement, at least by normalizing the blood phenylalanine. For these reasons synthetic analogues with high cofactor activity and more lipophilic to enter the brain (6-methyl-tetrahydropteridine, 6,6-dimethyl-tetrahydropteridine, 1'2'diacetyl-tetrahydropteridine) might appear an attractive route to cofactor therapy (31). Even with 6-methyl-tetrahydropteridine, variable response has been observed (29, Leupold et al this volume). Moreover, a precise evaluation of possible side effect has to be conducted; in one patient (TD), a 9 month administration of 6-methyl-tetrahydropteridine was accompanied by an elevation of serum transaminase levels, indicating a possible liver toxicity of the drug. In DHPR deficiency, the cofactor therapy has not been extensively investigated. A trial reported by Smith et al (52) demonstrated a poor response to THB.

Except for few cases, and considering those with a sufficient follow-up, it can be concluded that the general estimation of the therapy effectiveness is more pessimistic than initially expected.

If some neurological signs can be reversed by the therapy, the gross development remained delayed and the best results lead to children attending a "special school". In most of the cases some improvement was noted on muscular tone, seizures and alertness, but often after a period of initial improvement no more progress and even deterioration was noted.

Thus the success of the treatment still appears modest and variable. In order to avoid any subjective interpretation of the therapeutic results, tables VII and VIII summarize observation for each patient with the words used by their respective physician.

The variation in the therapeutic success might be due to late start of the specific therapy, possible fetal cerebral damage, difficulties in obtaining a balanced neurotransmitter replacement therapy, variation in the intensity of the metabolic block, or need of adjuvent therapy (ie: folate supplementation). The latter possibility has been documented in DHPR deficient patients following the demonstration that 8 out of 9 patients have low serum, erythrocyte and/or CSF levels of folic acid and that folinic acid administration has led to significant improvement of clinical signs in 3 cases. The reason of the folate deficiency remains obscur. DHPR may play a role in keeping folate in the tetrahydro form and dihydrobiopterin accumulation may competitively interfere with folate metabolism (30,52).

Table VII : Follow-up of biopterin-synthetase deficient patients on specific treatment

initial	treated at age	last record	result
WC (44)	2w	-	THB 10 mg/kg/d: improvement in sucking, sucking, + neurotransmitters: improvement of tone
YO (41)	2w	3y	THB monotherapy: 10 mg/kg/d: normal mental and physical development
BE (18)	2m	2y6m	neurotransmitters: good control of neurological signs but hypotonia, hypomobility (DQ: retarded by about 6-8 m) did not well respond to a high dose of THB (10 and 20 mg/kg/d), effect of THB alone lasted 1 d only; resistance to 1'2'diacetyl-THB
KK (8)	3m	deceased at 9m	neurotransmitters + THB 5 mg /kg/d: no improvement, tendency to opisthotonus, myoclonic jerks (severe cerebral damage)

MB (56)	5m	5y7m	neurotransmitters: no neurological signs, no convulsion normal EEG, low normal IQ (90)
AA	5m	1y	neurotransmitters + low dose of THB: better social contact, improvement of tone
RB (36)	7m	4y	neurotransmitters: improved neurological development until age 25m then remained hypertonic and choreoathetotic (DQ < 1y) 6 MPH4 8-38 mg/kg/d: no persistent effect were apparent although she had definite periods of increased alertness
PN (34)	8m	2y	neurotransmitters: EEG normalised, decreased hypertonia psycho-motor improvement (DQ: 14 m)
RM (37)	8m	1y1m	neurotransmitters: vomiting and night cry have disappeared, but neck not fixed
LL (4)	6m	1y6m	THB 40 mg/kg/d: normal EEG, improvement but myoclonic jerks of the extremities. + neurotransmitters: neurologically normal, delayed mental development but still making rapid progress
ML (50)	10m	1y9m	THB monotherapy: 150 mg/d: improvement of neurological status: decreased lethargy, improved tone but still irritable and remained retarded (deterioration if THB was stopped for a week)
KD	11m	10y	neurotransmitters: improvement of neurological signs special school (IQ: 65)
MG (20)	11m	2y4m	neurotransmitters: improvement, no further convulsions (DQ: 90 at 20 m). At 2y1m: + THB 5mg/kg/d then 1.25 mg/kg/d: more active (if stopped: short temper, tired, stridor) but neurotransmitters still needed (IQ: 118, but motor development: 1y6m)
KM (3)	1y	2y3m	neurotransmitters: disappearance of hyperpyretic crisis, myocloni, hypertonia and hypersalivation
MS (55)	1y2m	6y8m	neurotransmitters: progressive improvement of neurological signs, but incomplete (DQ: 52 at 5y6m), At 5y11m: with THB 5mg/kg/d then 1.25 mg/kg/d more active and temper improved but neurotransmitters still needed (normal EEG, IQ: 61)

FCP (21)	2y	4y	neurotransmitters: improvement in cognitive performances but little improvement in motor development (DQ: 8 m)
TAP	2y	2y11m	THB 2.5 mg/kg/d + neurotransmitters: improvement in the control of crisis, in speaking and in walking (DQ: 18 m)
SK	3y	5y5m	neurotransmitters: some improvement, but still retarded mental development: 2 y, motor: 1 y
CA (45)	3y4m	-	neurotransmitters: poor improvement
SM (24)	3y6m	-	neurotransmitters: more responsive to his environment, seizures have almost disappeared but there has been no essential change in hypotonia or the marked delay in speech and motor development
TSM (21)	4y	6y	neurotransmitters: improvement in cognitive performances, activity and speaking (DQ: 18 m)
YZ (41)	5y	8y	THB monotherapy (neurotransmitters inconstantly given at the beginning) 5 mg THB/kg/d : special school
TD (29)	7y	11y6m	neurotransmitters: developmental progress at a delayed rate. At 11 y: THB 20 mg/kg/d led to a pronounced improvement (tone, coordination, drooling, eye-rolling) 6-MPH4 12 mg/d led to increased motor ability, progress in speech, decreased drooling (but elevation of hepatic enzymes) reintroduction of neurotransmitters: loss of effectiveness (IQ 50)
TP	8y	9y	THB 2.5 mg/kg/d + neurotransmitters: improvement in seizure control and in hyperkinetic movements, special school

Table VII : Follow-up of dihydropteridine-reductase deficient patients on specific treatment

initial	treated at age	last record	result
DP (30)	3m	5y	neurotransmitters: good response but episodes of drug intolerance. At 3y9m: normal neurological status but poor intellectual performance and lack of spontaneous activity. At 4y3m: +folinic acid, B12: definite improvement (DQ: 4y)
NF (52)	3m	3y6m	neurotransmitters + THB (50 mg/d): symptoms disappeared, IQ at 18 m: 84, then progress slowed down. At 33m: + folic acid : symptoms worsened, + folinic acid: dramatic improvement, IQ at 42m: 64
WO (5)	6m	4y3m	neurotransmitters: speaks fluent arabic, walking alone since 2½ y, variable muscle tone
LM (22)	6m	1y6m	neurotransmitters: decreased spasticity of the limbs, walks without support, normal tone and reflexes, normal EEG
NV	8m	deceased at 11m	neurotransmitters: seizures more spaced, hypertonia has regressed leaving in its place a marked hypotonia. +folic acid and B12: some improvement (tone, psycho- motor development)
NC (5)	8m	8y	neurotransmitters: profound and continuing neurological deterioration
GC (33)	11m	-	neurotransmitters: without satisfactory results, frequent crisis of absence, tetraparesis, mental retardation
KA (32)	1y	3y10m	neurotransmitters: good response, resolution of status epilepticus. Mental ability: 14 m, motor development: 2 y. It exists definite neurological abnormalities (pyramidal lesions?)
KH	1y4m	2y1m	neurotransmitters: definite deterioration (epilepsy, blind, unresponsive, spastic) IQ: non testable

DG (33)	1y5m	-	neurotransmitters: initial improvement, at 21m major seizure activity was noticed, psychomotor retardation and neurological picture have progressively worsened
MO (5)	1y5m	-	neurotransmitters: no change, parents requested discontinuation of the treatment
HY	1y8m	2y1m	neurotransmitters: without convulsions, but no mental progress, social contact looks better
AN	1y9m	-	neurotransmitters: more alert, tone improved; lost for follow-up
SF (6)	2y	3y5m	neurotransmitters: muscular tone had improved considerably, IQ: 60
ETMH	2y	4y5m	neurotransmitters: socially more responsive, remains severely retarded, says 3 or 4 single words
KS	11y	13y	neurotransmitters: ataxic gait and convulsions disappeared, IQ: 24
MS	16y	18y	neurotransmitters: ataxic gait and convulsions disappeared, IQ: 25
NP (57)	18y	20y	neurotransmitters: no benefit
DS		13m	only PKU diet since d15, good result if blood Phe < 6 mg/dl
DA		3y	only PKU diet since d15, good result if blood Phe < 6 mg/dl hypotonia, hypersalivation if Phe 6

CONCLUSION

Tetrahydrobiopterin deficiency is a rare cause of hyperphenylalaninemia. Diagnostic tools are now available for the recognition of these variants among hyperphenylalaninemic babies. However, the description of atypical forms of the disease has risen new problems. Indeed, a great heterogeneity has been observed in the clinical presentation, the biochemical expression and the response to the treatment.

It is obvious that a residual THB synthetic activity exist even in "typical" forms, since significant amounts of neurotransmitter metabolites are present in CSF and these children usually tolerate relatively large amount of phenylalanine. However, efforts have to be directed toward a better characterisation of individual residual capacity to synthesise tetrahydrobiopterin in order to solve the problem in deciding whether or not to treat.

The most important question, whatever infants can achieved normal development if early treated, remains to be answered. A better definition of protocols for the follow-up of neurotransmitter replacement therapy and the development of new therapeutic ways may change the general prognosis of the disease.

Partial, peripheral and transient forms do not require a specific treatment, and because of their good prognosis investigations are refrained. However, the difficulties encountered in the diagnosis of some THB deficient patients plead for more uniform and complete study of each patient.

ACKNOWLEDGEMENT: We are indebted to Dr. S.Berlow, B.Blehova, S.Brown, B.Cohen, R.G.H.Cotton, D.M.Danks, J.P.Farriaux, R.Giugliani, A.Green, L.G.Greeves, F.Güttler, K.J.Hsiao, J.Hyanek, R.J.Leeming, B.Leroux, H.L.Levy, A.Lipson, I.Lyon, G.Malpuech, R.Matalon, N.Matsuo, J.F.Mattei, M.E.O'Flynn, B.Radeva, R.A.Roesel, C.Romano, G.Rylance, C.R.Sriver, J.Schirrer, I.Smith, K.Tada, R.M.Veall, H.Von Theile for their collaboration.

References:

1 Bartholomé, K. 1974. Lancet 2, 1580

2 Bartholomé, K., Byrd, D.J. 1975. Lancet 2, 1042

3 Bartholomé, K., Byrd, D.J., Kaufman, S., Milstien, S. 1977. Pediatrics 59, 757

4 Beck, B., Brandt, N.J., Christensen, E., Niederwieser, A. Pedersen, P.S. 1983. Acta Paediatr.Scand. 72, 449

5 Butler, I.J., O'Flynn, M.E., Seifert, W.E., Howell, R.R. 1981. J.Pediatr. 98, 729

6 Cerone, R., Scalisi, S., Cotellessa, M., Schiaffino, M.C., Caruso, U., Romano, C. 1986. J.Inher.Metab.Dis. 9 (suppl.2), 244

7 Clemens, P., Gruttner, R. 1985. J.Pediatr. 107, 312

8 Cohen, B.E., Szeinberg, A., Quint, J., Normand, M., Blonder, J., Peled, I. 1985. Isr.J.Med.Sci. 21, 520

9 Cotton, R.G.H., Jennings, I., Bracco, G., Ponzone, A., Guardamagna O. 1986. J.Inher.Metab.Dis. 9, 239

10 Curtius, H-Ch., Niederwieser, A., Viscontini, M., Otten, A., Schaub, J., Scheibenreiter, S., Schmidt, H. 1979. Clin.Chim.Acta 93, 251

11 Danks, D.M., Bartholomé, K., Clayton, B.E., Curius, H-Ch, Gröbe, H, Kaufman, S., Leeming, R., Pfleiderer, W., Rembold, H., Rey, F. 1978. J.Inher.Metab.Dis. 1, 49

12 Dhondt, J.L., Delcroix, M., Farriaux, J.P. 1982. Clin.Chim.Acta 121, 33

13 Dhondt, J.L., Farriaux, J.P., Leroux, B., Lyon, I.C.T. 1983. In : Biochemical and Clinical Aspects of Pteridines. (H.Ch. Curtius, W. Pfleiderer & H.Wachter eds.) WdeGruyter, p. 318

14 Dhondt, J.L., Leroux, B., Farriaux, J.P., Largillière, C., Leeming, R.J. 1983. Eur.J.Pediatr. 141, 92

15 Dhondt, J.L. 1984. J.Pediatr. 104, 501

16 Dhondt, J.L. 1986. In: Chemistry and Biology of Pteridines. (B.A.Cooper, V.M.Whitehead eds.) WdeGruyter, Berlin, p 385

17 Dhondt, J.L., Farriaux, J.P., Hayte, J.M. 1986. Arch.Fr.Pediatr. 43, 785

18 Endres, W., Niederwieser, A., Curtius, H.Ch., Wang, M., Ohrt, B., Schaub, J. 1982. Helv.paediat.Acta 37, 489

19 Firgaira, F.A., Choo, K.H., Cotton, R.G.H., Danks, D.M. 1981. Biochem.J. 198, 677

20 Fukuda, K., Tanaka, T., Hyodo, S., Kobayashi, Y., Usui, T. 1985. J.Inher.Metab.Dis. 8, 49

21 Giugliani, R., Costa da Costa, J., Dutra-Filho, C.S., Nierderwieser, A. 1983. Rev.Brasil.Genet. 6, 557

22 Gröbe, H., Bartholomé, K., Milstien, S., Kaufman, S. 1978. Eur.J.Pediatr. 129, 93

23 Güttler, F., Lou, H., Lykkelund, C., Niederwieser, A. 1984. Eur.J.Pediatr. 142, 126

24 Hase, Y., Shintaku, H., Turuhara, T., Oura, T. 1982. J.Inher.Metab.Dis. 5, 81

25 Hoganson, G., Berlow, S., Kaufman, S., Milstien, S., Schuett, V., Matalon, R., Naylor, E., Seifert, W. 1984. Pediatrics 74, 1004

26 Hreidarsson, S., Valle, D., Holtzman, N., Coyle, J., Singer, H., Kapatos, G., Kaufman, S. 1982. Pediatr.Res. 16, 192 A

27 Kaufman, S., Holtzman, N.A., Mielstien, S., Butler, I.J., Krumholz, A. 1975. N.Engl.J.Med. 293, 785

28 Kaufman, S., Berlow, S., Summer, G.K., Milstien, S., Schulman, J.D., Orloff, S., Spielberg, S., Pueschel, S. 1978. N.Engl.J.Med. 299, 673

29 Kaufman, S. Kapatos, G., McInnes, R.R., Schulman, J.D., Rizzo, W.B. 1982. Pediatrics 70, 376

30 Leeming, R.J., Harpey, J.P., Brown, S.M., Blair, J.A. 1982. J.ment.Defic.Res. 26, 21

31 Levine, R.A., Zoephel, G.P., Neiderwieser, A., Curtius, H-Ch., Traub, H., Pfleiderer, W. 1985. In: Biochemical and Clinical Aspects of Pteridines. Vol.4. (Wachter H., Curtius H-Ch., Pfleiderer W. Eds) WdeGruyter, Berlin, p 269

32 Lipson, A., Yu, J., O'Halloran, M., Potter, M., Wilken, B. 1984. J.Inher.metab.Dis. 7, 69

33 Longhi, R., Riva, E., Valsasina, R., Paccanelli, S., Giovannini, M. 1985. J.Inher.Metab.Dis. 8 (suppl.2), 97

34 Malpuech, G., Guyon, A., Demeocq, F., Piton, A., Boespflug ,O., Vanlieferinghen, Ph. 1984. Arch.Fr.Pediatr. 41, 5

35 Matalon, R. 1984. J.Pediatr. 104, 579

36 McInnes, R.R., Kaufman, S., Warsh, J.J., Van Loon, G.R., Milstien, S., Soldin, S., Walsh, P., MacGregor, D., Hanley, W.B. 1984. J.Clin.Invest. 73, 458

37 Milstien, S., Holtzman, N.A., O'Flynn, M.E., Thomas, G.H., Butler, I.J., Kaufman, S. 1976. J.Pediatr. 89, 763

38 Milstien, S., Kaufman, S., Summer G.K. 1980. Pediatrics 65, 806

39 Niedewieser, A., Curtius, H-Ch., Viscontini, M., Schaub, J., Schmidt, H. Lancet 1979, 1, 550

40 Niederwieser, A., Staudenmann W., Wang M., Curtius H-Ch., Joller P., Atarés M., Cardesa-Garcia J. 1982. Eur.J.Pediatr. 138, 97

41 Niederwieser, A., Curtius, H.Ch., Wang, M., Leupold, D. 1982. Eur.J.Pediatr. 138, 110

42 Niederwieser, A., Blau N., Wang M., Joller P., Atarés M., Cardesa-Garcia J. 1984. Eur.J.Pediatr. 141, 208

43 Niederwieser, A., Ponzone, A., Curtius, H.Ch. 1985. J.Inher.Metab.Dis. 8 (suppl.1), 34

44 Niederwieser, A., Leimbacher, W., Curtius, H.Ch., Ponzone, A., Rey, F., Leupold, D. 1985. Eur.J.Pediatr. 144, 13

45 Rey, F., Harpey, J.P., Leeming, R.J., Blair, J.A., Aicardi, J., Rey, J. 1977. Arch.Fr.Pediatr. 34, CIX

46 Rey, F., Saudubray, J.M., Leeming, R.J., Niederwieser,A., Curtius, H.Ch., Rey, J. 1983. Arch.Fr.Pediatr. 40, 237

47 Sahoda, Y., Leeming, R.J., Blair, J.A., Armstrong, R.A., Green, A., Cohen, B.E. 1986. J.Inher.Metab.Dis. 9 (suppl.2), 247

48 Sawada, Y., Shintaku, H., Isshiki, G., Hase, Y., Tsuruhara, T., Oura, T. 1984. Acta Paediatr.Jpn. 26, 561

49 Seifert, W.E., Fox, J.L., Butler, I.J. 1980. Ann.Neurol. 8, 38

50 Scriver, C.R., Clow, C.L., Kaplan, P., Watters, G.V., Laberge, C., Mohyuddin, F., Milstien, S. 1986. In: Chemistry and biology of pteridines. (Cooper B.A. and Whitehead V.M. eds.) W.de Gruyter, Berlin, p 403

51 Smith, I., Clayton, B.E., Wolff, O.H. 1975. Lancet 1, 1108

52 Smith, I., Hyland, K., Kendall, B., Leeming, R.J. 1985. J.Inher.Metab.Dis. 8 (suppl.1), 34

53 Smith, I., Dhondt, J.L. 1985. 1985. Lancet 1, 818

54 Takahashi, T., Kodama, S., Nishio, H., Takumi, T., Matsuo, T., Hase, Y., Sawada, Y. 1985. J.Inher.Metab.Dis. 8, 105

55 Tanaka, T., Aihara, K., Iwai, K., Kohashi, M., Tomita, K., Narisawa, K., Arai, N., Yoshida, H., Usui, Y. 1981. Eur.J.Pediatr. 136, 275

56 Theile, Von H., Bührdel, P., Sandig, K.R., Mücke, J., Bigl, V., Grimm, U. 1983. Dt.Gesundh.-Wesen 38, 1657

57 Young, J.H., Walker, V., Tippett, P.A., Clayton, B.E., Veall, R.M. 1983. J.Inher.Metab.Dis. 6, 111

PRENATAL DIAGNOSIS OF TETRAHYDROBIOPTERIN DEFICIENCY

N. Blau, A. Niederwieser[+], H. Ch. Curtius, W. Leimbacher, L. Kierat, A. Matasovic, and W. Staudenmann

Division of Clinical Chemistry, Department of Pediatrics, University of Zürich, CH-8032 Zürich, Switzerland

I. Ozalp

Institute of Child Health, Department of Pediatric Nutrition & Metabolism, Hacettepe University, Ankara, Turkey

INTRODUCTION

Three inborn errors of metabolism (Figure 1) are known to cause tetrahydrobiopterin (BH_4) deficiency (1-3,8): a defect of GTP cyclohydrolase I (GTPCH), 6-pyruvoyl-tetrahydropterin synthase (PPH_4S), and dihydropteridine reductase (DHPR). The first two enzymes are involved in the biosynthesis, the last one in the regeneration of BH_4. PPH_4S was designated in the past as "dihydrobiopterin synthetase" but it must be assumed that this enzyme actually does not exist. BH_4 is the cofactor of phenylalanine 4-hydroxylase, tyrosine 3-hydroxylase, and tryptophan 5-hydroxylase, and defects in its biosynthesis leads to hyperphenylalaninemia with progressive neurological disorders (4). The first symptoms like hypertonia of the extremities and hypotonia of the trunc often become apparent only at the age of 3-8 months. Patients usually do not respond clinically to a low phenylalanine diet, however they can be treated with the neurotransmitter precursors L-Dopa and 5-hydroxytryptophan (5), and with BH_4 (6,7).

Dedicated to Prof. A. Niederwieser who passed away on 5.2.87

Unconjugated Pterins and Related Biogenic Amines

Without treatment the infants usually have no social contact and die during the first years of life. Approximately 1-2% of the newborns with hyperphenylalaninemia are deficient in the BH_4 cofactor and therefore selective screening of cases detected by Guthrie mass screening should be performed systematically. In addition it is of great value to make a prenatal diagnosis of all three variants of BH_4 deficiency in families at risk. However, in the past a prenatal diagnosis was only possible for DHPR deficiency. Prenatal diagnosis of BH_4 metabolic disorders have been achieved mainly by measurement of metabolites in amniotic fluid and of enzyme activities in cultured amniotic fluid cells. In some cases fetal blood samples and fetal liver biopsy have proven useful when the activity of the enzyme involved was not expressed in amniotic fluid cells. Chorionic villi biopsy may be besides suitable in the prenatal diagnosis of DHPR deficiency. Rapid development of recombinant DNA technology will enable more sensitive and selective prenatal diagnosis of a variety of genetic disorders of BH_4 biosynthesis in the future.

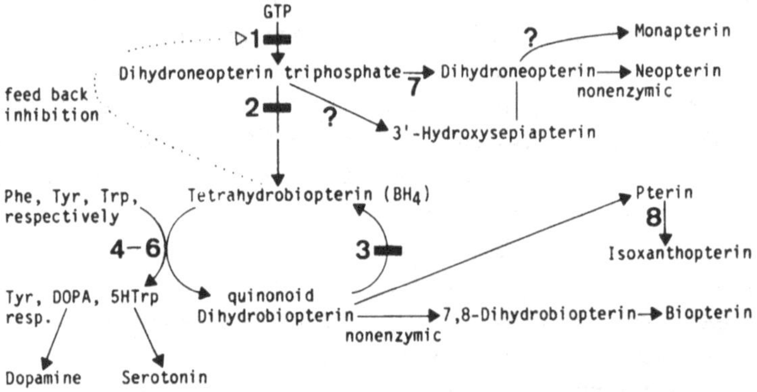

Fig.1 Possible metabolic defects in hyperphenylalaninemia. 1:GTPCH, 2:PPH_4S, 3:DHPR, 4:Phe-4-hydroxylese, 5:Tyr-3-hydroxylase, 6:Trp-5-hydroxylase, 7:alk. phosphatase, 8:xanthine oxidase

GTP CYCLOHYDROLASE I DEFICIENCY

In GTPCH deficiency, the enzyme catalyzing the first step in the biosynthesis of BH_4 by forming dihydroneopterin triphosphate from GTP is deficient. Since these patients are unable to form dihydroneopterin triphosphate and its metabolites no neopterin, biopterin, pterin, or isoxanthopterin can be detected in urine. Although this variant has been detected by Niederwieser et al. only recently (9,10), four patients are already known (11-13). Because GTPCH activity is not expressed in cultured amniotic fluid cells (14), prenatal diagnosis is possible only by measurement of pterins in amniotic fluid (15), and probably by measurement of GTPCH activity in stimulated mononuclear blood cells (16) of the fetus.

TABLE 1

GTP CYCLOHYDROLASE I ACTIVITY IN LIVER BIOPSIES AND IN STIMULATED LYMPHOCYTES (18)

Subjects	Relative GTPCH activity (%)	
	in liver homogenate	in stimulated lymphocytes
Patient	<4	<3
Father	-	30
Mother	-	46
Controls	100 n=15	100 n=4

-not measured

Assay of GTPCH activity in liver biopsy (17) of fetus might occasionally also be used. Table 1 shows that no GTPCH activity could be detected in PHA stimulated lymphocytes of the patient with proven GTPCH deficiency in liver. The samples from the father and mother showed 30 and 46%, respectively, of the mean enzyme activity of controls. These results demonstrate that heterozygotes of GTPCH deficiency can be recognized by measurement of the enzyme activity in stimulated lymphocytes.

Recently, Dhondt (personal communication) reported on the first prenatal diagnosis of GTPCH deficiency by measurement of pterins in amniotic fluid in a pregnancy at risk. However, because of abnormally low pterin concentrations discrimination between homozygotes and heterozygotes was not possible .

6-PYRUVOYL-TETRAHYDROPTERIN SYNTHASE DEFICIENCY

Niederwieser et al. showed recently (19) that in so called "dihydrobiopterin synthetase" deficiency the defective enzyme is 6-pyruvoyl-tetrahydropterin synthase. PPH_4S catalyzes the conversion of dihydroneopterin triphosphate to 6-pyruvoyl-tetrahydropterin through the sequence of triphosphate elimination and intramolecular redox reaction (20). Typical cases of PPH_4S deficiency show high levels of neopterin and abnormally low levels of biopterin in urine. Concentrations of homovanillic acid and 5-hydroxyindoleacetic acid in CSF are pathologically low even under low phenylalanine diet. Particulary PPH_4S deficiency seems to be heterogeneous and one should distinguish between classical and peripheral or partial defects (21-23). In the peripheral or partial form of PPH_4S deficiency normal levels of neurotransmitter metabolites in CSF were measured. In any case patients respond well to the BH_4 loading test. PPH_4S deficiency can be prenatally screened by measurement of neopterin and biopterin in amniotic fluid as shown by Niederwieser et al. (24). The percentage of biopterin of the sum of neopterin and biopterin was about 1 % (control range

19-43) in one case.

TABLE 2

6-PYRUVOYL-TETRAHYDROPTERIN SYNTHASE ACTIVITY IN ERYTHROCYTES OF PATIENTS AND CONTROLS

Case		Age	PPH_4S activity, % of control
Patient	A	5m	23
	B	2y	10
	C1	2y	8
	C2	5y	7
	D	15m	4
	E	5y	3
Fetus	E	19wg	2
Father	A	25y	97
Mother	A	25y	22
Father	B	34y	49
Mother	B	34y	32
Father	C	35y	19
Mother	C	35y	19
Father	D	-	20
Mother	D	-	19
Father	E	31y	23
Mother	E	29y	20
Controls		25y	100
n=9			(65-165)
n=2		18+23wg	195/295
Heterozygotes		25y	18
n=16			(8-31)
Homozygotes		1m-8y	7
n=11			(0-20)

Because PPH_4S is like GTPCH not expressed in amniotic cells, the enzyme defect can be confirmed only by measurement of activity in fetal liver (19) or fetal erythrocytes (24). Table 2 shows that a rest activity of up to 20% can be measured in erythrocytes of homozygotes, and that the values for obligate heterozygotes are lower than the expected 50%. The fetal controls showed 2 to 3 fold higher PPH_4S activities than adult controls. A few children with partial PPH_4S deficiency were also investigated and showed activities below 20%. By two-dimensional presentation of normalized values of PPH_4S activity versus normalized percentage of urinary biopterin, differentiation between homozygotes and obligate heterozygotes is possible (25).

DIHYDROPTERIDINE REDUCTASE DEFICIENCY

DHPR deficiency, a defect of regeneration of the cofactor, was detected by Kaufman (26). Because there is no feedback inhibition of GTPCH due to the lack of BH_4, biosynthesis of BH_4 is activated. Patients with DHPR deficiency showed very high biopterin excretion and percentage of biopterin was always higher than 80 (2,27). Biopterins are excreted as inactive 7,8-dihydrobiopterin and biopterin formed from q-dihydrobiopterin (Fig. 1). However, newborns with DHPR deficiency can be missed by measurement of pterins in urine, and measurement of DHPR activity in blood (Guthrie card) according to Arai et al. (28) is recommended. DHPR deficient patients failed in some cases to respond to the BH_4 oral loading test (8), and heterogeneity of the molecular defect was demontsrated in these patients (29). Because DHPR activity is present in normal fibroblasts and amniotic fluid cells (30,31), as well as in blood cells (28,32), and chorionic villi (13), DHPR deficiency can be diagnosed prenatally. The first prenatal determination of DHPR in a fetus at risk was described by Firgaira et al. (33). Normal results were obtained when cultured amniocytes from a pregnancy at risk were assayed for DHPR activity. Table 3 shows the data of prenatal determination of DHPR in a family with DHPR deficiency.Measurement of pterins in amniotic fluid obtained at 21 week of gestation showed slightly higher neopterin and normal biopterin (% biopterin = 19). By measurement of DHPR activity in cultured amniotic fibroblasts the fetus was diagnosed as heterozygote.

TABLE 3

PRENATAL DIAGNOSIS AND HETEROZYGOTE DETECTION IN FAMILY WITH DHPR DEFICIENCY

Case	DHPR activity, % of controls	
	Blood spots	Amniocytes
Fetus at risk	-	48
Baby at risk	54	-
Brother	<10	-
Mother	46	-
Father	49	-
Paternal uncle	103	-
Paternal aunt	89	-
Controls	100 n=29(28)	100 n=5(34)

An assay of DHPR activity in dried blood spots obtained after the birth of this child confirmed the amniotic cells results. The healthy, non-consanguineous parents as well as the baby showed the expected 50% activity when compared with the controls.

CONCLUSIONS

According to current experience, the prenatal diagnosis of BH_4 deficiency has to be based on the measurement of pterin metabolites in amniotic fluid, which actually reflect the levels in the fetus. Nevertheless, the corresponding enzymes should be measured in blood or in amniotic fluid cells.

ACKNOWLEDGEMENT

The auhors are grateful to Dr.A.Ponzone, Dr.W.Endres, Dr.H.Lehmann, Dr.H.Schmidt, Dr.H.Shintaku, Dr.O.Guardamagna, and Dr.D.Leupold. This work was supported by a grant from the Swiss National Science Foundation, Project No. 3.395-0.86.

REFERENCES

1. S.Kaufman. 1985. J.Inherited Metab.Dis. 8 Suppl 1,20-27
2. A.Niederwieser, A.Ponzone, H.Ch.Curtius. 1985. J.Inherited Metab.Dis. 8 Suppl 1,34-38
3. A.Niederwieser. 1987 In: Unconjugated Pterins in Neurobiology: Basic and Clinical Aspects (W.Lowenberg, R.A.Levine eds.). Taylor & Francis, London, p.107-118
4. D.M.Danks, K.Bartholome, B.E.Clayton, H.Ch.Curtius, H.Grobe, S.Kaufman, R.Lemming, W.Pfleiderer, H.Rembold, R.Rey. 1978. J.Inherited Metab.Dis. 1,49-53
5. K.Bartholome, D.J.Byrd. 1975. Lancet II,1042
6. A.Niederwieser, H.Ch.Curtius, O.Bettoni, J.Biri, B.Schircks, M.Viscontini, J.Schaub. 1979. Lancet I,131-133
7. H.Ch.Curtius, A.Niederwieser, M.Viscontini, A.Otten, J.Schaub, S.Scheibenreiter, H.Schmidt. 1979. Clin.Chim. Acta 93,251-265
8. J.L.Dhondt. 1984. J.Pediatr. 104,501-508
9. A.Niederwieser, W.Staudenmann. M.Wang, H.Ch.Curtius, M.Atares, J.Cardesa-Garcia. 1982. Eur.J.Pediatr 138,97
10. A.Niederwieser, N.Blau, M.Wang, P.Joller, M.Atares, J.Cardesa-Garcia. 1984. Eur.J.Pediatr 141,208-214
11. R.Matalon. 1984. Ped.Research 18,223A
12. J.L.Dhondt, J.P.Farriaux, A.Bondha, C.Largilliere, J. Ringel, M.M.Roger, R.J.Leeming. 1985. J.Pediatr. 106,954-956
13. E.W.Naylor, D.Ennis, A.G.F.Davidson, L.T.K.Wong, D.A.Appelgarth, A.Niederwieser. Pediatrics 79,374-378

14. J.Ferre, E.W.Naylor. 1986. In: Chemistry and Biology of Pteridines. (B.A.Cooper, V.M.Whitehead eds.) Walter de Gruyter, Berlin, p.309-313
15. J.L.Dhondt, J.M.Hayte, G.Forzy, M.Delcroix, J.P.Farriaux 1986. Clin.Chim.Acta 161,269-273
16. N.Blau, P.Joller, M.Atares, J.Cardesa-Garcia, A. Niederwieser. 1985. Clin.Chim.Acta 148,47-52
17. N.Blau, A.Niederwieser. 1983. Anal.Biochem. 128,446-452
18. N.Blau, A.Niederwieser. 1985. J.Clin.Chem.Clin.Biochem. 23,169-176
19. A.Niederwieser, W.Leimbacher, H.Ch.Curtius, A.Ponzone, F.Rey, D.Leupold. 1985. Eur.J.Pediatr. 144,13-16
20. D.Heintel, S.Ghisla, H.Ch.Curtius. 1987. In: Unconjugated Pterines in Neurobiology: Basic and Clinical Aspects (W.Lowenberg, R.A.Levine eds.). Taylor & Francis, London, p. 173-200
21. A.Niederwieser, H.Shintaku, W.Leimbacher, H.Ch.Curtius, J.Hyanek, J.Zeman, W.Endres. 1987. Eur.J.Pediatr. 146, 228-232
22. S.Hreidarsson, D.Vale, N.Kapatos, S.Kaufman. 1982. Pediatr.Res. 16,192A
23. J.L.Dhondt. 1987. In: Unconjugated Pterines and Related Biogenic Amines. (H.Ch.Curtius, N.Blau eds.), this volume
24. A.Niederwieser, H.Shintaku, Th.Hasler, H.Ch.Curtius, H.Lehmann, O.Guardamagna, H.Schmidt. 1986. Eur.J.Pediatr. 145,176-178
25. C.R.Scriver, C.L.Clow, P.Kaplan, A.Niederwieser. 1987. J.Pediatr. (in press)
26. S.Kaufman, N.A.Holtzman, S.Milstein, I.J.Butler, A.Krumholz. 1975. N.Engl.J.Med. 293,785-790
27. J.L.Dhondt, C.Largilliere, P.Ardouin, J.P.Farriaux, M.Dautrevaux. 1981. Clin.Chim.Acta 110,205-214
28. N.Arai, K.Narisawa, H.Hayakawa, K.Tada. 1982. Pediatrics 70,426-430
29. F.A.Firgaira, K.H.Choo, R.G.H.Cotton, D.M.Danks. 1981. Biochem.J. 198,677-682
30. F.A.Firgaira, R.G.H.Cotton, D.M.Danks. 1979. Clin.Chim.Acta 95,47-59

31. S.Milstein, N.G.A.Holzman, M.E.O'Flynn, G.H.Thomas, I.J.Butler, S.Kaufman. 1976. J.Pediatr. 87,763-766
32. F.A.Firgaira, R.G.H.Cotton, D.M.Danks. 1979. Lancet II,1260-1263
33. F.A.Firgaira, R.G.H.Cotton, D.M.Danks, K.Fowler, A.Lipson, J.S.Yu. 1983. Prenat.Diagn. 3,7-11

SEPIAPTERIN REDUCTASE IN HUMAN AMNIOCYTES, SKIN FIBROBLASTS, CHORIONIC VILLI AND STIMULATED MONONUCLEAR BLOOD CELLS.

J. Ferré

Department of Genetics, Faculty of Biology, University of Valencia, Burjasot (Valencia), Spain

E.W. Naylor

Department of Medical Genetics, Western Pennsylvania Hospital, Pittsburgh, PA 15224, U.S.A.

Introduction

Because of the important role of tetrahydrobiopterin (BH_4) in aromatic amino acid metabolism and neurotransmitter biosynthesis in mammals, the study of its biosynthesis has attracted the interest of many groups and is currently being intensively investigated. Two different pathways are known to exist that lead to BH_4 synthesis: the salvage and the *de novo* pathways (1). Sepiapterin reductase was first discovered as an enzyme that converted sepiapterin into 7,8-dihydrobiopterin (BH_2) in the salvage pathway, and more recently, it has been shown that it is also involved in the synthesis of BH_4 in the *de novo* pathway (2, 3).
Since 1975, several inborn errors of BH_4 metabolism in man have been discovered. These have included defects of dihydropteridine reductase (4), GTP cyclohydrolase (5), and 6-pyruvoyl-H_4-pterin synthase (6). Up to date, no case of sepiapterin reductase deficiency has been demonstrated. However, a defect of this enzyme is likely to be found in the future. In order to be able to screen for sepiapterin reductase deficiency, we have looked at several tissues of generalized use in diagnosis, including those used in prenatal diagnosis.

Unconjugated Pterins and Related Biogenic Amines

Materials and Methods

Sample preparation

Amniocentesis was performed at 14-18 weeks gestation, and amniotic fibroblasts obtained by culturing amniocytes in Chang Medium. When cultures were confluent (in 75 cm^2 flasks), cells were pooled, washed, and suspended in 2 ml of 0.1 M Tris-HCl buffer (pH 7.8) containing 0.15 M KCl, 5% glycerol and 2.5 mM EDTA ("storage buffer"). The same protocol was used to obtain fibroblasts from skin biopsies (from newborns). Cell membranes were disrupted by freezing and thawing in an acetone-dry ice bath. After centrifugation at 15,000 g for 2 min, the supernatant was stored at -40°C until analyzed.

Chorionic villi samples obtained from spontaneous abortions at 10 and 18 weeks gestation were stored at -40°C until used. Approximately 100 mg of fetal tissue was dissected, washed in 0.9% NaCl, and suspended in 0.5 ml of the "storage buffer". After homogenization in a glass homogenizer, the sample was centrifuged at 15,000 g for 2 min and the supernatant passed through a small Sephadex G-25 column.

Blood samples were obtained from adult donors. Hemolysis of red blood cells was performed as described by Katoh et al. (7). Mononuclear blood cells were separated from whole blood (4-8 ml) in Ficoll-Paque (Pharmacia). Stimulation of mononuclear blood cells with phytohemagglutinin (PHA) was performed following Blau et al.'s procedure with small modifications (8). Cells were harvested after 5 days, washed with 0.9% NaCl, suspended in 0.5 ml of the "storage buffer", and then proceeded as described for fibroblasts. Prior to analysis, plasma, red cell hemolysates and mononuclear blood cells lysates were passed through a small Sephadex G-25 column.

Sepiapterin reductase assay

The assay was carried out as follows: 29 µl of sample was incubated in 0.1 M potassium phosphate buffer (pH 6.8), 100 µM NADPH and 50 µM sepiapterin (50 µl final volume) at

37°C for 30 min (except where otherwise indicated) in the dark. The reaction was stopped with 25 µl of 1% I_2/2% KI in 1 M HCl. After 10 min at room temperature, precipitated proteins were separated by centrifugation at 15,000 g for 3 min, and 25 µl of 1% ascorbic acid added. The resulting biopterin was assayed by high-performance liquid chromatography (HPLC) injecting 50 µl of the mixture, in a reverse-phase column using 5% methanol in water as the mobile phase. Detection of biopterin was carried out by fluorescence (excitation at 362 nm, emission at 435 nm).

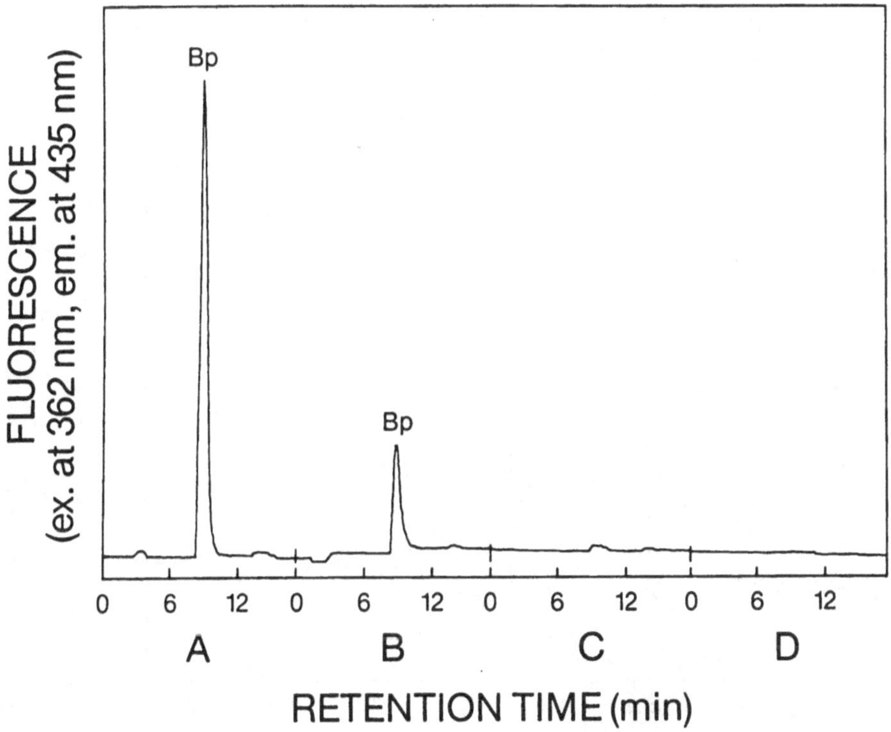

Fig. 1.- HPLC analysis of the sepiapterin reductase reaction product in stimulated mononuclear blood cells. (A) Complete reaction mixture after I_2 oxidation, (B) complete reaction mixture without I_2 oxidation, (C) and (D) reaction mixture without NADPH, with and without I_2 oxidation respectively. Bp = Biopterin.

Results

Sepiapterin reductase activity was demonstrated by the disappearance of sepiapterin (followed by measuring the decrease of absorbance at 420 nm) and the characterization of biopterin after oxidation with iodine and HPLC separation (Fig. 1)(prior to oxidation, some biopterin can be detected because of the spontaneous oxidation of BH_2). The biopterin yield from the reacted sepiapterin was found to be of approximately 95% in the reaction conditions. Sepiapterin reductase activity was linear up to 45 min of incubation at 37°C (Fig. 2), and proportional to the concentration of protein in the reaction at least up to 0.5 mg/ml. The reaction was inhibited by

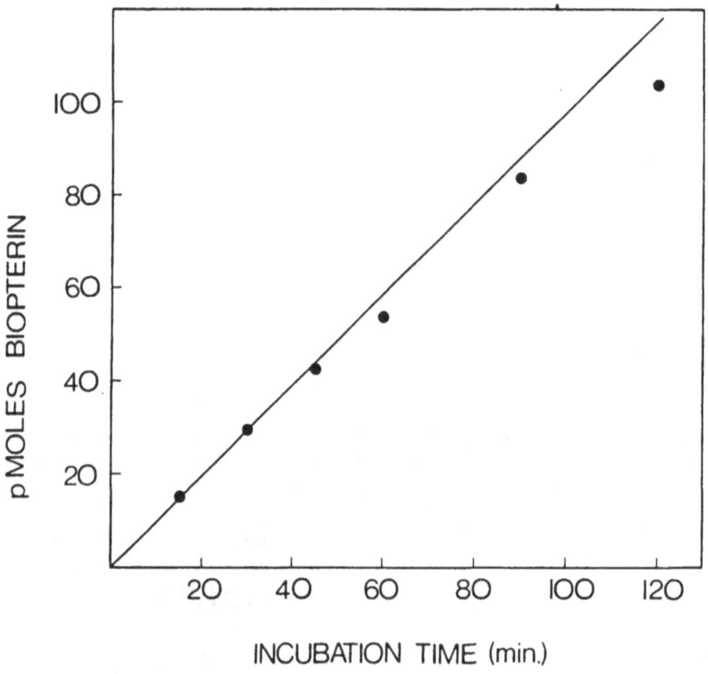

Fig. 2.- Sepiapterin reductase activity vs. incubation time in a sample from amniotic fibroblasts. The values are the means of two determinations.

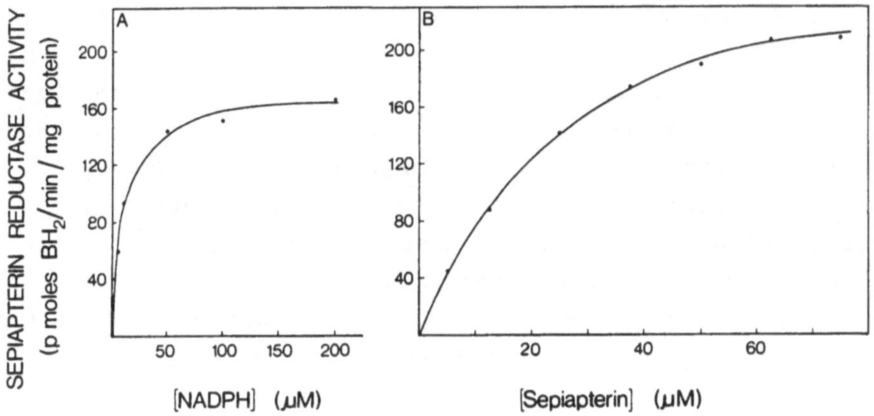

Fig. 3.- Dependence of sepiapterin reductase activity from amniocytes on the concentration of NADPH and sepiapterin in the assay. The reaction mixture was incubated for 10 min using (A) different concentrations of NADPH and 50 μM sepiapterin, and (B) different concentrations of sepiapterin and 100 μM NADPH.

N-acetylserotonin, a specific inhibitor of sepiapterin reductase (9). The reaction was also inhibited by $NADP^+$, thus care was taken to use fresh preparations of NADPH. Sepiapterin was purified in C_{18} Sep-Pak (Waters Assoc.) prior to use (10) to eliminate blue fluorescent impurities that would interfere with the reaction. In Fig. 3 is shown the dependence of sepiapterin reductase on NADPH and sepiapterin concentrations. The effect of the "storage buffer" on the reaction was checked comparing the activity in two samples of amniotic fibroblasts from which an aliquot had been harvested in 0.1 M potassium phosphate buffer (pH 6.8) and another aliquot in the "storage buffer". No significative difference in the specific activity of sepiapterin reductase in the two samples was found.

Sepiapterin reductase activity in different tissues and blood fractions

The sepiapterin reductase activity found in samples from different sources is summarized in Table 1. The specific activity in skin fibroblasts falls within the range obtained in amniotic fibroblasts, but with a mean value a little higher. Sepiapterin reductase activity was also detected in chorionic villi, although analysis of a larger number of samples will be required before a reliable mean value for the specific activity could be obtained.

Table 1.- Sepiapterin Reductase Activity in Fibroblasts, Chorionic Villi and Blood Fractions.

Sample	Number of Samples Analyzed	Specific Activity (pmoles BH_2/min/mg protein)		
		mean	S.D.*	range
Amniotic Fibroblasts	21	170	76	95-315
Skin Fibroblasts	6	201	48	120-263
Chorionic Villi	2	60	24	43-77
Plasma	10	0.08	0.04	0.04-0.16
Red Blood Cells	10	0.46	0.08	0.38-0.64
Mononuclear Blood Cells	17	30	35	0-91
Stimulated Mononuclear Blood Cells	17	118	23	83-161

* Standard Deviation

Sepiapterin reductase activity in plasma was extremely low and close to the detection limits of the assay. Red blood cells hemolysates showed a mean activity of 0.46±0.08 pmoles BH_2 /min/mg protein (mean ± S.D.). Non-stimulated mononuclear blood cells showed a very variable sepiapterin reductase activity, ranging from non-detectable to 91 pmoles BH_2/min/mg protein, with some samples with no measurable activity. However, when stimulated with PHA for 5 days, the specific activity increased to 118±23 pmoles BH_2/min/mg protein (mean ±S.D.), and the values from the different samples became more uniform.

Stability in frozen samples

The stability of sepiapterin reductase in homogenates prepared in the "storage buffer" was checked. Aliquots of the same sample were stored at -40°C and analyzed at different times. Samples of amniotic fibroblasts after 1 month practically had the same activity as when freshly prepared. The activity in skin fibroblasts homogenates was also found to be stable, at least up to 15 days. In chorionic villi homogenates the activity decreased rapidly in samples kept at -40°C; approximately 70% of the activity was lost after one week, and 90% after two weeks. In stimulated mononuclear blood cells, we found that approximately 14% of the initial sepiapterin reductase activity was lost upon freezing, but thereafter it remained unchanged for at least two weeks.

Discussion

Sepiapterin reductase activity has already been reported in several mammalian tissues, including T-cells (11), adrenal medulla and cortex, brain, liver, spleen and erythrocytes. In this paper we demonstrate this enzyme activity in amniotic and skin fibroblasts, chorionic villi and stimulated mononuclear blood cells. The presence of sepiapterin reductase in these tissues will make it possible the screening for plausible sepiapterin reductase deficiencies, including prenatal diagnosis.

Compared to GTP cyclohydrolase and pyruvoyl-H_4-pterin (PTP) synthase, the two enzymes in the BH_4 pathway for which inborn errors have been found, sepiapterin reductase seems to be much more ubiquitous. GTP cyclohydrolase deficiency can only be confirmed by liver biopsy or, with reserves, in stimulated mononuclear blood cells (8). This enzyme activity could not be found in amniotic fibroblasts or chorionic villi samples (12). Diagnosis of PTP synthase deficiency is more accessible, since this enzyme is measurable in red blood cells and prenatal diagnosis has already been performed with fetal blood (13). Sepiapterin reductase has been measured in fetal erythrocytes (13), and we have demonstrated that it is also measurable in amniotic fibroblasts and chorionic villi for prenatal diagnosis.

Screening for sepiapterin reductase deficiency can be performed either in skin biopsies or, more easily, in blood samples. Although we have found activity in plasma, this is extremely low, and we cannot discard the possibility of a contamination with enzyme from blood cells lysed during the separation procedure.

We have found that the activity in mononuclear blood cells that had not been subjected to stimulation was very variable, including samples that showed no detectable activity at all. However, upon PHA stimulation, the specific activity increased to a mean value 256 times higher than the one found in erythrocytes, and with a comparable standard deviation from the mean. Therefore, it is possible to use the sepiapterin reductase assay in stimulated mononuclear blood cells as confirmatory or complementary to the one carried out in erythrocytes. The variable sepiapterin reductase activity found in fresh mononuclear blood cells samples may be due to the heterogeneous type of cells in the preparation, or to the immunological condition of the donor. However, stimulation for 5 days with PHA gives rise to a more homogeneous final population of cells.

Recently, several papers have appeared demonstrating a relationship between pteridine synthesis and lymphocyte activation. The participation of BH_4 in the regulation of lymphocyte proliferation has been reported (11, 14, 15, 16) and it has also been shown that stimulation of human mononuclear blood cells produces an increase in the synthesis of pteridines (11, 17). Our results regarding the increase of sepiapterin reductase activity during mononuclear blood cells stimulation suggest that this enzyme may be involved in the regulation of lymphocyte activation. Since GTP cyclohydrolase activation in stimulated human mononuclear blood cells has also been reported (8), it seems likely that the synthesis of the enzymes of the BH_4 pathway could be triggered during the PHA induction of lymphocyte proliferation.

Acknowledgments

This work was supported in part by a grant from the US Department of Health and Human Services, Bureau of Health Care Delivery and Assistance (Project MCJ 9049) and by a grant from the Board of Trustees of the Magee-Womens Hospital. Support for J.F. was provided by a grant from the "Conselleria de Cultura, Educació i Ciència de la Generalitat Valenciana" (Spain). We also wish to acknowledge M. Guetthoff for her valuable technical assistance and D.C. Ennis for the HPLC analyses.

References

1. Nichol, C.A., C.L. Lee, M.P. Edelstein, J.Y. Chao and D.S. Duch. 1983. Proc. Natl. Acad. Sci. USA 80, 1546-1550.
2. Milstien, S. and S. Kaufman. 1983. Biochem. Biophys. Res. Commun. 115, 888-893.
3. Milstien, S. and S. Kaufman. 1986. In: Chemistry and Biology of Pteridines (B.A. Cooper and V.M. Whitehead, eds.). Walter de Gruyter, pp. 169-181.

4. Kaufman, S., N. Holtzman, S. Milstien, I.J. Butler and A. Krumholz. 1975. N. Engl. J. Med. 293, 785-790.

5. Niederwieser, A., N. Blau, M. Wang, P. Joller, M. Atarés and J. Cardesa-García. 1984. Eur. J. Pediatr. 141, 208-214.

6. Niederwieser, A., W. Leimbacher, H.Ch. Curtius, A. Ponzone, F. Rey and D. Leupold. 1985. Eur. J. Pediatr. 144, 13-16.

7. Katoh, S., Y. Arai, T. Taketani and S. Yamada. 1974. Biochim. Biophys. Acta 370, 378-388.

8. Blau, N., P. Joller, M. Atarés, J. Cardesa-García and A. Niederwieser. 1985. Clin. Chim. Acta 148, 47-52.

9. Katoh, S., T. Sueoka and S. Yamada. 1982. Biochem. Biophys. Res. Commun. 105, 75-81.

10. Ferré, J. and K.B. Jacobson. 1985. J. Chromatogr. 350, 389-398.

11. Ziegler, I. J. Ellwart and U. Schwulera. 1986. In: Chemistry and Biology of Pteridines (B.A. Cooper and V.M. Whitehead, eds.). Walter de Gruyter, pp. 209-212.

12. Ferré, J. and E.W. Naylor. 1986. In: Chemistry and Biology of Pteridines (B.A. Cooper and V.M. Whitehead, eds.). Walter de Gruyter, pp. 309-313.

13. Niederwieser, A., H. Shintaku, Th. Hasler, H.Ch. Curtius, H. Lehmann, O. Guardamagna and H. Schmidt. 1986. Eur. J. Pediatr. 145, 176-178.

14. Ziegler, I., U. Hamm and J. Berndt. 1983. Cancer Res. 43, 5356-5359.

15. Ziegler, I., U. Schwulera, H.H. Sonneborn and W.J.P. Müller. 1985. Naturwiss. 72, 330-331.

16. Webler, S. and D.L. Jaye. 1986. In: Chemistry and Biology of Pteridines (B.A. Cooper and V.M. Whitehead, eds.). Walter de Gruyter, pp. 235-238.

17. Ziegler, I. 1985. Biochem. Biophys. Res. Commun. 132, 404-411.

IMPAIRED BIOPTERIN SYNTHESIS IN A PATIENT WITH MILD HYPERPHENYLALANINEMIA. A NEW VARIANT ?

J.L. Dhondt, G. Forzy, J.M. Hayte

Laboratoire de Biochimie. Faculté Libre de Médecine de Lille, 56 rue du Port, F-59046 LILLE Cédex, France

P. Guibaud, M.O. Rolland, C. Dorche, S. Andre

Centre d'Etude des Maladies Métaboliques, Hôpital Debrousse, Lyon, France.

Introduction

Because the possible benefit of an early institution of treatment with pterins and neurotransmitter precursors in tetrahydrobiopterin deficiency, investigation of tetrahydrobiopterin metabolism has been recommended for all newly screened hyperphenylalaninemic babies. But systematic search of cofactor deficiency among hyperphenylalaninemic newborns has recently brought variant forms of biopterin-synthetase deficiency to light. These variants have been called transient, partial or peripheral form, but no clear diagnostic distinction exists between them (J.L.Dhondt this volume).

The present study deals with a case of mild hyperphenylalaninemia presumably caused by a defect on the tetrahydrobiopterin synthetic pathway different from defects previously reported.

Case report

Primael is the third boy from healthy and unrelated parents. Pregnancy and delivery were normal, Apgar score at 10 and birth weight of 4300 g. Tremors were observed, but blood glucose and calcium levels were normal. The Guthrie test performed at 5 days of age was positive. At 3 weeks of age, serum phenylalanine was 1260 µmol/l (21 mg/100 ml) and tyrosine 75 µmol/l while the child was on a diet providing 400 mg of phenylalanine per day. Clinical examination was normal except few tremors of upper limbs after stimulation and a moderate tendency to hypertonia. EEG examination revealed a few spikes and sharp waves in the frontal area. On a 500 mg phenylalanine/d diet, the blood phenylalanine dropped to 485 µmol/l. A phenylalanine load (75 mg/kg) resulted in an increase of phenylalanine (840 µmol/l 90 minutes after the load) with a slow decrease (660 µmol/l 24 hours after).

Unconjugated Pterins and Related Biogenic Amines

Table I. Blood phenylalanine levels and urinary pteridines excretion.

	age	blood phe µmol/l	neopterin	biopterin	N/B
			nmol/mmol of creatinine		
P. Primael					
	26d	680	18670	868	21.5
	37d	445	14363	979	14.7
untreated PKU children					
	(2w-1m)		7418 ±5682	3550 ±1980	2.3 ± 1.6
Controls					
	(2w-1m)		2036 ± 787	817 ± 341	2.8 ± 1.4

Pteridine analysis of urine samples obtained at 26 and 37 days of age showed increased neopterin and normal biopterin levels with an increased neopterin to biopterin ratio (N/B). However, the biopterin excretion was low when compared to values in untreated PKU infants (Table I).

At 2 months of age, the neurologic examination confirmed a degree of hypertonia and minor disturbances of EEG tracing were still observed. CSF pterins and monoamine metabolites were measured (Table II). A tetrahydrobiopterin loading test (8.5 mg/kg/d) was performed while the child received 450 mg of phenylalanine per day. The blood phenylalanine was 360 µmol/l and decreased to 105 and 85 µmol/l at 4 and 8 hours respectively after the tetrahydrobiopterin challenge. In the urine, an increase of biopterin excretion attested for the good absorption of the oral tetrahydrobiopterin and neopterin continuously decreased after the tetrahydrobiopterin administration.

Since no major perturbation in monoamine neurotransmitters metabolism was noted in CSF, no specific treatment was decided. The dietary phenylalanine tolerance progressively increased to 1000 mg/d. At 7 months of age, irritability and sleep perturbations were noted; these troubles were related to an elevation of blood phenylalanine (670 µmol/l) due to an increase of phenylalanine intake to 1500 mg/d. On a 1200 mg/d diet, these troubles disappeared and blood phenylalanine decreased to 335 µmol/l. At 9 months of age, clinical examination and EEG were normal.

Methods

Tetrahydrobiopterin tablets and pteridine standards were obtained from Dr.B.Schircks (Jona, Switzerland).

Routine analysis of pterins in urine and CSF was

performed by the method previously described (1, J.L.Dhondt this volume).

Dihydropteridine reductase (DHPR), 6-pyruvoyl-tetrahydro-pterin synthase (6-PT synthase) and sepiapterin reductase have been measured in erythrocytes by Dr. A.Niederwieser (Zurich, Switzerland).

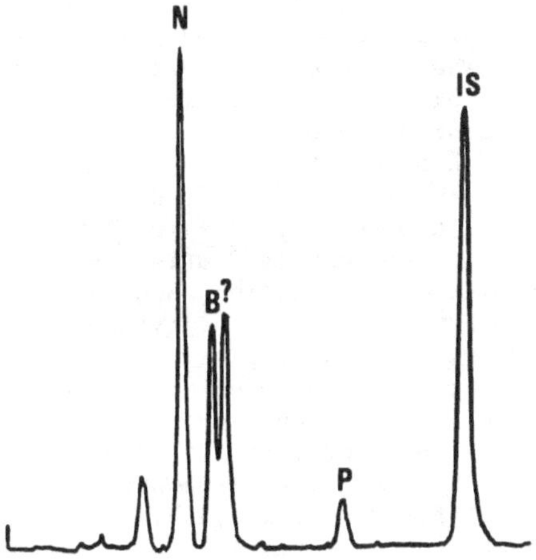

Figure 1 : HPLC separation of urinary pteridines in the patient. N: neopterin, B: biopterin, ?: unknown peak, P: pterin, IS: internal standard (6-methyl-pteridine).

Table II. CSF concentrations of pteridines and monoamine metabolites (nmol/l)

	Primael			controls (n=76)
at age:	2m	6m	9m	
neopterin	42.8	11.9	117.0	22.2 ± 10.2
biopterin	31.3	15.9	33.6	28.3 ± 10.1
unknown peak	22.4	6.6	18.6 *	
N/B ratio	1.4	0.8	3.5	0.8 ± 0.4
5-HIAA	315	170	178	225 ± 107
HVA	393	135	369	364 ± 139

* calculated as biopterin equivalent

Results

The HPLC profile of urinary pteridines revealed the existence of an abnormal peak (Fig.1) which has not been seen in the 7200 chromatograms already performed in the laboratory, and particularly in urine from other typical or partial forms of biopterin synthetase deficiency. This peak was found present in all the samples collected between 1 month and 9 months of age, that ruled out the possibility of an interfering substance. The ratio of this peak to that of biopterin was found to be relatively stable in 23 urine samples collected during a 48 hours period: 1.44 ± 0.21 . This peak was well separated on a SCX column but coeluted with biopterin on a reverse-phase column eluted with disodium phosphate 0.02 M, citric acid 0.04 M (pH 3) and 5% methanol. The peak was also observed in non-oxidised urine and was found to be labile under alkaline oxidation. None of the reference compounds which have been tested (erythro- and threo-biopterin, xanthopterin, isoxanthopterin, 6-formylpterin, 6-hydroxymethylpterin, pterin-6-carboxylic acid, monapterin, lumazine, biolumazine, sepiapterin, pterin) coeluted with that peak. Thin layer chromatography on cellulose with K_2HPO_4 0.01 M gave a good separation of that compound (Rf = 0.60 vs 0.54 for biopterin), which was found to have a blue fluorescence under UV similar to that of biopterin.

That peak was also found in CSF; however, biopterin concentration was within normal range and N/B ratio was not increased as in urine, except at 9 months of age.

Pteridine excretion in the members of the family is reported in table III. Biopterin and neopterin excretion was found normal except the presence of the unknown peak in both the parents and one of the two brothers, but at a lower concentration.

The erythrocyte activity of DHPR, 6-pyruvoyl-tetrahydropterin synthetase and sepiapterin reductase was normal.

Discussion

The diagnosis of tetrahydrobiopterin synthesis deficiency is usually based on the demonstration of a rapid normalisation of blood phenylalanine level following an oral tetrahydrobiopterin load, decreased levels of monoamine metabolites in CSF, an abnormal profile of pteridines in biological fluids with neopterin accumulation and trace amount of biopterin resulting in a high neopterin to biopterin (N/B) ratio (1), and the absence of activity of the biopterin-synthetase system, likely due to a lack of the 6-pyruvoyl-tetrahydropterin synthase component (2).

As in other screening programs for metabolic diseases, the screening of patients with hyperphenylalaninemia for

Table III. Urinary excretion of pteridines in parents and siblings (nmol/mmol of creatinine)

subject	age	neopterin	biopterin	unknown peak *	N/B
Mother	31 y	624	856	130	0.73
Father	43 y	298	695	82	0.43
controls (adults)		297 ± 88	679 ± 186	-	0.46 ±0.17
Philippe	7 y	404	879	165	0.46
Emmanuel	9 y	409	766	nd	0.53
controls (7-9 y)		607 ± 307	893 ± 230	-	0.79 ±0.65
Primael	9 m	10644	1835	2275	5.80

* calculated as biopterin equivalent

disorders of biopterin metabolism can be expected to detect a heterogeneous group of patients. Indeed, several patients have been described with ill-defined forms of biopterin-synthetase deficiency. The apparent absence of CNS complication did distinguish these forms from other "typical" biopterin-synthetase deficiency. However, if such atypical forms are certain, the distinction between peripheral, partial and transient forms is not clear (J.L.Dhondt this volume).

In the present case, the following data suggest that the infant's hyperphenylalaninemia resulted from a defective synthesis of tetrahydrobiopterin:

- the N/B ratio at the neonatal period was higher than that observed in normal or PKU neonates. Moreover, the biopterin excretion can be considered as a low value. Indeed, if its absolute value was within the normal range for age, the value was low when compared to that in hyperphenylalaninemic babies, since an increase of blood phenylalanine is known to increase biopterin excretion (3). Moreover, the significant decrease of blood phenylalanine following tetrahydrobiopterin loading suggested that the in vivo phenylalanine hydroxylation was limited without cofactor supplementation. The progressive increase of the dietary phenylalanine tolerance (Fig.2) suggested that the metabolism of phenylalanine had improved with time, but remained limited since an increase of phenylalanine intake to 1500 mg/d resulted in an increase of blood phenylalanine.
- the chromatographic profile of urinary pteridines revealed the existence of an abnormal peak which has many of the characteristics of a pteridine. This peak was also found in

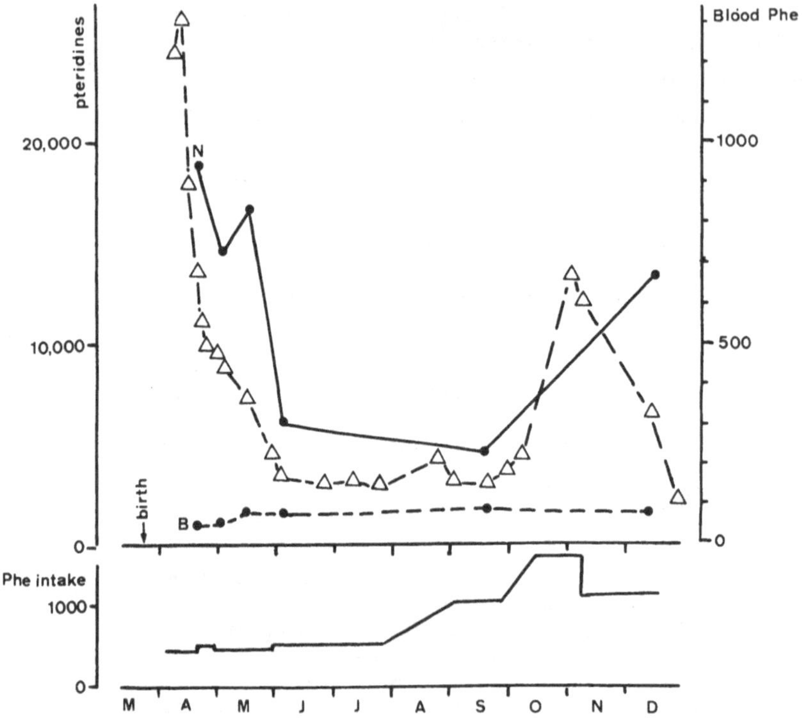

Figure 2 : Evolution of blood phenylalanine levels (△ µmol/l), urinary pteridine excretion (• nmol/mmol creatinine, N: neopterin, B: biopterin), and dietary phenylalanine intake (mg/d) in patient PP.

the CSF. Such a profile has never been observed in other typical or partial forms of biopterin synthetase deficiency nor described in the literature. The peak was labile to alkaline oxidation, that suggested a tetrahydro-form of the original pterin. Definite characterisation of this compound is in progress.

- the minor clinical and EEG abnormalities noticed at the neonatal period are not specific enough to define CNS involvement. However, the observation of sleeping difficulties at 7 months of age, when the blood phenylalanine concentration was no longer controlled, perhaps demonstrated a toxic effect of high phenylalanine level. In two case of partial DHPR deficiency (B. Leroux, personal communication), an increase of blood phenylalanine was responsible for behaviour and neurological alterations which were controlled by a phenylalanine restricted diet without the need of neurotransmitters supplementation. In a DHPR deficient patient on neurotransmitter precursors recently reported by Ponzone (4), CSF HVA and HIAA concentrations were found higher when the blood phenylalanine level was strictly controlled by a PKU diet. Two patients with partial biopterin-synthetase deficiency (5) were reported to be rather sleepy after an oral phenylalanine load. It seems that in patients with limited tetrahydrobiopterin synthesis neurological signs can be induced by an excess of phenylalanine at a concentration (about 10 mg/dl) lower than that generally believed to be harmful in classical PKU (above 20 mg/dl).

The enzymatic investigations performed in all the members of the family were negative (data not shown) as far as the enzymes which have been measured are the only ones involved in the metabolism of tetrahydrobiopterin. Other metabolic routes for tetrahydrobiopterin synthesis have been speculated, it has been proposed that a specific 6-pyruvoyl-tetrahydropterin reductase can also convert 6-pyruvoyl-tetrahydropterin into tetrahydrobiopterin but at a slower rate than does sepiapterin reductase. A defect in this reductase or in an other reductase may cause a mild HPA with practically normal brain metabolism like in the present case.

The presence of the unknown peak in the urine of parents suggests a recessive autosomal inheritance of the metabolic defect. The characterisation of the unknown peak still appeared as the only way to establish which metabolic route of biopterin synthesis is altered.

Nevertheless, this observation points out the necessity of a careful evaluation of tetrahydrobiopterin metabolism in each babies with neonatal hyperphenyl-alaninemia, even mild or transient.

REFERENCES

1. Dhondt, J.L., Largillière, C., Ardouin, P., Farriaux, J.P., Dautrevaux, M. 1981. Clin.Chim.Acta 110, 205

2. Niederwieser, A., Leimbacher, W., Curtius, H-Ch., Ponzone, A., Rey, F., Leupold, D. 1985. Eur.J.Pediatr. 144, 13

3. Dhondt, J.L., Ardouin, P., Hayte, J.M., Farriaux, J.P. 1981. Clin.Chim.Acta 116, 143

4. Ponzone, A., Guardamagna, O., Ferraris, S., Biasetti, S., Bracco, G., Niederwieser, A. 1987. Eur.J.Pediatr. 146, 93

5. Rey, F., Saudubray, J.M., Leeming, R.J., Niederwieser, A., Curtius, H-Ch., Rey, J. 1983. Arch.Fr.Pediatr. 40, 237

DIHYDROPTERIDINE REDUCTASE ACTIVITY IN FETAL TISSUES

G. Bracco, A. Iavarone, S. Pagliardini, F. Levis
Laboratorio Analisi Chimico-cliniche, Ospedale Infantile Regina Margherita, 10126 Torino, Italy

O. Guardamagna, S. Ferraris, A. Ponzone
Clinica Pediatrica dell'Università, 10126 Torino, Italy

Introduction

Dihydropteridine reductase (DHPR) (E.C. 1.6.99.7) is an enzyme which catalyzes the transformation of dihydropteridine into its active form tetrahydrobiopterin (BH4), cofactor for phenylalanine-, tyrosine-, tryptophan-hydroxylases.

DHPR deficiency (McKusick 26163) is the effect of at least two different autosomal recessive mutations, leading either to lack of DHPR protein (DHPR- CRM-) or to production of an immunologically reactive but enzymatically inefficient protein (DHPR- CRM+)(1). These two different defects possibly cause two different kinds of clinical presentation, more severe in the DHPR- CRM+ instance (2). DHPR deficiency causes lack of BH4 and accumulation of dihydropteridine and its pterin precursors. Impaired activity of aromatic amino acid hydroxylases results in hyperphenylalaninemia and neurotransmitter deficiency. As previously reported (3), DHPR deficiency seems to be more frequent in Southern Italian populations than in other ethnic groups.

DHPR deficiency can be diagnosed by urinary pterin analysis, and

Unconjugated Pterins and Related Biogenic Amines

by direct enzyme activity measurement on spotted blood, lymphocytes, fibroblasts.

Surveillance of at risk pregnancies for genetic metabolic disorders is now a reality by means of new and safe procedures of fetal tissue sampling.

BH4 deficiencies are disorders which meet the criteria for prenatal diagnosis, because they are severely debilitating and life-threatening, and therapy is difficult and still has an uncertain outcome. Unfortunately, heterozygote detection is so far only possible for DHPR deficiency: opportunity of prenatal diagnosis of other BH4 deficiencies (pyruvoyltetrahydropterin synthase or PTPS deficiency, guanosine triphosphate cyclohydrolase I or GTP-CH I deficiency) can only arise from the unlucky event of a previous affected child in the offspring.

Prenatal diagnosis of PTPS deficiency can be accomplished by measuring enzyme activity in fetal red blood cells and by amniotic fluid pterin analysis (4).

DHPR is a ubiquitarious enzyme, which is present with different degrees of activity in whole blood (5), lymphocytes and platelets (6), fibroblasts (7), amniotic fluid cells (8).

Prenatal diagnosis of DHPR deficiency can be obtained by means of enzyme measurement on cultured amniotic cells (9).

The procedure of culturing amniotic fluid cells is time consuming, and genetic counselling is rather delayed. Our laboraty has access to fetal blood samples for prenatal diagnosis of thalassemia and to chorionic villi for prenatal diagnosis of metabolic disease and from voluntary pregnancy termination.

Owing to these reasons, DHPR activity has been determined on fetal blood and on chorionic villi.

Materials and Methods

DHPR activity has been measured on 497 subjects, as indicated in table I.

Table I - Tissues Examined for DHPR Activity in Different Study Groups.

STUDY GROUP	Number	TISSUE
Normal Controls		
Newborns	65	blood spot
Subjects over 1 year of age with no major pathology	118	blood spot
Selected patients		
Children with unrelated disease	4	fibroblasts
Children with neurological disease	64	blood spot
PKU: homozygotes and relatives	175	blood spot
PTPS deficiency: homozygotes and relatives	5	blood spot
DHPR deficiency: homozygotes and relatives	49	blood spot
	3 *	fibroblasts
Pregnancies at risk for thalassemia (19-21 weeks)	13	fetal blood spot
Terminating pregnancies (8-12 weeks)	4	chorionic villi

* Also examined by blood spot DHPR measurement.

Of the four DHPR deficiencies examined, one has an immunologically inefficient DHPR (DHPR- CRM+), one has no immunological evidence of DHPR protein (DHPR- CRM-), and two are now being studied for immunological characterization. All four have also been diagnosed by means of urinary pterin analysis.
Termination of pregnancy was under a voluntary basis, and signed, informed consent was obtained by each pregnant woman in order to sample chorionic villi immediately before termination.

DHPR activity of each sample was determined at least twice. Blood spot DHPR activity was measured according to Arai et al. (5) and fibroblast DHPR according to Firgaira et al. (7) . The same procedure was adjusted for chorionic villi.

Results

Mean DHPR activity shows no significant differences among the following classes: subjects over one year of age with no major pathology and PKU (homozygotes and relatives). This group will now be referred to as the general population. DHPR activity in newborns is higher than in the general population. DHPR activity in fetal tissues is lower than in control groups (see Table II).

Table II - DHPR Activity in Normal Subjects

Group	Number	DHPR	S.D.
General Population (blood spot)	328	3.88	1.22
Newborns (blood spot)	65	5.91	1.64
Fetal blood (blood spot)	13	3.04	0.77
Children with unrelated disease (fibroblasts)	4	33.0	3.2
Chorionic villi	4	7.92	3.2

Blood spot DHPR activity is nmol ferrocytochrome/min/5 mm disc. Fibroblast and chorionic villi DHPR activity is nmol NADH/min/mg of protein.

No DHPR activity was found in blood spots from 2 previously described children (2), nor from two more children with clinical signs of malignant hyperphenylalaninemia. All four had also been diagnosed by urinary pterin analysis. Examined fibroblasts of one of them also show no activity.

A reduction of blood spot activity to about 50% of the general population was found in:

7 obligate hetezygotes from 4 families, and 6 relatives from three of these families; similar reduction was observed in fibroblasts from two of the 7 parents examined;

2 brothers, both affected by classic PKU;

1 parent of a classic PKU child (not related to the former);

1 PTPS deficient patient and his father;

3 subjects over one year of age with no major pathology (one child and two adults);

1 child with neurological involvement.

Appearently frequent linkage of low DHPR activity with PKU families must be further evaluated as it has so far no explanation.

Table III summarizes our results with DHPR measurements on DHPR deficiencies, obligate heterozygotes and DHPR relatives with reduced activity, subjects suspected of being heterozygotes because of low activity, and controls.

Discussion

As previously suspected (3), DHPR deficiency seems confirmed as a rather frequent genetic disorder in Southern Italy. In fact, 8 subjects (from 6 families) out of the 328 of the general population (i. e. not suspected of being DHPR deficient), show a decrease in activity to about 50% of controls, the same activity found in obligate heterozygotes. All 8 cases and the 4 affected by total DHPR deficiency are of Southern Italian ancestry, no one being from North of Neaples.

TABLE III - DHPR Activity in Blood Spots, Fibroblasts, Fetal Blood Spots, and Chorionic Villi.

	DHPR-homozyg. (N. 4)	DHPR-heteroz. (N. 13)	DHPR-suspected heteroz. (N.8)	Controls (N.344)
Blood Spots : Number	4	13	8	328
Mean DHPR	0	1.38	1.35	3.88
S.D.	-	0.32	0.34	1.22
Fibroblasts: Number	1	2	-	4
Mean DHPR	0	16.4	-	33.0
S.D.	-	-	-	3.2
Fetal Blood: Number	-	-	-	13
Mean DHPR	-	-	-	3.04
S.D.	-	-	-	0.77
Chorionic Villi: Number	-	-	-	4
DHPR (8 weeks)				4.1
(9 weeks)				6.1
(9 weeks)				8.9
(10 weeks)				12.6

Blood spot DHPR activity is nmol ferrocytochrome/min/5 mm disc. Fibroblast and chorionic villi DHPR is nmol NADH/min/mg of protein.

Low DHPR activity of one of the 64 children with neurological involvement is not here taken into consideration: diagnoses of each case must be carefully evaluated before drawing conclusions, also keeping in mind that reduced DHPR activity has recently been associated with X-linked mental retardation (10).

DHPR activity in fetal blood samples is slightly lower than that of the general population. DHPR activity of chorionic villi at 8-10 weeks of gestation yields low values, possibly because of the cellular heterogeneity of the sample. More chorionic villi have to be examined to demonstrate whether it is necessary to culture villi in order to obtain higher activities. Low chorionic villi activity can stand for low DHPR activity in fetal tissues, as demonstrated by lower activity in fetal blood (19-21 weeks) than in newborn blood. Increase of chorionic villi activity with increase of pregnancy time (see Table III) cannot be evaluated on a small series.

Chorionic villi and fetal blood are now easy and safe fetal tissues to be sampled in order to measure DHPR activity. If the blood sample is a mixture of fetal and maternal blood, selective lysis of maternal erythrocytes can be obtained. DHPR activity was greatly reduced but still measurable in one pure fetal blood sample examined also after such lysis (2.05 nmol/min/disc prior to treatment, 0.62 nmol after treatment).

Prenatal diagnosis for DHPR deficiency is of great interest, but it has so far not been demonstrated as possible by means of amniotic fluid pterin determination. DHPR activity has previously been measured on cultured amniotic cells (8,9), and on fetal liver (11). Only amniotic cell cultures are suitable for prenatal diagnosis: but this method is time consuming and genetic counselling is delayed. Data here presented are preliminary, but demonstrate that DHPR can also be determined in other fetal tissues, namely chorionic villi and fetal blood, by the same procedures used for spotted blood and cultured fibroblasts, so possibly allowing a simple and early test for prenatal diagnosis.

Acknowledgements

The authors are indebted to Prof. Vilma Gabutti (Clinica Pediatrica dell'Università, Torino), Dr. Ines Pelaggi (Ospedale G. Ciaccio, Catanzaro), Prof. M. Massobrio (Ospedale Mauriziano, Torino) and Prof. F. Ravetto (Clinica Neuropsichiatrica Infantile dell'Università, Torino), who kindly provided blood spot and fetal tissue samples.

1. F.A. Firgaira, R.G.H. Cotton, D.M. Danks. 1983. Dihydropteridine reductase (DHPR) deficiency. Demonstration of DHPR- CRM+ and DHPR- CRM- mutants. In: Chemistry and Biochemistry of Pteridines (J.A.Blair ed.). Walter de Gruyter, Berlin New York.

2. R.G.H. Cotton, I. Jennings, G. Bracco, A. Ponzone, O. Guardamagna.1986 Tetrahydrobiopterin non-responsiveness in dihydropteridine reductase deficiency is associated with the presence of a mutant protein. J.Inher. Metab. Dis. 9,239-243.

3. A. Ponzone, V. Ricca, S. Ferraris, O. Guardamagna, G. Bracco, S. Pagliardini, F. Levis, M. Giovannini, E. Riva, R. Longhi.1984. DHPR deficiency in Italy. J. Pediatr. 105, 1008.

4. A. Niederwieser, H. Shintaku, Th. Hasler, H.Ch. Curtius, O. Guardamagna, A. Ponzone, H. Schmidt. 1986. Prenatal diagnosis of 'dihydropteridine synthetase' deficiency, a variant form of phenylketonuria.Eur. J. Pediatr. 145, 176-178.

5. N. Arai, K. Narisawa, M.H. Kawa, K. Tada.1982. Hyperphenylalaninemia due to dihydropteridine reductase deficiency: diagnosis by enzyme assay on dried blood spots. Pediatrics 70, 426-430.

6. F.A. Firgaira, R.G.H. Cotton, D.M. Danks. 1979. Dihydropteridine reductase deficiency diagnosis by assay on peripheral blood cells.Lancet ii, 1260-1263.

7. F.A. Firgaira, R.G.H. Cotton, D.M. Danks. 1979. Human dihydropteridine reductase. A method for the measurement of activity in cultured cells, and its application to malignant hyperphenylalaninemia. Clin. Chim. Acta 95, 46-59.

8. S. Milstien, N.A. Holtzman, M.E. O'Flynn, G.H. Thomas, I.J. Butler,S. Kaufman. 1976. Hyperphenylalaninemia due to dihydropteridine reductase deficiency. J. Pediatr. 89, 763-766.

9. F.A. Firgaira, R.G.H. Cotton, D.M. Danks, K. Fowler, A. Lipson, J.S. Yu. 1983. Prenatal determination of dihydropteridine reductase in a normal fetus at risk for malignant hyperphenylalaninemia. Prenat. Diagn. 3, 7-11.

10. R.A. Armstrong, A. Sahota, J.A. Blair, B.E. Cohen. 1986. Genetic analysis of partial dihydropteridine reductase deficiency in families with mental retardation. J. Inher. Metab. Dis. 9, 400-401.

11. L. Sanchez-Urretia, C.E. Pierce, O. Greengard. 1978. Dihydropteridine reductase activity in adults, fetal and neoplastic tissues. Enzyme 23, 346-352.

MOLECULAR ANALYSIS OF HUMAN DIHYDROPTERIDINE REDUCTASE AND DIHYDROPTERIDINE REDUCTASE DEFICIENCY

R.G.H. Cotton, W.M. Hutchison, S. Wake, I.G. Jennings, W.J. McAdam, D.M. Danks, and H.-H.M. Dahl.

Murdoch Institute, Royal Children's Hospital, Melbourne, Australia, 3052.

Introduction

Malignant hyperphenylalaninaemia is caused by defects in the biosynthetic pathway of tetrahydrobiopterin (BH_4) (1-5). Dihydropteridine reductase (hDHPR) deficiency was the first enzyme defect recognised to cause this condition (3), and it accounts for approximately 30% of the cases. As BH_4 is an essential co-factor for the three aromatic amino acid hydroxylases (phenylalanine, tryptophan and tyrosine hydroxylase) (6), hDHPR deficiency leads not only to hyperphenylalaninaemia, but also to severe neurological disturbances, and is generally lethal if untreated (1). We have identified and isolated a cDNA clone containing the full coding region of hDHPR in order to study hDHPR further at the molecular level, and to establish screening methods for hDHPR deficiency by detection of restriction enzyme fragment length polymorphisms (RFLPs).

Results and Discussion

Human liver hDHPR, which appears to be a dimer composed of identical subunits with a molecular weight of 26,000 (7), was purified to apparent homogeneity (Fig. 1A). The purification steps have been described (7), except that a FPLC step was introduced instead of DEAE Sephadex chromatography. Polyclonal antibodies had previously been raised in rabbit

Unconjugated Pterins and Related Biogenic Amines

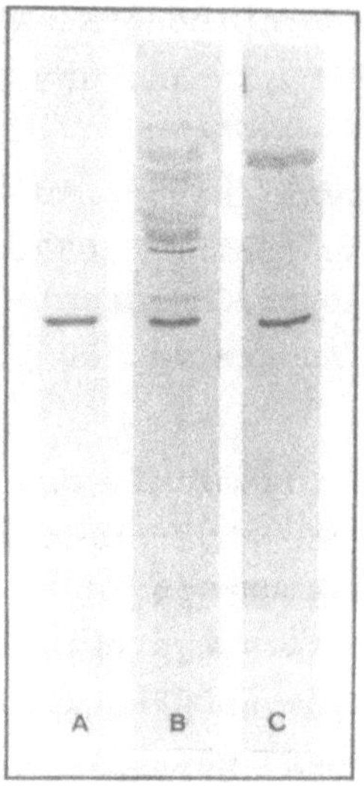

Figure 1. Analysis of purified human liver hDHPR and hDHPR antibodies. Lane A: polyacrylamide gel electrophoresis of purified hDHPR stained with comassie brilliant blue. Lane B: Western blot of human liver extract probed with the polyclonal antibodies to hDHPR. Lane C: Western blot of human liver extract probed with the affinity purified antibodies to hDHPR.

using hDHPR purified as published (7). When used to probe a protein blot of human liver extract, these antibodies not only recognised hDHPR, but also a number of other proteins (Fig. 1B). The antibodies were therefore further purified by affinity chromatography. 1 mg hDHPR (Fig. 1A) was bound to CNBr-activated Sepharose 4B and the antibodies passed over the column (equilibrated in 10 mM Tris/HCl, pH 7.4; 0.9% NaCl). After washing, the affinity purified antibodies were eluted in 0.1 M glycine, pH 2.5 and neutralised immediately with 0.1 M Tris/HCl, pH 8.5. Protein blot analysis of human liver extract with the affinity purified antibodies showed that only two bands were recognised (Fig. 1C). Using the affinity purified antibodies, we screened a human cDNA library constructed in the expression vector λgt11 (Clontech Inc.

U.S.A.). Potential positive hDHPR clones were identified using the ProtoBlot Immunoscreening System (Promega Biotech, U.S.A.).

Purified hDHPR was also subjected to trypsin and CNBr-degradation, and the peptides purified by HPLC-chromatography. The amino acid sequence of selected peptides was determined and degenerate DNA probes synthesised, based on the sequence information from two of these peptides.

Positive clones identified by immunoscreening were purified and probed with the kinase labelled oligonucleotide probes. Only clones positive with both screening methods were further analysed. This led to the identification of a 1240 bp cDNA clone (hDHPR 19) (8).

The nucleotide sequence of this cDNA clone was determined by established methods (9) and the hDHPR amino acid sequence deduced (Fig. 2).
The evidence that this is a hDHPR clone is as follows: the β-galactosidase fusion protein reacts strongly with affinity purified antibodies to hDHPR and the predicted molecular weight of hDHPR (25.700) is very close to those reported.

```
         10         20         30         40         50         60         70
MAAAAAAGEA RRVLVYGGRG ALGSRCVQAF RARNWWVASV DVVENEEASA SIIVKMTDSF TEQADQVTAE

         80         90        100        110        120        130        140
VGKLLGEEKV DAILCVAGGW AGGNAKSKSL FKNCDLMWKQ SIWTSTISSH LATKHLKEGG LLTLAGAKAA

        150        160        170        180        190        200        210
LDGTPGMIGY GMAKGAVHQL CQSLAGKNSG MPPGAAAIAV LPVTLDTPMN RKSMPEADFS SWTPLEFLVE

        220        230        240        250        260        270        280
TFHDWITGKN RPSSGSLIQV VTTEGRTELT PAYF*
```

Figure 2. Amirio acid sequence of hDHPR precursor predicted from the cDNA sequence.

	Human Liver		Sheep Liver		Bovine Liver	
Amino Acid	Dahl et al (cDNA) (8)	Firgaira et al (7)	Craine et al (10)	Cheema et al (11)	Hasegawa (12)	Korri et al (13)
Asp	9	16	18	19	15	17
Asn	7					
Thr	17	17	15	18	18	15
Ser	19	18	21	20	22	18
Glu	14	22	26	26	21	23
Gln	7					
Pro	9	9	10	9	9	10
Gly	25	26	26	24	22	20
Ala	33	31	30	32	32	26
Val	18	17	17	17	17	16
Met	8	4	4	5	5	5
Ile	9	8	6	7	6	7
Leu	21	21	20	25	22	21
Tyr	3	3	2	3	2	3
Phe	7	7	6	7	6	6
His	4	5	5	5	4	5
Lys	14	14	15	14	14	14
Arg	9	9	8	10	9	8
Cys	4	3	2	3	1	9
Trp	7	5	2	3	1	3

Table 1. Amino acid composition of hDHPR. The amino acid composition determined by Dahl et al (8) is based on the protein sequence presented in Fig. 2.

Furthermore, all of the tryptic and CNBr peptide sequences that were determined by amino acid sequencing are present in the deduced sequence and the amino acid composition of the deduced hDHPR agrees well with published data (Table 1).

The hDHPR amino acid sequence presented in Figure 2 is that of the precursor. The N-terminus of the mature hDHPR is blocked, presumably by acetylation (13). Hence we have not yet been able to determine the N-terminal amino acid of the mature protein although it is likely to be one of the six alanine residues following the initiating methionine.

The predicted amino acid sequence shows that all nine proline residues are present in the C-terminal half of the protein. Protein structure and hydrophobicity prediction based on the Chou and Fassman (14) algorithm is shown in Fig. 3.

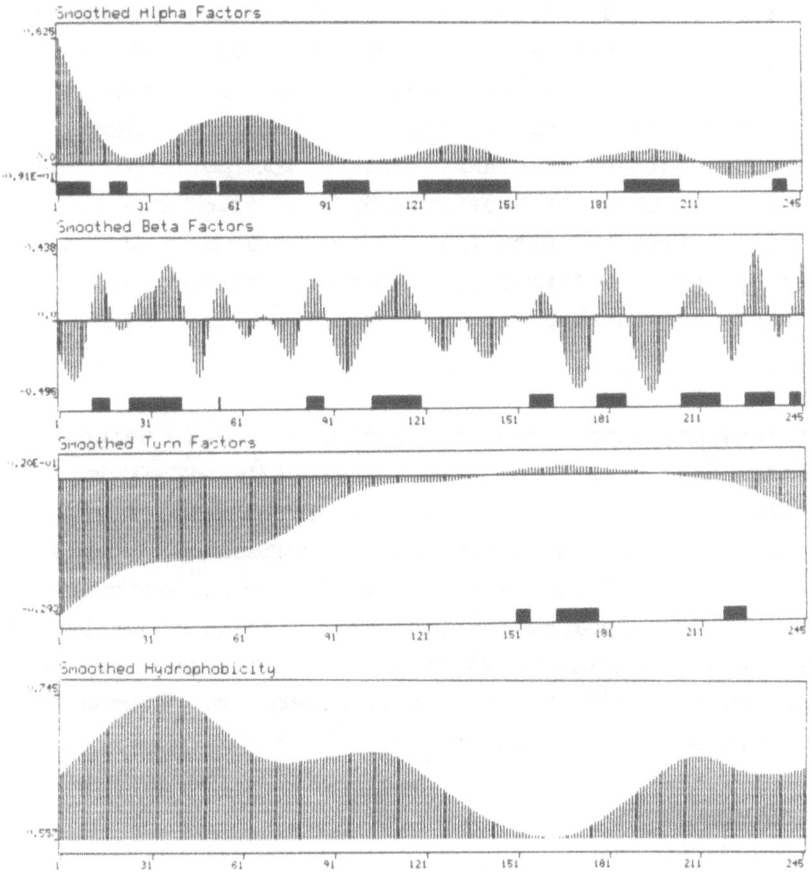

Figure 3. Prediction of the secondary structure of hDHPR according to the algorithm of Chou and Fasman (14).

The hDHPR cDNA probe has been used to study RFLPs generated with a variety of restriction enzymes. So far we have detected five RFLPs with the restriction endonucleases Ava II, Msp I and Nco I. Although these RFLPs are not in total equilibrium, they should be useful in prenatal diagnosis in families with hDHPR deficiency.

We have also examined patient cell lines for the presence of antibody cross-reacting material (CRM) and hDHPR mRNA (8). From those studies it is clear that at least three groups of patients exist, namely mRNA negative/CRM negative, mRNA positive/CRM negative, and mRNA positive/CRM positive. These results suggest that hDHPR deficiency - as expected - is caused by a variety of mutations. This conclusion is supported by the observation that we have not been able to show linkage between specific RFLP haplotypes and hDHPR deficiency.

Computer assisted comparison of hDHPR with protein sequences in the data bases Newat (Doolittle), NBRF-PIR, PSD-Kyoto (Ooi/Nakashima) and P.G. Trans did not reveal significant homologies with other published protein sequences. When a direct comparison of hDHPR was done with human dihydrofolate reductase (hDHFR), regions centred around amino acids 102-108 in hDHPR and amino acids 18-24 in hDHFR showed some homology. Comparison with 5,10-methylene tetrahydrofolate reductase from E.coli showed no significant homology.

Acknowledgements.

We thank Dr. F.J. Morgan for help with the protein sequence analysis, and Dr.N. Hoogenrath and Ms L. Graf for oligonucleotide probes.

References

1. Danks, D.M., Bartholome, K., Clayton, B.E., Curtius, H., Grobe, H., Kaufman, S., Leeming, R., Pfleiderer, W., Rembold, H. and Rey, F. 1978. Malignant hyperphenylalaninaemia - current status (June 1977). J. Inher. Metab. Dis. 1, 49-54.

2. Smith, I., Clayton, B.F. and Wolff. O.H. 1975. New variant of phenylketonuria with progressive neurological illness unresponsive to phenylalanine restriction. Lancet i, 1108-1111.

3. Kaufman, S., Holtzman, N.A., Milstein, S., Butler, I.J. and Krumholz, A. 1975. Phenylketonuria due to a deficiency of dihydropteridine reductase. New Engl. J. Med. 293, 785-790.

4. Bartholome, K. 1974. A new moleuclar defect in phenylketonuria. Lancet, ii, 1580.

5. Danks, D.M., Cotton, R.G.H. and Schlesinger, P. 1975. Tetrahydrobiopterin treatment of variant form of phenylketonuria. Lancet ii, 1043

6. Kaufman, S. and Fisher, D.B. 1974. In: Molecular Mechanism of Oxygen Activation (O. Hayaishi, ed.) Academic Press, New York, 285-369.

7. Firgaira, F.A., Cotton, R.G.H. and Danks, D.M. 1981. Isolation and Characterisation of dihydropteridine reductase from human liver. Biochem. J. 197, 31-43.

8. Dahl, H.-H.M., Hutchison, W., McAdam, W., Wake, S., Morgan, F.J. and Cotton R.G.H. 1987. Human dihydropteridine reductase: Characterisation of a cDNA clone and its use in analysis of patients with dihydropteridine reductase deficiency. Nucl. Acids. Res., in press.

9. Sanger, F., Nicklen, S. and Coulson, A.R. 1977. DNA Sequencing with Chain-terminating Inhibitors. Proc. Natl. Acad. Sci. U.S.A. 74, 5463-5467.

10. Craine, J.E., Hall, E.S. and Kaufman, S. 1972. The Isolation and Characterization of Dihydropteridine Reductase from Sheep Liver. J. Biol. Chem. 247, 6082-6091.

11. Cheema, S., Soldin, S.J. Knapp. A., Hoffman, T. and Scrimgeour, K.G. 1973. Properties of purified Quinumoid dihydropteridine reductase. Can. J. Biochem. 51;1229-1239.

12. Hasegawa, H. 1977. Dihydropteridine Reductase from Bovine Liver. J. Biochem (Tokyo) 81, 169-177.

13. Korri, K.K., Chippel, D., Chauvin, M.M., Tirpak, A., and Scrimgeour, K.G. 1977. Quinonoid Dihydropteridine Reductase from Beef Liver. Can. J. Biochem. 55, 1145-1152.

14. Chou. P.Y. and Fasman, G.D. 1978. Prediction of the secondary structure of proteins from the amino acid sequence. Adv. Enzymol. 47, 45-148.

FATE OF PERIPHERALLY ADMINISTERED TETRAHYDROBIOPTERIN IN CONGENITAL TETRAHYDROBIOPTERIN DEFICIENCY

S. Ferraris, O. Guardamagna, G. Bonetti, G. Bracco, L. Leone, A. Ponzone.

Department of Pediatrics, University of Turin, Regina Margherita Children's Hospital, 10126 TORINO, Italy.

A. Niederwieser

Department of Pediatrics, Division of Clinical Chemistry, 8032 Zurich, Switzerland.

Introduction

Tetrahydrobiopterin (BH_4) administration for both diagnosis and treatment of BH_4 deficient patients was first attempted by Danks et al. in 1975 (1,2), when the lack of a pure product appeared to be the main limitation to its use. In time, with improved availability, animal esperiments (3,4) and clinical studies (5) revealed that the poor brain penetration and the lack of recycling in dihydropteridine reductase deficiency (6,7) could represent the most limiting factors for BH_4 therapy. Replacement of the missing cofactor constitutes in theory the most rational approach to the therapy of BH_4 deficiency, also because all its functions are not known (8).

Unconjugated Pterins and Related Biogenic Amines

Only a few patients, unfortunately, and specifically only some of those that carry a synthetic defect, show a good outcome when treated with BH_4 monotherapy (9,10). Synthetic, lipid-soluble, analogues gave, on the other hand, few results despite their advantage of crossing more easily the blood-brain barrier (11,12,13). At the clinical level, our understanding of the pharmacological characteristics of BH_4 are on the whole very limited, due to the few studies so far performed.

We were surprised by examining the results of an in vivo study we performed, which largely differ from the currently accepted information.

Patient and Methods

A PTPS deficient patient (14), who was shown to be only partially responsive to the BH_4 monotherapy, at the age of two years had discontinued cofactor and neurotransmitter therapy because of difficulties in drug supply. Neurological symptoms had reappeared and serum phenylalanine (Phe) rose to about 20 mg% when he was orally loaded with a single BH_4 ($BH_4 \cdot 2HCl$, Dr Schirks Laboratory, Wettswil a A., Switzerland) dose of 20 mg/kg after overnight fasting. Serum, cerebrospinal fluid (CSF) and urine samples were taken at different intervals and the patient was kept on a free diet without neurotransmitter precursor administration during the study. Pterins and neurotransmitter metabolites homovanillic acid (HVA) and 5-hydroxyndole acetic acid (5-HIAA) were measured by HPLC (15), amino acids by an automatic amino acid analyzer.

Pharmacokinetic Analysis

Times of peak (t_{max}) and peak concentrations (C_{max}) were derived directly from experimental values. Elimination rate constant (ß) of biopterin (B) was calculated by linear regression from the log serum concentration versus time plot during the terminal phase of the curve (Fig. 1).
Excretion rate constant (K_e) was calculated from the log urinary excretion rate versus midpoint collection time (Fig. 2), according to standard pharmacokinetic methods (16) Half-lives ($t\frac{1}{2}$) were calculated as 0.693/ß. Area under the serum (and CSF) concentration curve (AUC) was measured by the

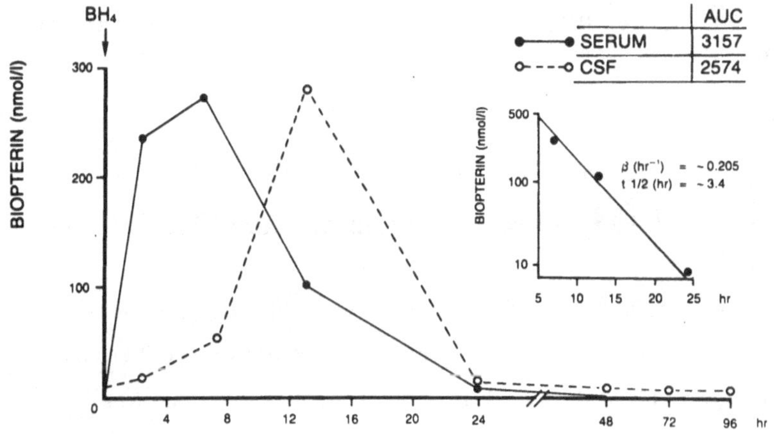

Fig. 1. Serum and CSF biopterin concentrations following an oral load dose of 20 mg/kg BH_4 in a PTPS deficient patient. Area under curve (AUC), elimination rate constant (ß), and half-life ($t\frac{1}{2}$) are shown.

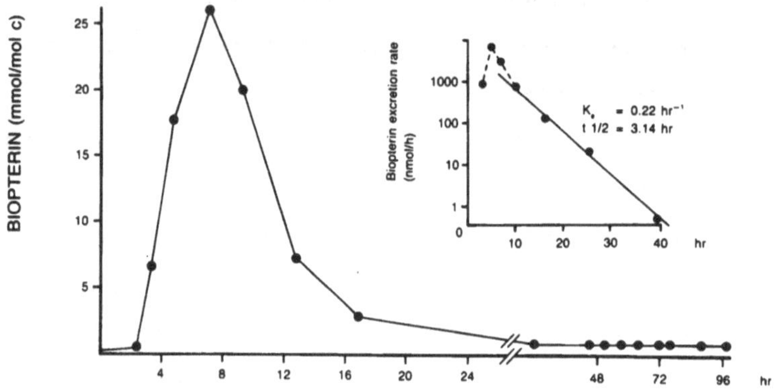

Fig. 2. Urine biopterin concentrations following an oral load dose of 20 mg/kg BH_4 in a PTPS deficient patient. Urinary elimination constant (K_e) and half-life ($t\frac{1}{2}$) are shown.

trapezoidal rule and extrapolated to infinity. B renal clearance (Clr) was calculated as $\frac{\text{amount excreted in urine}}{\text{AUC}}$.

Time courses of B levels in serum and CSF are reported in Fig. 1. C_{max} are closely similar in the two biological fluids, t_{max} being delayed in CSF by about 6 hrs. Within 24 hrs after the load, B levels return to basal values both in serum and in CSF. Also the elimination rate constant (ß) might be similar in serum and in CSF, though in the latter it was not calculated because of the scarcity of available data.

Time course of B levels in urine is shown in Fig. 2. t_{max} are slightly delayed with respect to t_{max} of serum B, but again all B is cleared within 24 hrs. B K_e, as well as B t $\frac{1}{2}$, are in the same order of magnitude of serum B ß and t $\frac{1}{2}$.

Urinary recovery of the administered BH_4, as calculated from the sum of its main excreted metabolites B, pterin and isoxanthopterin, accounts only for 2,7% of the load dose (Fig. 3). Time courses of neopterin (N) levels in serum, CSF and urine appears in Fig. 4. In all biological fluids, N concentrations significantly increase following the load, and decrease under basal values within 6-24 hrs. This effect lasts for some days. These data indicate that BH_4 can easily cross the blood-brain barrier. The low urinary recovery, in absence of other extrarenal excretion pathways, such as bile or intestine, suggests a low bioavailability of the orally administered cofactor. This

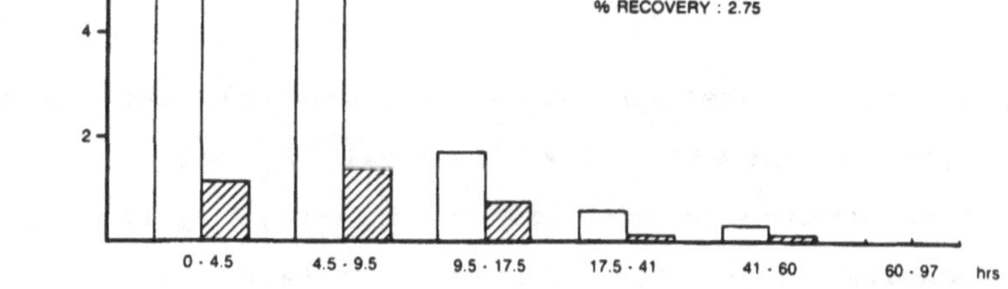

Fig. 3. Time course of urinary excretion and final urinary recovery of an orally administered dose of 20 mg/kg BH_4 in a PTPS deficient patient.

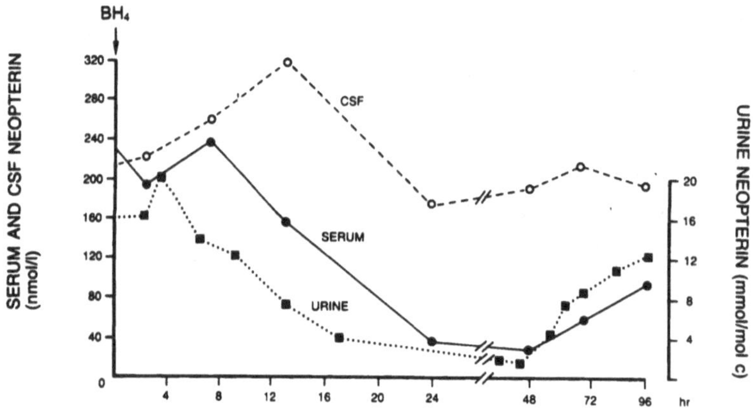

Fig. 4. CSF, serum and urine neopterin concentrations following an oral load dose of 20 mg/kg BH_4 in a PTPS deficient patient.

well agrees with the observations of Kaufman (8), who compared serum B concentrations after i.v. and oral BH_4 administration. Though no pharmacokinetic analysis was attempted by the author, from his results it is possible to calculate (AUC oral/AUC i.v.) an absolute bioavailability value of about 6%. This could indicate that the main feature that accounts for the low B concentrations reached in CSF, as well as in serum, is the low availability of orally administered BH_4 .

Pharmacodynamic Analysis

Peripheral effect of BH_4 therapy has been sufficiently shown, since easier to investigate (9,10,11). It has been demonstrated that also at the central level orally administered BH_4 can variably enhance the hydroxylating activity, seen by the improvement in neurological symptoms and by the increase in neurotransmitter production (5,10,12,13).

By comparing the time course of B concentration to that of amino acids and neutransmitter metabolites concentrations in serum and in CSF (Fig. 5), we can show that the pharmacodynamic effect is evident for about three days in both compartments. It begins as soon as B concentration increases, hence earlier in periphery, and lasts for about two days after C_{max} has been reached. In this patient, a doubling of basal CSF neurotransmitter metabolites HVA and 5-HIAA values was obtained with a single BH_4 administration.

Discussion

PTPS deficiency, leading to the absence of endogenous *de novo* synthesis of BH_4, makes the patient a good experimental model for this kind of study. The most striking finding we had, refers to the fact that BH_4 appeared to readily cross the blood-brain barrier. Previous studies in humans could be misleading because the times chosen for sampling did not take account of the kinetic properties of BH_4 and of the differences between peripheral and central compartments (5,13,17).

The lack of clinical improvement and of biochemical correction observed in many patients may not be due to the BH_4 scarce brain penetration, but possibly to poor gut absorption or to insufficient dosage, and definitely to the own central unresponsiveness or limited responsiveness. In such patients, the rate limiting factor for hydrolases dependent monoamine synthesis could be represented, as suggested by McInnes et al. (13), not by BH_4 availability in the brain (18), but by the ontogenetic underdevelopment of brain monoaminergic target areas. Even increasing the dose of exogenous BH_4 or administering high lipid-soluble analogues (8,12,13), neurotransmitter production remains at a low level, as in the

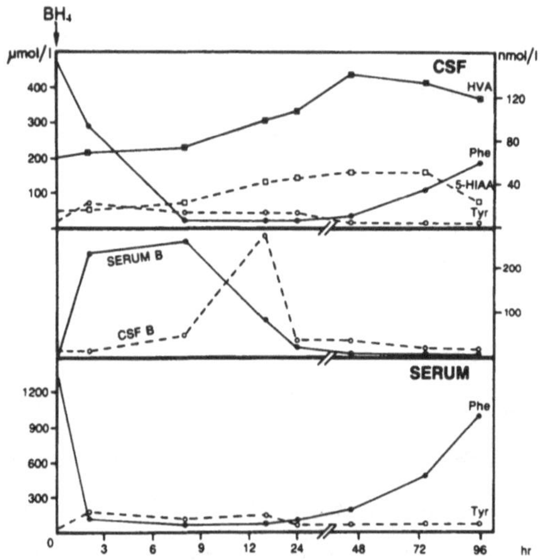

Fig. 5. Peripheral and central pharmacodynamic effect following a single dose of 20 mg/kg BH_4 in a PTPS deficient patient. Time course of CSF amino acids, 5-hydroxyndole acetic acid and homovanillic acid concentrations and of serum amino acid concentrations is compared to that of serum and CSF biopterin concentrations.

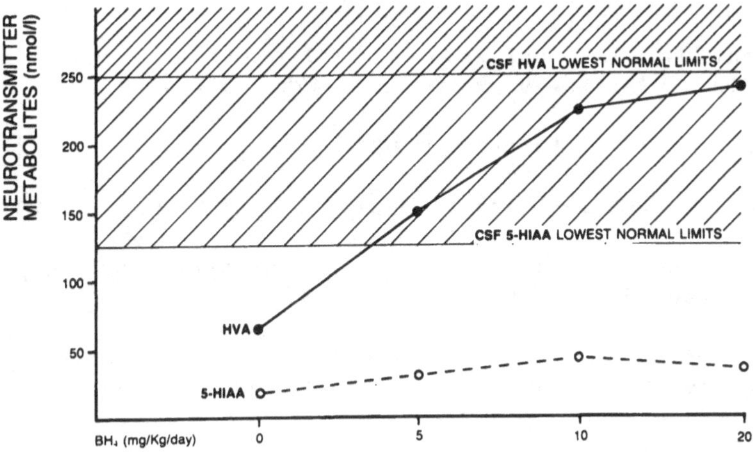

Fig. 6. CSF neurotransmitter metabolites during BH_4 monotherapy in a PTPS deficient patient: dose/effect relationships.

present case of PTPS deficiency (Fig. 6), or it is almost absent, as in another case of PTPS deficiency with neonatal presentation of low birthweight and microcephaly.

References

1. Danks, D.M., R.G.H. Cotton, P. Schlesinger. 1975. Lancet 2, 1043.

2. Danks, D.M., R.G.H. Cotton, P. Schlesinger. 1976. Lancet 2, 1236.

3. Gal, E.M., G. Hanson, A. Sherman. 1976. Neurochem. Res. 1, 511-523.

4. Kaufman, S., G. Kapatos. 1981. Science 212, 955-956.

5. Kaufman, S., G. Kapatos, R.R. McInnes, J.D. Schulman, B. Rizzo. 1982. Pediatrics 70, 376-380.

6. Kaufman, S., N.A. Holtzman, S. Milstien, I.J. Butler, A. Krumholz. 1975. New Engl. J. Med. 293, 785-790.

7. Kaufman, S. 1975. Lancet 2, 767.

8. Kaufman, S. 1985. J. Inher. Metab. Dis. 8, Suppl. 1, 20-27.

9. Niederwieser, A., H.-Ch. Curtius, M. Wang, D. Leupold. 1982. Europ. J. Pediatr. 138, 110-112.

10. Endres, W. 1985. In: Inherited Diseases of Amino Acid Metabolism (H. Bickel, V. Wachtel, eds.). Georg Thieme Verlag, Stuttgart, pp. 124-130.

11. Endres, W., A. Niederwieser, H.-Ch. Curtius, M. Wang, B. Ohrt, J. Schaub. 1982. Helv. paediat. Acta 37, 489-498.

12. Kaufman, S., G. Kapatos, W.B. Rizzo, J.D. Schulman, L. Tamarkin, G.R. Van Loon. 1983. Ann. Neurol. 14, 308-315.

13. McInnes, R.R., S. Kaufman, J.J. Warsh, G.R. Van Loon, S. Milstien, G. Kapatos, S. Soldin, P. Walsh, D. Mac Gregor, W.B. Hanley. 1984. J. Clin. Invest. 73, 458-469.

14. Niederwieser,A., W.Leimbacher, H. Ch. Curtius, A. Ponzone, F. Rey, D. Leupold. 1985. Eur. J. Pediatr. 144, 13-16.

15. Niederwieser, A., W. Staudenmann, E. Wetzel. 1984. J. Chromatogr. 290, 237-246.

16. Gibaldi, M., D. Perrier. 1975. Pharmacokinetics. Marcel Dekker, Inc., New York, pp. 6-17.

17. Danks, D.M., P. Schlesinger, F. Firgaira, R.G.H. Cotton, B.M. Watson, H. Rembold, G. Hennings. 1979. Pediat. Res. 13, 1150-1155

18. Kettler, R., G. Bartholini, A. Pletscher. 1974. Nature 249, 476.

6-PYRUVOYL-TETRAHYDROPTERIN SYNTHASE DEFICIENCY: THERAPEUTIC TRIAL WITH TWO DIFFERENT SYNTHETIC PTERIN ANALOGUES IN THREE PATIENTS

D. Leupold, H. Lehmann

Department of Pediatrics, University of Ulm
D-7900 Ulm/Do., Germany

H.Ch. Curtius, A. Niederwieser

Department of Pediatrics, University of Zürich
CH-8032 Zürich, Switzerland

Introduction

Inborn errors of tetrahydrobiopterin (BH_4) synthesis have been detected in 1 - 3 % of patients with hyperphenylalaninemia (1). Two enzyme defects are known in the biosynthetic pathway of BH_4: GTP-cyclohydrolase I (2) and a phosphate eliminating enzyme leading to 6-pyruvoyltetrahydropterin. For defects of this enzyme the term 6-pyruvoyl-tetrahydropterin synthase (PTPS) deficiency is proposed.

BH_4 is the cofactor not only for phenylalanine hydroxylase but also for the hydroxylation of tyrosine (3) and tryptophan (4). BH_4 deficiency results in a severe neurological disorder due to impaired synthesis of the neurotransmitters dopamin and serotonin. Therefore the treatment consists in substitution of the neurotransmitter precursors L-dopa and 5-hydroxytryptophan (5-HTP) and Carbidopa in order to overcome the central nervous system defect and of BH_4 to keep the plasma phenylalanine level in the normal range. It is, however, not known whether a life long treatment with neurotransmitter precursors might not produce severe side effects (5, 6). Therefore, since it has been reported (7, 8) that BH_4 is capable of penetrating the

Unconjugated Pterins and Related Biogenic Amines

blood-brain-barrier in up to 1 per cent of the applicated dose, several attempts have been made to find BH_4 analogues with the same cofactor activity but better penetration into the brain (9).

In this paper we will report clinical and biochemical data from a trial with 6-methyltetrahydropterin (6-MPH_4) in two siblings and of 6 -hydroxymethyltetrahydropterin (6-$HMPH_4$) in another child with PTPS deficiency.

Clinical data

1. 6-MPH_4

Y.Z. is treated successfully with BH_4 monotherapy since 1981(10) Y.Ö., her younger brother, the third child of this family, is treated in the same way from 18 days of life. Both children are physically and neurologically normal but slightly mentally retarded.

After informed consent was obtained from the parents BH_4 (5mg/kg bw in one single dose)was substituted by 6-MPH_4, given each morning for four days. About 36 hours after the last application of BH_4 behaviour and neurological state of the patients changed. Both got restless and irritable and had marked muscular hypotonia. The five years old girl developed all signs of a crisis seen after BH_4 withdrawal (11), i.e. pallor, ptosis, hypersalivation, ataxia and sleepiness. The 12 months old boy refused to eat, cried with a high-pitched voice and lost social contact. The fourth dose of 6-MPH_4, given 96 hours after the last application of BH_4, was doubled to 10 mg/kg bw, but when 5 hours later no clinical improvement occurred, the trial was stopped and 10 mg/kg bw BH_4 was given. 5 hours later both children had recovered completely.

2. 6-$HMPH_4$

W.C. is the second child of unrelated german parents. He was born at the 36. week of gestation with a birth weight of 1700 gms. PTPS deficiency was diagnosed at the 8th day of life and BH_4 monotherapy was tried without success. Only after substitution with neurotransmitter precursors his clinical symptoms (hypothermia, severe muscular hypotonia and failure to thrive) disappeared. He is below the third centile for heighth and weight but mentally appropriate for age.

At the age of 26 months, after informed consent of the parents, neurotransmitter precursors and BH_4 were stopped for 12 hours. 6-$HMPH_4$ was given orally in a dose of 20 mg/kg bw. (This high dose has been chosen, because it was known from the newborn period, that this amount of BH_4 was insufficient to rise neurotransmitter metabolites in the CSF). Within three hours the boys vigilance changed from normal to somnolence and after two more hours he was in a deep coma and responded to painful stimuli only. Another 20 mg/kg bw was applicated. 4 hours afterwards he shortly was awake and responded to words but suddenly his condition worsened. He developed marked ataxia and tremor of the limbs, got unconscious within two hours and had a cerebral convulsion lasting for two minutes. His temperature had constantly rosen during the trial and was 39 °C at the time of the convulsion (i.e. 24 hours after the last dose of neurotransmitter precursors).

The trial was stopped and L-dopa and 5-hydroxytryptophan were given. The boy regained conciousness within 2 hours, responded adequately to words but was sleepy. His temperature fell to normal values without antipyretic drugs. In order to test the effect of 6-$HMPH_4$ on liver phenylalanine hydroxylase it was continued at a dose of 2 and 4 mg/kg bw for four days - in addition to the normal neurotransmitter precursors L-dopa/Carbidopa and 5-hydroxytryptophan. The boy was clinically and neurologically normal.

Laboratory data

1. 6-MPH_4

In the patients Y.Z. and Y.Ö. BH_4 monotherapy is given in one single dose in the morning. The effect lasts at least for 24

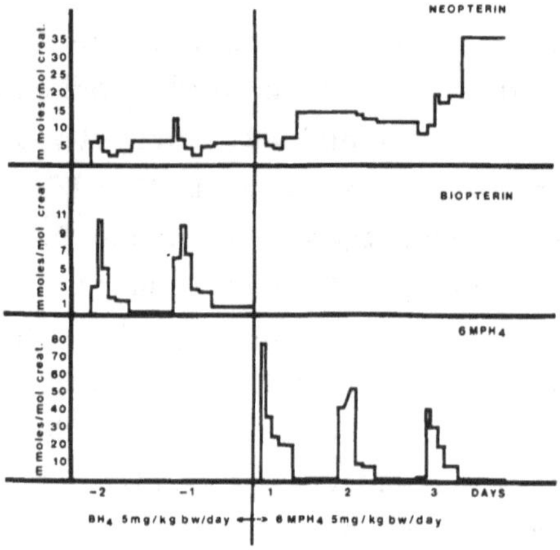

Figure 1: Excretion of pterins after loading with 6-MPH_4 in a patient with PTPS deficiency

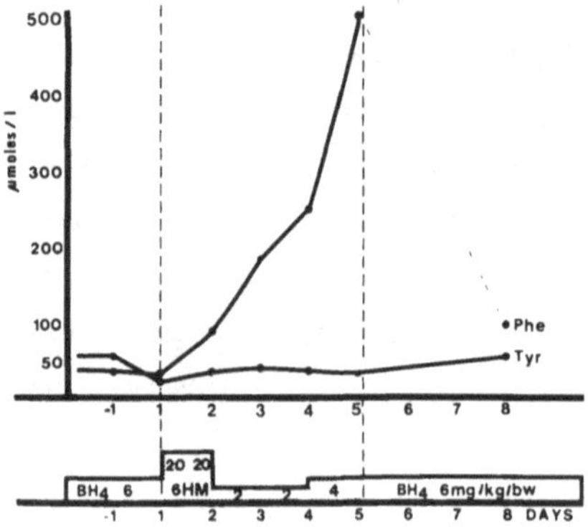

Figure 2: Effect of 6-$HMPH_4$ on phenylalanine metabolism in a patient with PTPS deficiency (6 HM = 6 -$HMPH_4$)

hours. Plasma phenylalanine concentration estimated before application was usually below 50 µM.

24 hours after 6-MPH_4, however, plasma phenylalanine concentration was increased eightfold and rose further. After 72 hours it was 1080 µmoles/l in Y.Ö. and 470 in Y.Z.

In figure 1 pterin metabolite excretion in the urine is shown. Neopterin is generally higher with 6-MPH_4 and the decrease during the first eight hours after application is not as notable as with BH_4. The excretion of 6-methylpterin clearly indicates that 6-MPH_4 must have been absorbed in the gut and is cleared very effectively.The neurotransmitter metabolites homovanillic acid (HVA) and 5-hydroxyindole acetic acid (5-HIAA) in CSF were not altered after 76 hours of 6-MPH_4 administration.

2. 6-$HMPH_4$

In figure 2 the plasma phenylalanine concentration in patient W.C. is demonstrated. Inspite of the extremely high dose of 2 x 20 mg/kg bw given within 5 hours, phenylalanine had doubled after 24 hours and increased up to 518 µmoles/l during the next days without any change of tyrosine.

Urinary excretion increases from 4.77 mmoles/mol creatinine at 0 - 2 hours to 8.08 at 24 - 36 hours.

In table I changes in CSF are shown.

Whereas during the normal medication of BH_4, L-dopa/Carbidopa and 5-HTP the neurotransmitter metabolites HVA and 5-HIAA were in the normal range for age, a sharp decrease was found 24 hours after withdrawal of neurotransmitter precursors. Neopterin rose from 54.1 to 72.2 nmoles/l. 6-$HMPH_4$ concentration was very high. The most unexpected finding, however, was an extremely high glycine concentration of 289 µmoles/l, which is about 50 times the normal value. Serine concentration was normal. Plasma glycine at the same time was in the lower normal range.

TABLE I

Effect of 6-$HMPH_4$ on CSF aminoacids, pterins, and neurotransmitter metabolites.

Date	Medication (mg/kg bw)		HVA (nmoles / l)	5-HIAA (nmoles / l)	N (nmoles / l)	B (nmoles / l)	6-$HMPH_4$ (nmoles / l)	Phe (µmoles/ l)	Tyr (µmoles/ l)	Gly (µmoles/ l)	Ser (µmoles/ l)	Plasma Gly (µmoles/ l)
Oct. 20.	BH_4	6	358,9	217,6	54,1	n.d.	n.d.	6,87	10,16	4,38	23,84	205
	L-DOPA	7,25										
	5-HT	6,6										
Oct. 21.	6-$HMPH_4$	2x20	50,4	98,5	72,2	n.d.	65,50	8,29	5,91	289,0	26,97	118
Oct. 28.	BH_4	6										
	L-DOPA	7,25										
	5-HT	6,6	239,6	155,6	79,9	n.d.	n.d.	12,03	7,94	5,75	24,16	232

n.d. = not detectable

Discussion

High-dose BH_4 monotherapy has been shown to be an effective treatment in some patients with PTPS deficiency (10, 12) but not in others (13, 14). In contrast to the good effect of the cofactor analogue 6-MPH_4 reported by Kaufman (15) the results were discouraging in two of our patients and in another from Vienna (14).
The cofactor activity on liver phenylalanine hydroxylase apparently is negligible, since plasma phenylalanine continuously increases. In Y.Z. some effect might be supposed, because four hours after the third dose phenylalanine decreased from 344 to 247 µM and a small increase of tyrosine was observed. In Y.Ö. only an increase of tyrosine occurred. This is in a sharp contrast to the effect of BH_4. Clinically both children developed severe neurological symptoms indicating neurotransmitter deficiency. The actual CSF concentration of HVA and 5-HIAA, however, which was in the lower normal range for age with BH_4 had not changed significantly after 4 days on 6-MPH_4. For explanation of this discrepancy one may speculate, that CSF neurotransmitter metabolites do not exactly mirror the actual concentration of dopamine and serotonin at the synaptic cleft.

The reason for the trial with 6-$HMPH_4$ in our third patient, W.C., were experimental data, which showed a 3 to 6 fold better penetration into rat brain and an in vivo cofactor activity for phenylalanine hydroxylase 50 % of that of BH_4 (16).
In contrast, in patient W.C. there seems to be no cofactor activity at all, either for phenylalanine hydroxylase (see figure 2) or for cerebral neurotransmitter synthesis.
The severe neurological symptoms with coma and convulsion are not explained by neurotransmitter deficiency, since before start of treatment in the newborn period the CSF concentration of HVA and 5-HIAA had been much lower. It also is unlikely, that the abrupt withdrawal of neurotransmitter precursors caused these symptoms. On a former occasion, when L-dopa and 5-HT

were withheld for 24 hours, the decrease in CSF neurotransmitter metabolites was in the same order of magnitude as now, but only mild neurological impairment was observed.
The high CSF concentration of 6-$HMPH_4$ showes, that the substance penetrated the blood brain barrier effectively. This is in contrast to BH_4, where practically no biopterin was found after a dose of 20 mg/kg bw (Data from the newborn period). Whether the excessively elevated CSF glycine concentration is an indicator for an acute toxic effect of 6-$HMPH_4$ on the glycine-cleavage system in brain remains speculative. In fact, the CSF glycine concentration observed in W.C. is higher than in patients with non-ketotic hyperglycinemia, an inborn error of glycine metabolism, leading to severe cerebral damage.

Further investigations are necessary to solve these question.

References

1. Danks, D.M., K. Bartholomé, B.E. Clayton et al. 1978.J.Inher. Metab. Dis. 1, 49 - 53

2. Niederwieser, A., W. Staudenmann, M. Wang, H.Ch. Curtius, M. Atares, J. Cardesa-Garcia. 1982. Eur.J.Pediatr. 138, 97

3. Shiman, R., M. Akino, S. Kaufman. 1971. J.Biol. Chem. 246, 1330-1340

4. Friedman, P.A., A.H. Kappelman, S. Kaufman. 1972. J. Biol. Chem. 247, 4165 - 4173

5. Sternberg, E.M., M.H. van Woert, S.N. Young, J. Magnussen, H. Baker, S. Gauthier, K. Osterland. 1980. New Engl.J. Med. 303, 782 - 787

6. Shaw, K.M., A.J. Lees, G.M. Stern.1980. Q.J. Med. 49, 283 - 293

7. Gal, E.M., G. Hanson, A. Sherman.1976.Neurochem.Res.1, 511-523

8. Kapatos, G., S. Kaufman.1981.Science, 212, 955 - 956

9. Levine, R.A., G.P. Zoephel, A. Niederwieser, H.Ch. Curtius. 1985. In: Biochemical and Clinical Aspects of Pteridines. (H.Wachter et al eds) Vol. 4, 269

10. Niederwieser, A., H.Ch. Curtius, M. Wang, D. Leupold.1982. Eur.J. Pediatr. 138, 110 - 112

11. Leupold, D., M. Wang, A. Niederwieser. 1982.In: Biochemical and Clinical Aspects of Pteridines (H.Wachter, H.Ch. Curtius, W. Pfleiderer, eds) de Gruyter, Berlin,Vol. 1, p. 307

12. Kaufman, S., G. Kapatos, R. Mc Innes, J. Schulman, W. Rizzo. 1982.Pediatrics 70(3), 376 - 380

13. Endres, W., A. Niederwieser, H.Ch. Curtius, M. Wang, B. Ohrt, J. Schaub. 1982. Helv. Paed.Acta 37, 489 - 498

14. Scheibenreiter, S. pers. communication

15. Kaufman, S., G. Kapatos, W.B. Rizzo.1983.Amer.Neurol.14, 308 - 315

16. Levine, R. et al to be published.

PLACENTAL BARRIER IN MOTHER-TO-FETUS TRANSFER OF TETRAHYDROBIOPTERIN IN HUMANS

A.Ponzone, S.Ferraris, S.Biasetti, I.Dianzani, O.Guardamagna
Department of Pediatrics, Regina Margherita Children's Hospital, 10126 Turin, Italy

L.Kierat, A.Niederwieser
Department of Pediatrics, University of Zürich, 8032 Zürich, Switzerland

Introduction

Metabolic homeostasis of the fetus results from the interaction between the maternal environment and its own developing homeostatic mechanisms. It is mediated by the placenta, alternatively blocking or allowing mono- or bi-directional transfer of substances whose concentration may not be normal. The ensuing development of the fetus, in presence of intact placental function, may be unaffected or variably affected by both maternal or its own disordered homeostasis. Final outcome will depend on the extent of the metabolic capability of the feto-placental unit to neutralize the potentially altered maternal or fetal environment. The underlying result of these interactions is already known for several inherited metabolic diseases (Table 1).

Fetal Homeostasis in Tetrahydrobiopterin (BH4) Deficiency

In the case of BH4 deficiency, the homozygous newborn can appear either clinically normal, or carrying the specific neurological symptoms associated with low birthweight and microcephaly. This issue has already

Unconjugated Pterins and Related Biogenic Amines

Table 1. Fetal Outcome in Pregnancies with Maternal or Fetal Metabolic Disorders (1,2).

Maternal metabolic disorder	fetus damaged	PKU type II tyrosinemia homocystinuria (?) diabetes IgG mediated autoimmunity
	fetus not damaged	Hartnup disease isovaleric acidemia hyperlysinemia hyperornithinemia hyperprolinemia I hypertryptophanemia hydroxyprolinemia hystidinemia hypertriglyceridemia I galactosemia xantine oxydase deficiency conjugated hyperbilirubinemia
Fetal metabolic disorder	fetus damaged	lysosomal storage diseases lipid storage diseases alpha-1-antitrypsin deficiency conjugated hyperbilirubinemia
	BH4 deficiency?	
	fetus not damaged	amino acidopathies organic acidurias urea cycle disorders glu-6-phosphatase deficiency unconjugated hyperbilirubinemia

been discussed in a previous Symposium (3) in physiological terms, i.e., if maternal BH4 does cover the fetal requirement by crossing the placenta, and in medical terms, i.e., if a pregnancy in which the fetus is found by prenatal diagnosis to have cofactor deficiency should be treated antenatally by administering BH4 to the mother.

Placental transfer of naturally occurring or exogenous substances has been studied over a long time, but not fully understood in humans, mainly because of ethical constraints. From a general standpoint, it has been ascertained to be a species-specific phenomenon, controlled by facilitating or inhibiting mechanisms, and depending to some extent on the physico-chemical properties of the substance. Some excellent reviews have appeared on this subject (4,5,6).

On a physico-chemical basis, BH4 belongs to the class of intermediate-sized (molecular weight 241), water-soluble, lipid-insoluble compounds. They do not cross or slowly cross to the fetus. If they cross, they can be detected in amniotic fluid (AF) 20 minutes after administration to the mother and after some hours they reach peak levels that are higher than maternal serum levels (6).

As regards our understanding of BH4 placental transfer, some observations we made recently may be useful.

Pterins in AF

Analysis of pterins was carried out by HPLC in AF (8) in a group of pregnant women at 14-20 weeks and in a group at 37-40 weeks (Fig.1).

Pterins were already detectable in amniotic fluid at the beginning of the second trimester, in higher concentrations than in maternal serum, and their concentrations still rose by the end of gestation. Similar results have been recently reported by Dhondt et al. in a larger series of pregnancies (7).

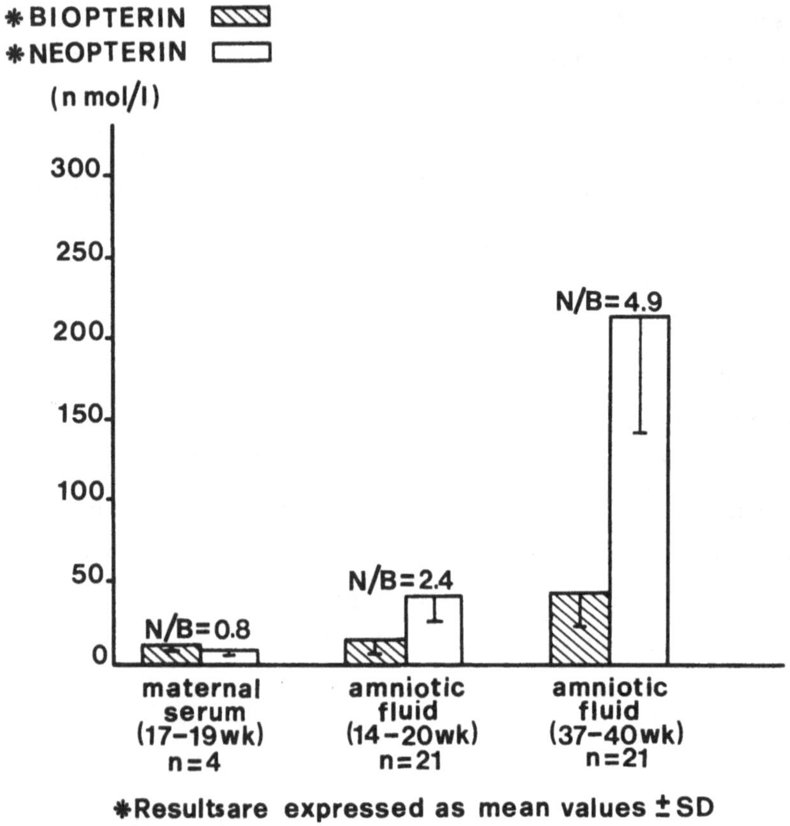

Fig.1: Maternal serum and AF pterins during pregnancy.

Such a pterin accumulation in AF against a transplacental gradient could be expression alternatively of active transfer from the mother or of the fetus' own synthesis, or of both. It also makes a back diffusion process unlikely. As the production of fetal urine progressively increases after the first trimester of gestation, the high pterin values found near term could be due to increased renal excretion rather than to increased fetal synthesis or placental transfer. The high neopterin/biopterin (N/B) ratio in AF, above that in maternal serum, could be indicative of active fetal syntesis as well as of preferential placental transfer.

BH4 Load in Pregnant Women

To solve the above questions, 4 women, at 17-19 weeks of pregnancy, before undergoing prostaglandin abortion (Prostin F2α , Upjohn S.p.A., Milano, Italia; 40 mg in a single intraamniotic administration), were examined after informed consent. Three were orally loaded with 20 mg/Kg (6RS)-L-erythrotetrahydrobiopterin dihydrochloride (BH4·2HCl, Dr.Schirks Laboratory, Wettswil a.A., Switzerland), and the non-loaded woman was the control. Before loading, and at time intervals after loading, serum, urine and AF pterins were measured.

BH4 absorption was shown by the fast increase in serum and urine biopterin levels (Fig.2). In AF B levels were still within the normal range for that gestational age 4-12 hours after loading. Several hours after loading AF B levels were slightly increased in three cases, including the control, possibly following the AF infusion of prostaglandins (Fig.2).

Serum, urine and AF N concentrations showed no significant alterations after the load, with the exception of a fast increase in serum N in all loaded cases and of a late increase in AF N in the control (Fig.3).

Discussion

Reported data indicate that placenta has a barrier function for BH4 in humans. Being placenta impermeable to BH4, AF pterins must be of fetal origin. Thus, the human fetus can be considered autotrophous for its BH4

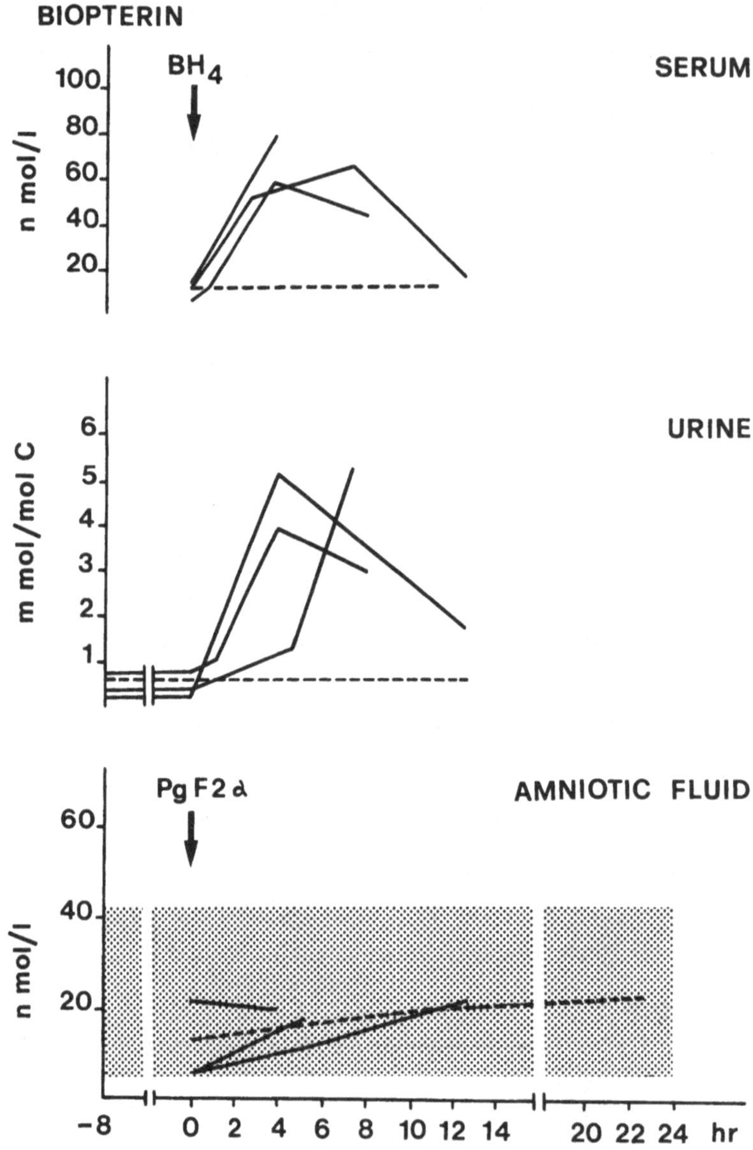

Fig.2: Time course of serum, urine, AF biopterin levels following an oral loading dose of 20 mg/Kg BH4 in three pregnant women during prostaglandin abortion. Control was a non-loaded woman who underwent prostaglandin abortion (dotted line).

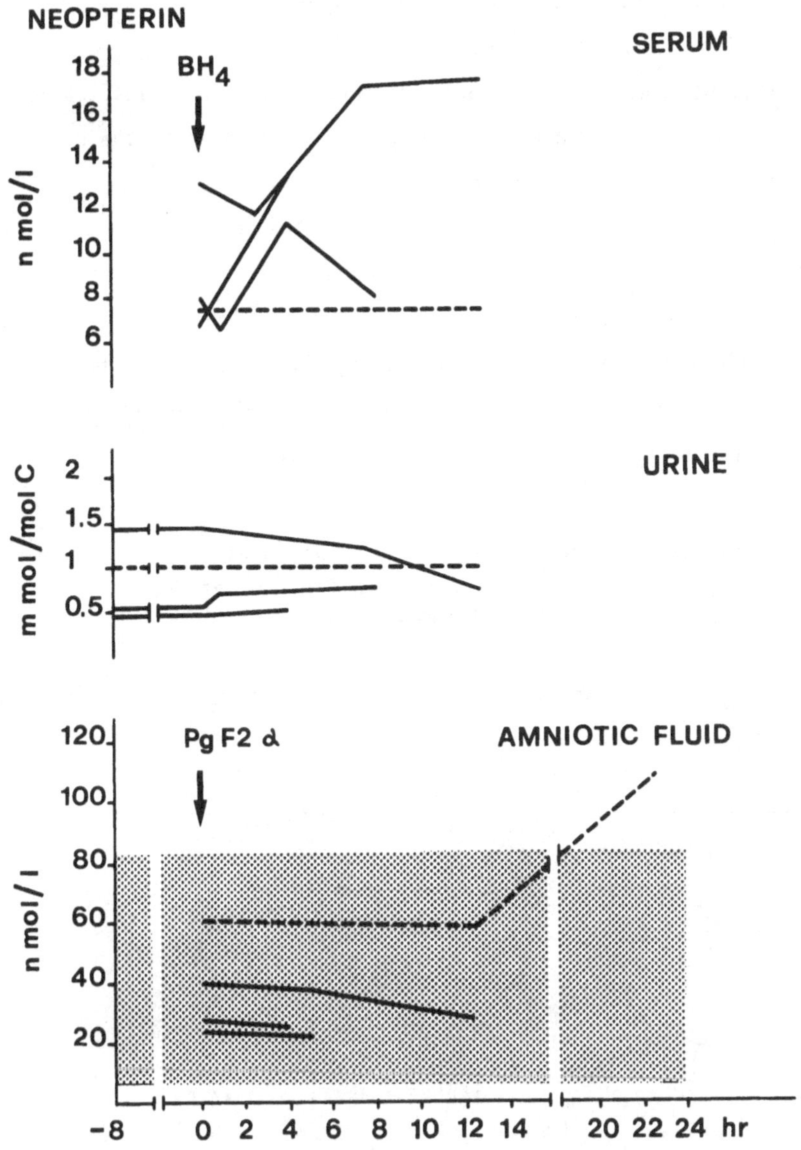

Fig.3: Time course of serum, urine, AF neopterin levels following an oral loading dose of 20 mg/kg BH4 in three pregnant women during prostaglandin abortion. Control was a non-loaded woman who underwent prostaglandin abortion (dotted line).

requirements, in keeping with early expression and high activity of the key enzymes of this metabolic pathway (8,9).

Additional evidence has been recently derived from the investigations carried out for prenatal diagnosis in two pregnancies in which the parents were heterozygous for 6-pyruvoyl tetrahydropterin synthase (PTPS) deficiency (8,10). In the instance of the fetus being affected, as revealed from low PTPS activity in its erythrocytes, the pattern of AF pterins reflected the synthetic block with high N and low B concentration (Table 2). Following this evidence, it is likely, as it will be later discussed, that prenatal diagnosis of other BH4 deficiencies will be possible through the very simple analysis of AF pterins.

Table 2. Pattern of AF Pterins and PTPS Activity in Fetal Erythrocytes assayed during two Pregnancies for Prenatal Diagnosis of PTPS Deficiency.

fetus	wk of gestation	fetal blood PTPS activity (μU/gHb)	%	amniotic fluid biopterin (nmol/l)	neopterin	% B*
female, affected	19	1.4	1.6	1.8	136	1.3
female, normal	19	108.7	128.7	15	33	31.2
controls	14-20					
mean		84.4	100	17	42.7	29.2
range		47.3-132	56-168	6-43	14-83	15-47
		(n=4)			(n=21)	

$$*B= \frac{\text{biopterin} \cdot 100}{\text{biopterin + neopterin}}$$

A still unsolved question is the variable outcome of the homozygous fetus. It may well depend on its residual enzyme activity, or, alternatively, on the supply of hydroxylated neurotransmitter precursors by the mother via placenta. It is noteworthy that, in one of the two PTPS deficient patients we observed, prenatal damage was associated with lower B and neurotrasmitter metabolite levels in cerebrospinal fluid (Table 3).
Due to the barrier function of the placenta, prenatal treatment of a BH4 deficient fetus appears to be not achievable through cofactor administration to the mother. Because of its kinetic properties, BH4 AF infusion, at least in synthetic defects, could be indicated.

Table 3. Cerebrospinal Fluid homovanillic acid (HVA), 5-hydroxyndole acetic acid (5-HIAA), Biopterin and Neopterin Concentrations in two PTPS deficient Patients with different clinical Status at Birth.

Patients' clinical status at birth	PTPS activity in erytrocytes (μU/gHb)	%	CSF B	N	HVA	5-HIAA
			(nmol/l)			
A.I.: low birthweight, microcephaly, tremors, limb hypertonicity	2	2.3	0	449	0	13
	PTPS activity in liver (μU/mg protein)	%				
M.G.: normal	0.03	3	4.3	131	134	38

Acknowledgements

We thank Prof.Mario Campogrande, Department of Gynecology, University of Torino, and Prof.V.Gabutti, Pediatric Clinic, University of Turin, for performing the amniocenteses.

References

1. Levy, H.L. 1985. In: Genetic and Metabolic Disease in Pediatrics (J.K. Lloyd and C.R. Scriver, eds.). Butterworths, pp.250-267.

2. Stacey, T.E. 1985. In: Genetic and Metabolic Disease in Pediatrics (J.K Lloyd and C.R. Scriver, eds.). Butterworths, pp.234-249.

3. Niederwieser, A., H.-Ch.Curtius. 1985. In: Inherited Diseases of Amino Acid Metabolism (H.Bickel and U. Wachtel, eds.). Georg Thieme Verlag, pp.104-123.

4. Boyd, R.D.H., F.C. Battaglia (eds.).1983. Perinatal Medicine. Butterworths.

5. Chamberlain, G.V.P., A.W. Wilkinson (eds.).1979. Placental transfer. Pitman Medical.

6. Seeds, A.E. 1980. Am.J.Obstet Gynecol. 138, 575-586.

7. Dhondt, J.L, J.M. Hayte, G. Forzy, M. Delcroix, J.P. Farriaux. 1986. Clin.Chim.Acta, 161, 269-273.

8. Niederwieser, A., H. Shintaku, Th. Hasler, H.-Ch.Curtius, H. Lehamann, O. Guardamagna, H. Schmidt. 1986. Eur. J. Pediatr. 145, 176-178.

9. Bracco,G., A.Iavarone, S.Pagliardini, O.Guardamagna, S.Ferraris, A.Ponzone. 1987. In: Unconjugated Pterins and Related Biogenic Amines (H.-Ch. Curtius, ed.), de Gruyter, pp.

10. Niederwieser, A., W. Leimbacher, A. Matasovic, L. Kierat, N.Blau. 1987. In: Unconjugated Pterins and Related Biogenic Amines (H.-Ch. Curtius, ed.) de Gruyter, pp.

SECTION F: MONOAMINE NEUROTRANSMITTERS AND ENZYMES

THERAPEUTIC POTENTIAL OF TETRAHYDROBIOPTERIN THERAPY IN NEUROLOGICAL AND PSYCHIATRIC ILLNESS.

Robert A. Levine, Ph.D.
Laboratory of Molecular Neurobiology, Lafayette Clinic and Wayne State University, Detroit, Michigan 48207

Introduction

Tetrahydrobiopterin (BH_4), which is synthesized intracellularly from guanosine triphosphate (GTP), is the required cofactor for tyrosine hydroxylase and tryptophan hydroxylases (see figure 1). Many of the control mechanisms governing tyrosine and tryptophan hydroxylase activities appear to be dependent upon the tissue concentration of BH_4. BH_4 is thought to limit the reaction rates of tyrosine and tryptophan hydroxylases, which are the rate-limiting enzymes in dopamine and serotonin synthesis, respectively. Increasing the tissue concentration of BH_4 (or synthetic cofactors) by exogenous administration has been shown to increase dopamine and serotonin synthesis in brain (1,2), synaptosomes (3), brain slices (2,4) and cultured cells (5,6). These observations have contributed to the concept that certain neurological and psychiatric diseases thought to involve a biogenic amine deficiency in brain may be treated by BH_4 therapy. This chapter will review the status of therapeutic trials with BH_4 and discuss some of the current considerations for treating patients with BH_4.

Background

Alterations in BH_4 metabolism are responsible for the rare childhood disease of BH_4-deficient hyperphenylalaninemia, which is also referred to as atypical phenylketonuria (PKU). This disease is characterized by a lack of appropriate BH_4 levels in the brain and/or liver (7,8,9) due to a deficiency in one of the enzymes involved in BH_4 metabolism; the defect can

Unconjugated Pterins and Related Biogenic Amines

occur in one of the biosynthetic enzymes or in dihydropteridine reductase (DHPR), which is responsible for the regeneration of BH_4 from quinoid-dihydrobiopterin (q-BH_2) after a hydroxylation reaction. Some, although not all of the children with a lack of BH_4 in the CNS, can be treated with BH_4 given orally and do not require the neurotransmitter precursors, 5-hydroxytryptophan (5-HTP) and L-dihydroxyphenylalanine (L-DOPA) (9,10). Others can be treated with BH_4 for the peripheral defect, but they also require 5-HTP and L-DOPA for amelioration of CNS problems (11,12). Still other newborns are totally unresponsive to BH_4 or neurotransmitter precursor therapy and continue to exhibit CNS disturbances (11).

Changes in the metabolism of BH_4 and biogenic amines have also been observed in a variety of neurological and psychiatric disorders. The cerebrospinal fluid (CSF) concentration of BH_4 is decreased in Parkinson disease (13), familial dystonia (14,15), Alzheimer's disease (16,17), and other neurological diseases (16); a decreased level of BH_4 in CSF with increasing age has also been reported (18,19). In vitro BH_4 biosynthesis from GTP in post mortem Alzheimer's cortical tissue has been shown to be decreased from control (20). Whether diminished BH_4 levels are the result of biogenic amine neuronal loss or due to altered BH_4 metabolism is unknown. However, any involvement of altered biogenic amine metabolism in disease may be the result of defects in BH_4 metabolism.

Figure 1 summarizes in schematic form the involvement of BH_4 in dopamine and serotonin synthesis in the CNS. BH_4 biosynthesis occurs in both dopamine and serotonin neurons, as depicted by the overlapping area in the middle of Figure 1. In BH_4 biosynthesis, guanosine triphosphate (GTP) is converted enzymatically to dihydroneopterin triphosphate (NH_2P_3) by GTP cyclohydrolase. NH_2P_3 is converted to 6-pyruvoyl-tetrahydropterin (6-PPH_4) by 6-PPH_4 synthase. Sepiapterin reductase (SR) is responsible for at least one and possibly both side-chain reductions that are required to make BH_4 from 6-PPH_4 (21,22,23), which has two keto functions in the side-chain. Another keto reductase has recently been recognized (for review, see 24) that may catalyze the reduction of one keto function, although the physiological contribution of this other reductase has not been studied. After BH_4 is formed, it can donate an electron in the process of aromatic amino acid hydroxylation, which yields quinoid dihydrobiopterin (q-BH_2); q-BH_2 is reduced back to BH_4 by (quinoid) dihydropteridine reductase (DHPR), which requires NADH as cofactor.

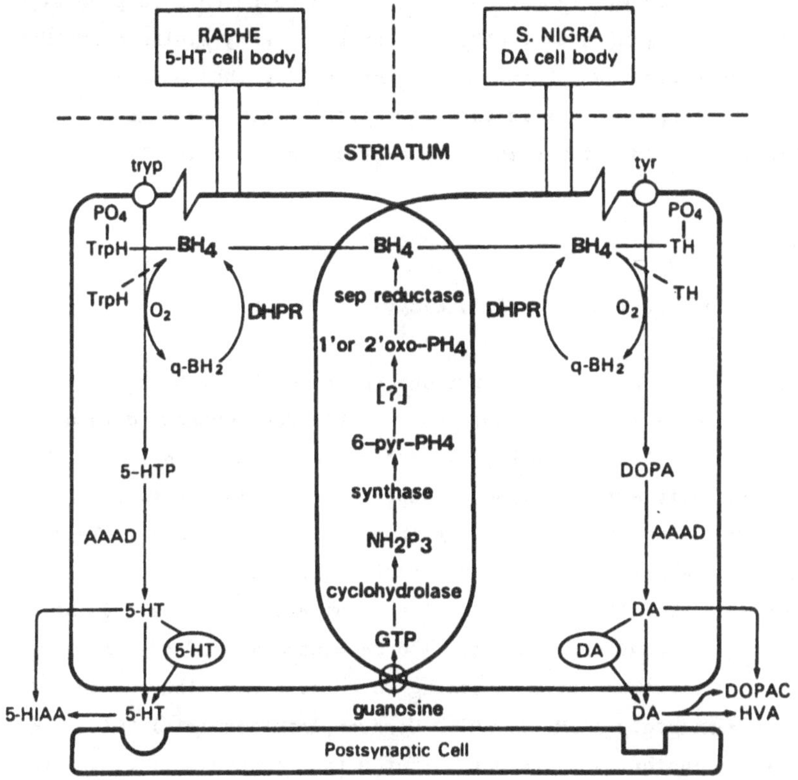

Figure 1. BH_4 and biogenic amine metabolism. Abbreviations are in alphabetical order: AAAD- aromatic amino acid decarboxylase; DA- dopamine; DHPR- dihydropteridine reductase; DOPA- dihydroxyphenylalanine; DOPAC- dihydroxyphenylacetic acid; 5-HT- serotonin; 5-HIAA- 5-hydroxy-indoleacetic acid; HVA- homovanillic acid; PH_4-tetrahydropterin; pyr-pyruvoyl; q-BH_2- quinoid-dihydrobiopterin; sep- sepiapterin; tryp- tryptophan; TrpH- tryptophan hydroxylase (PO_4-phosphorylated); tyr- tyrosine; TH- tyrosine hydroxylase (PO_4-phosphorylated).

Only recently has it become clear that human BH_4 biosynthesis is regulated differently from nonprimates (25). In humans, it is likely that an accumulation of intracellular NH_2P_3 with subsequent dephosphorylation leads to the appearance of neopterin in the circulation, which is not normally found in significant amounts in nonprimates. Therefore, the rate of human BH_4 biosynthesis is probably controlled by an enzyme(s) after NH_2P_3, possibly 6-PPH_4 synthase or SR. It should be considered when studying BH_4 metabolism that nonprimate models may not be appropriate in certain instances for learning about BH_4 metabolism and its interaction with biogenic amine metabolism in humans.

Theoretical Considerations For BH_4 therapy

Although there have been only a few therapeutic trials with BH_4, they have yielded information that should be valuable for conducting future trials. The issue of how much BH_4 can enter the human brain and reach the appropriate site after peripheral administration is always of concern. It has been shown previously that BH_4 entrance into whole rat brain (26) and discrete areas (27,28) is limited, with whole brain content being doubled (26) and striatal content raised 30% (27,28) after an intraperitoneal dose of 20 mg/kg. The concentration of BH_4 in the striatum on a wet weight basis (assuming ubiquitous distribution in all cells) is approximately 1 uM (29,30). However, nigrostriatal 6-hydroxydopamine lesions in rats demonstrated that approximately 85% of all BH_4 and GTP cyclohydrolase are contained in dopamine neurons (31). Other evidence demonstrating the concentration of BH_4 in striatal dopamine terminals is that in postmortem parkinsonian striatal samples, where there is a dramatic loss of nigrostriatal dopamine cells, there is an 80% reduction of striatal BH_4 compared to control (32). If one assumes that BH_4 is only located in striatal dopamine terminals and these terminals occupy 1% of the total striatal volume, then the dopamine terminal concentration of BH_4 would be 100 uM. From this argument, it follows that raising the striatal BH_4 concentration by 30% (28) in all striatal cells would increase the dopamine terminal concentration of BH_4 by only about 0.3%, assuming administered BH_4 distributes ubiquitously in the striatum and is not concentrated in dopamine terminals. It does not seem likely that there is a concentrating mechanism for peripherally administered BH_4 or other tetrahydropterin derivatives, since the amount of BH_4 or other cofactors found in striatum, cerebellum, or hippocampus after peripheral dosing was similar, even though the concentration of biogenic amine terminals is probably much higher in the striatum than the other two areas.

Phosphorylation of tyrosine hydroxylase has been proposed as an important regulatory mechanism in controlling dopamine synthesis by elevating the activity of the enzyme; this results in a more efficient utilization of existing concentrations of BH_4 (K_m for BH_4 is lowered through phosphorylation). It has been proposed that intracellular BH_4 levels in the striatum (and possibly throughout the CNS) are subsaturating for the majority of tyrosine hydroxylase molecules, since these molecules may be in an activity state (i.e. nonphosphorylated; high K_m for BH_4) that is not very responsive to the existing intracellular BH_4 concentration (31). The remaining smaller percentage (estimated to be 20-35% based on kinetic studies) of enzyme molecules are likely to be highly active (due to phosphorylation) in the presence of normal intracellular BH_4 concentrations; they are thought to provide for the majority of catecholamine synthesis. Although not known, this same potential relationship between BH_4 and tyrosine hydroxylase may apply to BH_4 and tryptophan hydroxylase in serotonin neurons. Thus, the theoretical basis for successful BH_4 therapy involves raising the intracellular concentration of BH_4 high enough to elevate the activity of nonphosphorylated enzyme molecules, which may require higher concentrations of BH_4 for activity.

If only a small percentage of phosphorylated tyrosine hydroxylase molecules are providing for the majority of striatal dopamine synthesis and the remaining nonphosphorylated molecules are less active, this condition may not be conducive for successful BH_4 therapy. Since BH_4 is concentrated in striatal dopamine terminals, large amounts of BH_4 or synthetic analogues may have to enter the brain from the periphery in order to raise significantly the striatal dopamine terminal cofactor concentration. Also, if nonphosphorylated molecules are to contribute significantly to striatal dopamine synthesis, it may require that these molecules be elevated to a large degree along their substrate-velocity curve (with respect to BH_4). The concentration of BH_4 or synthetic cofactor required to achieve this elevation is not known. It is important to consider that most of the administered BH_4 or analogue will be present outside dopamine terminals, and cannot participate in dopamine synthesis. For these reasons, it may be necessary to achieve very high whole brain concentrations before the full potential of BH_4 therapy and be tested.

In studies on the entrance of synthetic cofactors into brain after peripheral administration, 6-methyltetrahydropterin (6-MPH_4) was shown to enter brain approximately 8 to 10 times better than BH_4 (26,28). In testing a whole series of derivatives of BH_4, it was shown that tetrahydropterin (no substituent group in the 6 position) entered the brain about 60 times better than BH_4, while larger and generally more lipophilic derivatives entered brain more poorly than BH_4 itself (28). None of the cofactors tested were able to increase brain content of biogenic amine neurotransmitters or metabolites. Since direct injection of BH_4 in the rat brain has been shown to elevate tyrosine and tryptophan hydroxylation and increase metabolite levels (2), it can be inferred that very high peripheral doses of BH_4 or other active analogues would be necessary to achieve an increase in hydroxylation in a "control" brain. However, it is possible that patients with a BH_4 deficiency in brain might need less cofactor to enter the brain to achieve an increase in hydroxylation. This could be because the majority of hydroxylase enzyme molecules might exist in an activated (phosphorylated) state; in the absence of proper amounts of BH_4, these activated enzyme molecules might be able to utilize smaller amounts of BH_4 to generate activity. It is also possible in a BH_4 deficiency that hydroxylase molecules are not surrounded by "normal" amounts of BH_4; these molecules might have a protective effect and sequester small amounts of BH_4 that might reach the appropriate site of action after administration, allowing for their continued use. This might explain why a single large dose of BH_4 can have therapeutic benefit for several days in certain cases of atypical PKU. Currently, there are no good animal models of BH_4 deficiency to test this hypothesis.

Given that a majority of BH_4 and cyclohydrolase are highly concentrated in nigrostriatal dopamine neurons that comprise a small percentage of the striatal volume, the question arises whether BH_4 is contained in other nonaminergic cells in the striatum. If BH_4 is contained in nonaminergic striatal cells, the concentration is probably much lower than in dopamine neurons. If there is another physiological role for BH_4 in the general metabolism of all cells in the striatum, a high concentration of BH_4 must not be required. Alternatively, some small percentage of striatal BH_4 may be sequestered in other specialized cells where the role of BH_4 would not be of a general nature. The potential for BH_4 to have other physiological roles remains very much an open question (see Milstien, this volume).

BH_4 Metabolism and Human Disease

The discovery of atypical PKU has led to our increased understanding of BH_4 metabolism and its critical role in biogenic amine synthesis. The successful treatment of some cases of atypical PKU indicates the potential therapeutic usefulness of BH_4 in both BH_4 deficiencies discovered at birth and in BH_4 or biogenic amine deficiencies in adults. Atypical PKU probably provides the best example of therapeutic benefit from BH_4 administration, where cofactor replacement can increase biogenic amine production toward normal in some patients, although not all. Therapeutic benefit from BH_4 administration has also been reported in familial dystonia (33), Parkinson's disease (34, 35) and endogenous depression (36), although others have reported a lack of success after BH_4 therapy in Parkinson disease (15) and depression (Dr. D. Pikar, NIMH, personal communication). In all the above trials, BH_4 was administered orally, except in familial dystonia, where i.v. dosing was used (discussed below). The oral route of administration could play a part in some of the studies with negative results due to the relatively poor entrance of BH_4 into brain from the periphery (see below).

One of the more exciting discoveries regarding atypical PKU has been the observation that certain cases appear to be restricted to the liver, since CNS BH_4 and biogenic amine metabolism appears normal as reflected in CSF markers, and the patients exhibit no neurological deficits (9). Since a BH_4 deficiency can be system-selective in the periphery (liver defect)(9), it may be possible for a BH_4 deficiency to involve only the brain and not the periphery, which has been referred to as a "CNS-selective" BH_4 deficiency (25). This possibility currently would not be detected at birth by classical PKU screening, since only peripheral defects leading to hyperphenylalaninemia are discovered. That genetic defects in BH_4 biosynthesis occur in newborns leads one to suspect that the etiology of certain adult neuropsychiatric illnesses might be the result of altered expression of BH_4 biosynthetic enzymes. Certainly, many of the regulatory mechanisms for tyrosine and tryptophan hydroxylases revolve around BH_4.

In the studies of Miwa et al. (2), the observation that 6-MPH_4 was much less effective than BH_4 in elevating biogenic amine metabolism may explain the conflicting results of cofactor therapeutic trials in newborns and young children with atypical phenylketonuria (PKU). In some cases of atypical PKU, BH_4 (10) and 6-MPH_4 (9) monotherapies have proven effective in alleviating clinical symptom in the periphery and CNS. However, there have been 2 other independent cases where 6-MPH_4 administration was ineffective. Dr. Leupold, FRG, reported that 6-MPH_4 was totally ineffective during the 4 days of treatment in an atypical PKU patient whose CNS and peripheral symptoms have been continuously alleviated by BH_4 monotherapy. Dr. Scheibenreiter, Austria, also has reported on an atypical PKU patient whose periphery normally responsive to BH_4 such that a low phenylalanine diet is not required. In this patient, 6-MPH_4 was unable to prevent a large rise in serum phenylalanine over the course of several days, which was subsequently reduced to normal by reinstating BH_4 therapy. It is unusual that 6-MPH_4 would exhibit different characteristics from BH_4 in certain patients in vivo, since BH_4 and $6\text{-}MPH_4$ are both active hydroxylase cofactors and exhibit similar kinetic profiles in vitro. This is another example of how results from in vitro experiments may not be predictive of in vivo function.

In most cases where BH_4 therapy has been attempted, the compound was administered by the oral route, although intravenous (i.v.) dosing has been employed infrequently. Even in atypical PKU where there is a demonstrated lack of BH_4 synthesis, responses to oral BH_4 have been variable, with the majority not responding to BH_4 and a few showing dramatic improvement. The issue is confounded since some BH_4 deficient newborns that cannot be treated with BH_4 alone will respond to the neurotransmitter precursors, 5-HTP and L-DOPA, either alone or in combination with BH_4. One could speculate that either not enough BH_4 is entering the brain after oral administration (see below) or that there may be an additional problem with one or both of the hydroxylase reactions, such as a defective hydroxylase or other factor(s) required for hydroxylation (25).

Muenzer and Milstien (personal communication) recently investigated various routes of administration of BH_4 to humans and measured blood BH_4 levels over several hours. Their data showed that subcutaneous injection of BH_4 in humans can achieve about a 40-fold higher plasma concentration of BH_4 than oral administration, and subcutaneous administration was nearly as effective as i.v. administration for raising blood levels. Although CSF levels were not measured after these tests, it seems likely that brain levels would be significantly higher after subcutaneous administration of BH_4 as compared to the oral route. It is hard to interpret the mixed findings in the literature concerning therapeutic efficacy of BH_4, especially since most trials have used the oral route of administration.

Recent studies by LeWitt and coworkers (33) demonstrated one of the most convincing examples of successful BH_4 therapy, where 3 out of 10 patients with familial dystonia responded in double blind fashion to BH_4 given up to 10 mg/kg i.v. in a single dose. When a dose response to BH_4

was carried out and CSF BH_4 levels measured, one patient showed no beneficial therapeutic response to 2.5 mg/kg i.v., but did respond to 5 and 10 mg/kg i.v.. Interestingly, BH_4 in the CSF taken 4 hours after each dose was only elevated 50% at 2.5 mg/kg, whereas after 5 and 10 mg/kg the BH_4 level was increased by 7 and 16 fold, respectively. One interpretation of this limited data suggests that there is a threshold dosing level which has to be exceeded before BH_4 levels in the CSF, and possibly brain, are elevated in a proportional dose-dependent fashion. Otherwise, if there was no threshold, the 2.5 mg/kg dose would might have been expected to increase CSF BH_4 by 3 to 4 fold (instead of only 50%) to be in line with the CSF BH_4 levels at the higher doses. The existence of a threshold dosage level would argue for the use of single high doses of BH_4 rather than divide doses or infusions in treating patients. It will be difficult to discover the full therapeutic potential of BH_4 in neuropsychiatric illnesses until high enough brain concentrations can be achieved either with BH_4 or synthetic derivatives

Acknowledgement

This chapter is dedicated to Alois Niederwieser, who will long be remembered for his dedicated hard work, prominent contributions to the field of pterins and the study of inborn errors of metabolism, and most of all, he will be remembered by all of those who he enlightened with his friendship.

References

1. Kettler, R., Bartholini, G., and Pletscher, A.: Nature <u>249</u>, 476-478 (1974).

2. Miwa, S., Watanabe, Y., and Hayaishi, O.: Archives of Biochem. & Biophy. 239, 155 (1985).

3. Patrick, R.L., and Barchas, J.D.: J. Pharm. Exp. Ther. 197, 97-104 (1976).

4. Galloway, M.P. and Levine, R.A.: In: Chemistry and Biology of Pteridines, ed. by B.A. Cooper and V.M. Whitehead; Walter de Gruyter, Berlin, New York, pp 347-50 (1986).

5. Abou-Donia, M.M., Wilson, S.P., Zimmerman, T.P., Nichol, C.A., and Viveros, O.H.: J. Neurochem. 46, 1190-1199 (1986).

6. Levine, R.A. and Kelner, K.L.: In: Chemistry and Biology of Pteridines, ed. by B.A. Cooper and V.M. Whitehead; Walter de Gruyter, Berlin, New York, pp 373-376 (1986).

7. Danks, D.M. et al.: J. Inherit Metab Dis. 1, 49-53 (1978).

8. Niederwieser, A. and Curtius, H.-Ch.: In: Inherited Diseases of Amino Acid Metabolism, ed. by Bickel and Wachtel; Thieme, Stuttgart, pp 104-121 (1985).

9. McInnes, R.R., Kaufman, S., Warsh, J.J., Van Loon, G.R., Milstein, S., Kapatos, G., Soldin, S., Walsh, P., MacGregor, D., and Hanley, W.B.: J. Clin. Investigation 73, 458 (1984).

10. Niederwieser, A., Curtius, H.-Ch., Wang, M., and Leupold, D.: Eur. J. Pediatr. 138, 110-112 (1982).

11. Endres, W.: In: Inherited Diseases of Amino Acid Metabolism, ed. by Bickel and Wachetel; Thieme, Stuttgart, pp 124-132 (1985).

12. Bartholome, K. and Byrd, D.J.: Lancet II, 1042-1043 (1975).

13. Lovenberg, W., Levine, R.A., Robinson, D.S., Ebert, M., Williams, A.C. and Calne, D.B.: Science 204, 624-626 (1979).

14. Williams, A.C., Eldridge, R., Levine, R.A., Lovenberg, W. and Paulson, G.: Lancet 2 8139, 410-411 (1979).

15. LeWitt, P., Miller, L., Levine, R., Lovenberg, Newman, R., Papavasiliou, A., Rayes, A., Eldridge, R., and Burns, R.: Adv. Neurol. in press (1986).

16. Williams, A.C., Levine, R.A., Calne, D.B., Chase, T.N.and Lovenberg, W.: J. Neurol. Neurosurgery and Psychiatry 43, 735-738 (1980).

17. Kay, A.D., Milstein, S., Kaufman, S., Creasey, H., Haxby, J.V., Cutler, N.R.,and Rapoport, S.I.: Arch. Neurol. 43, 996-999 (1986).

18. Williams, A., Ballenger, J., Levine, R. A., Lovenberg, W. and Calne, D.B.: Neurology 30, 1244-1246 (1980).

19. Levine, R.A., Kuhn, D.M., Williams, A.C. and Lovenberg, W.: In: Influence of Age on the Pharmacology of Psychoactive Drugs, ed. by Raskin, Robinson and Levine; Elsevier, New York, pp 37-46 (1981).

20. Barford, P.A., Blair, J,.A, Eggar, C.S, Hamon, C.G.B., and Whitburn, S.B.: J. Neurol. Neurosurg. Psychia. 47, 736 (1984).

21. Milstein, S., Kaufman, S.: Biochem. Biophys. Res. Comm. 115, 88-893 (1983).

22. Heintel, D., Ghisla, S., Curtius, H.-Ch., Niederwieser, A., and Levine, R.A.: Neurochem. Int. 6, 141 (1984).

23. Smith, G.K., and Nichol, C.A.: In: Biochemical and Clinical Aspects of Pteridines, Vol 2, ed. by Curtius, Pfleider, and Wachter; de Gruyter, Berlin, pp 123 (1983).

24. Milstien, S. and Kaufman, S.: In: Chemistry and Biology of Pteridines 1986, ed. by Cooper and Whitehead; de Gruyter, Berlin, New York, pp 169-181 (1986).

25. Levine, R.A., Heintel, D., Leimbacher, W., Niederwieser, A., Curtius, H.-Ch., and Ghisla, S.: In: Biochemical and Clinical Aspects of Pteridines, Vol. 2, ed. by Curtius, Pfleiderer, and Wachter, de Gruyter, Berlin, pp 325-337 (1983).

26. Kapatos, G. and Kaufman, S. Science 212, 955-956 (1981).

27. Levine, R.A., Zoephel, G.P., Niederwieser, A. and Curtius, H.-Ch.: In: Biochemical and Clinical Aspects of Pteridines, Vol. 4, ed. by Wachter, Curtius, and Pfleiderer; de Gruyter, Berlin and New York, pp 267-276 (1985).

28. Levine, R.A., Zophel, P., Niederwieser, A., and Curtius, H.-Ch.: J. Pharmacol. Exp. Ther., accepted for publication (1987).

29. Bullard, W.P., Guthrie, P.B., Russo, P.V., and Mandell, A.J.: J. Pharmacol. Exp. Ther. 206, 4-20 (1978).

30. Levine, R.A., Kuhn, D.M. and Lovenberg, W.: J. Neurochem. 32, 1575-1578 (1979).

31. Levine, R., Miller, L. and Lovenberg, W.: Science 214, 919-921 (1981).

32. Nagatsu, T., Yamaguchi, T., Kaio, T., Sugimoto, T., Matsuura, S., Akino, M., Nagatsu, I., Iisuka, R., and Naraboyashi, H.: Clin. Chim. Acta 109, 305-311 (1981).

33. LeWitt, P., Miller, L., Levine, R., Lovenberg, W., Newman,R., Papavasiliou, A., Rayes, A., Eldridge, R., and Burns, R.: Neurology 36, 760-764 (1986).

34. Narabayashi, H., Kondo, T., Nagatsu, T., Matsuura, S.: Proc. Japan Acad. 58, 283-287 (1982).

35. Curtius, H.-Ch., Muldner, H., and Niederwieser, A.: J. Neural Transmission 55, 301-318 (1982).

36. Curtius, H.-Ch., Niederwieser, A., Levine, R.A., Lovenberg, W., Woggon, B., and Angst, J.: Lancet 1, 657-658 (1983a).

ORGANIZATION OF THE CATALYTIC AND REGULATORY SITES OF RAT LIVER PHENYLALANINE HYDROXYLASE[1]

Michael A. Parniak

Lady Davis Institute for Medical Research, 3755 Cote Ste-Catherine Rd., Montreal, Quebec, CANADA H3T 1E2

Phenylalanine hydroxylase (PAH)[2] catalyzes the tetrahydropterin-dependent conversion of phenylalanine to tyrosine. The enzyme purified from rat liver is an oligomeric protein composed of M_r=50000 subunits which are identical in primary structure (1). Native PAH shows complex kinetics with respect to its substrate, phenylalanine, when assayed with the naturally-occurring cofactor, BH_4 (2). The positive cooperative response of PAH to phenylalanine is intrinsic to the native enzyme molecule, since sigmoidal saturation curves for the binding of phenylalanine are observed even in the absence of tetrahydropterin cofactor (3). Native PAH binds 1.5 phe/M_r=50000 subunit, whereas the enzyme which has been desensitized and activated by modification with NEM binds only 1 phe/subunit (3). These data implicate the presence of distinct catalytic and regulatory binding sites for phenylalanine on PAH, a concept strengthened by the observation of positive heterotropic effects on the hydroxylase by tryptophan and 3-phenylserine (4).

Homotropic and heterotropic effects on enzyme activity are noted primarily with oligomeric proteins (5) and result from ligand-induced conformational alterations in these enzymes. Kaufman and Fisher (6) reported that purified native PAH is a mixture of tetramers (M_r=200000) and dimers (M_r=100000). Although Døskeland et al. also reported that PAH exists as a mixture of tetramers and dimers (7), conflicting results have been

[1] a portion of this work has been presented in abstract form: M.A. Parniak, S. Kaufman (1985) Biochemistry 24:3379 (Abstr.).

[2] the abbreviations used are: PAH, phenylalanine hydroxylase; BH_4, tetrahydrobiopterin; $6MPH_4$, 6-methyltetrahydropterin; NEM, N-ethylmaleimide; HPGPC, high performance gel permeation chromatography; SDS-PAGE, sodium dodecyl sulfate polyacrylamide gel electrophoresis.

Unconjugated Pterins and Related Biogenic Amines

obtained by others. Thus Nakata and Fujisawa (8) and Shiman (9) stated that the enzyme exists solely as a tetramer, whereas Tourian (10) and Ayling et al. (11) have reported that the hydroxylase exists solely as a dimer. Phenylalanine-induced alterations in secondary and tertiary structure (12, 13) as well as changes in quaternary structure (10, 11) have been described. In this report we outline the use of HPGPC to investigate in greater detail the oligomeric structure of rat liver PAH, and present a model for the organization and role of the catalytic and regulatory binding sites for phenylalanine in the expression of activity of PAH.

Methods

Phenylalanine hydroxylase was purified from rat liver by the method of Kaufman and Fisher (6) or by the procedure of Shiman et al. (13), and was >95% pure as determined by SDS-PAGE. Hydroxylase activity was determined by either a spectrophotometric method (as described in ref. 3) or with a fixed-time assay (14).

HPGPC analyses were performed with a column (7.5 x 600 mm) of TSK 3000 SW (Altex) at ambient temperature (23-25°C) with 0.1M potassium phosphate, pH 6.8, as the mobile phase. The injection volume was 50 uL for all analyses. The column was calibrated with standard proteins of known molecular weight. Further experimental details can be found in the legends to the various Figures.

Results

Two protein peaks were observed when purified rat liver PAH was analyzed by HPGPC (Fig.1). The major peak corresponded to a molecular weight of 200000 and the smaller protein peak eluted as M_r=100000. Since PAH consists of identical subunits of M_r=50000 (1), these oligomeric forms of the hydroxylase are presumed to be the tetramer and dimer, respectively. Analysis of 10 different preparations of rat liver PAH, purified by either of two different procedures (6, 13), showed that the tetramer accounts for 78 +/- 6% of the total hydroxylase protein under these conditions. The relative amounts of the tetramer and the dimer were unchanged over a 100-fold range of protein concentration, from 0.5 to 50 mg/mL PAH injected for analysis (Fig.1, insert). At concentrations less than 0.5 mg/mL protein, the relative amount of dimer appeared to increase. Limitations of detector sensitivity prevented analysis of samples of less than 0.1 mg/mL hydroxylase

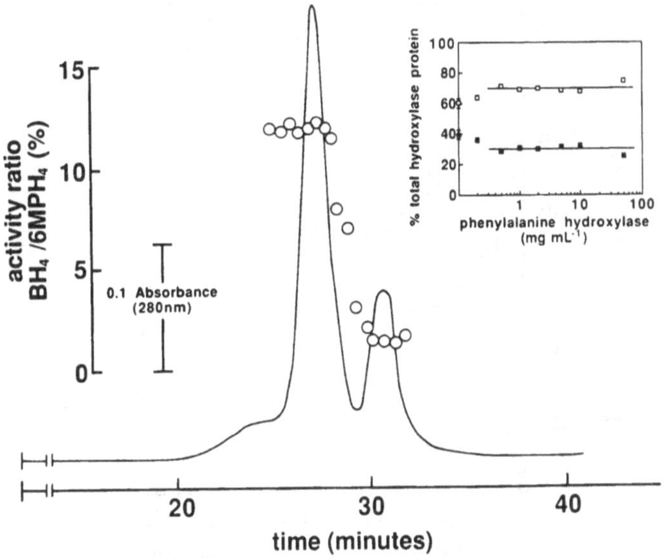

FIGURE 1: HPGPC of purified rat liver PAH. 50 uL of a 20 mg/mL solution of PAH was chromatographed, as described in Methods, at 0.5 mL/min, and 0.5 min fractions were collected. Aliquots of these fractions were assayed with 40 uM BH_4 and 2 mM phenylalanine, and with 300 uM $6MPH_4$ and 10 mM phenylalanine. PAH activity (O) is shown as the ratio % of BH_4-dependent specific activity to $6MPH_4$-dependent specific activity (14). Insert: Proportion of tetramer (□) and dimer (■) as a function of the concentration of PAH injected for analysis.

protein. No evidence for the presence of the monomeric M_r=50000 subunit was obtained under denaturing conditions. Preincubation of PAH with dithiothreitol prior to chromatography did not lead to the appearance of monomer. In addition, only the monomer was detected by SDS-PAGE in the absence of thiols. Thus it is unlikely that intersubunit disulfide bonds contribute to the oligomeric organization of PAH.

Aliquots of the separated tetramer and dimer were reanalyzed by HPGPC at various times after isolation. The tetramer slowly reverted over a 6h period to a mixture of tetramer and dimer in a proportion identical to that seen with the original sample of PAH. In contrast, the isolated dimer remained as a dimer under the same conditions. However, certain conditions can affect the dimer. Preincubation of native PAH with 2 mM phenylalanine and subsequent HPGPC using a mobile phase containing 2 mM phenylalanine

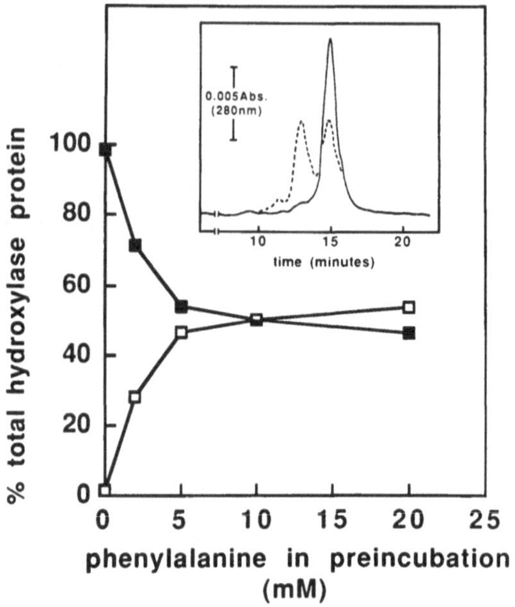

FIGURE 2: Effect of preincubation with phenylalanine on the conversion of dimer (■) to tetramer (□). Dimeric PAH was incubated for 15 min at 25°C in 0.1M phosphate, pH 6.8, containing the indicated concentrations of phenylalanine. Samples were then analyzed by HPGPC using a mobile phase of 0.1M phosphate, pH 6.8, containing 2 mM phenylalanine. Insert: HPGPC profile of dimeric PAH before (——) and after (----) preincubation with 10 mM phenylalanine.

showed PAH to be >98% tetramer (data not shown). Preincubation with phenylalanine did not affect the elution position of the isolated tetramer. However, similar treatment of the isolated dimer resulted in an increased formation of the tetramer with increasing phenylalanine concentration (Fig. 2). Complete conversion of the isolated dimer to the tetramer was not observed - this may be due in part to the low protein concentration in the isolated dimer pool (~0.3 mg/mL) (cf. insert to Fig. 1). PAH activated by phosphorylation has some of the same characteristics as the enzyme which has been activated by preincubation with phenylalanine (15). Analysis by HPGPC of PAH with a phosphate content of 0.6 mol P_i/mol M_r=50000 subunit showed that less than 5% of the hydroxylase protein eluted as the dimer (data not shown). This contrasts with the 25% dimer exhibited by the normal enzyme which contains 0.05-0.2 mol P_i/mol subunit (15).

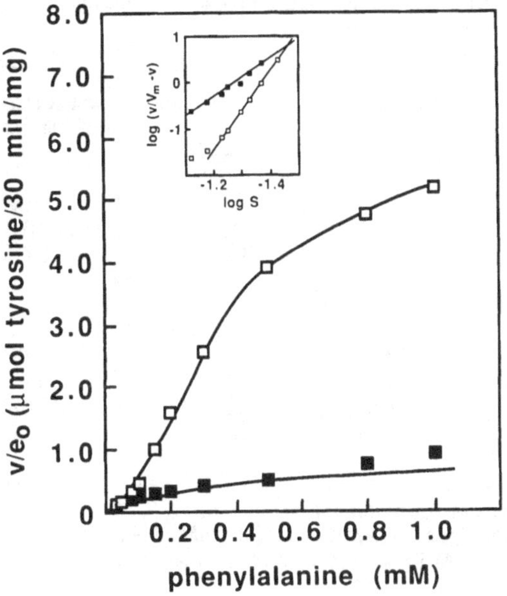

FIGURE 3: BH_4-dependent activity of isolated dimeric (■) and tetrameric (□) PAH as a function of phenylalanine concentration. The oligomeric forms of PAH were isolated as described in the legend to Fig. 1 and immediately assayed with 25 uM BH_4 and the indicated concentrations of phenylalanine. Insert: Hill plots of the data from the main Figure. n_H for the tetramer was 2.1; n_H for the dimer was 1.0.

Kinetics of the oligomeric forms of PAH

Both the tetramer and the dimer have identical specific activities when assayed with the synthetic cofactor, $6MPH_4$. However, the full expression of the regulatory properties and the various activation phenomena to which PAH is subject is noted only with the natural cofactor, BH_4. Substantial differences were observed in the BH_4-dependent specific activity of the two oligomeric forms of PAH (Fig. 1); the tetramer has a BH_4-dependent specific activity approximately 5 to 6 times that of the dimer. When the resolved oligomeric forms of PAH were assayed immediately after isolation, the BH_4-dependent activity of the tetramer showed a high degree of cooperativity (n_H = 2) in response to variations in phenylalanine concentration (Fig.3), similar to that given by native PAH. The activity of the dimer, on the other hand, was strictly hyperbolic up to about 0.5 mM phenylalanine. The

Table 1: Activation of BH_4-dependent activity of isolated dimeric and tetrameric PAH

Additions	Dimer initial rate (nmol/min/mg)	Dimer -fold activation	Tetramer initial rate (nmol/min/mg)	Tetramer -fold activation
Experiment 1				
None	13	-	62	-
0.2mM phe	25	2	1080	17
0.2mM lysolecithin	1106	88	1175	19
Experiment 2				
None	56	-	240	-
2mM phe	760	14	2510	11
5mM phe	1120	20	2640	11
20mM phe	1140	20	2510	11

Aliquots of samples were preincubated for 10 min at 25°C in 0.1M phosphate, pH 6.8, with the indicated addition, then assayed with 50 uM BH_4 and 0.2 mM phenylalanine (Experiment 1) or with 38 uM BH_4 and 2 mM phenylalanine (Experiment 2).

BH_4-dependent activity of the dimer was however greater than that predicted by the calculated kinetic constants, at higher phenylalanine levels. We feel that this may be the result of the formation of some tetrameric PAH promoted by these higher levels of phenylalanine (cf. Fig. 2).

Preincubation of PAH results in 10 to 20-fold increases in BH_4-dependent activity (13, 16). The apparent half-maximal concentration of phenylalanine necessary for this activation at pH 6.8 and 25°C is about 0.1 mM (4, 16). As shown in Table 1, preincubation with 0.2 mM phenylalanine gave a 20-fold increase in the activity of the tetramer, but only a 2-fold increase with the dimer. The extent of activation of the dimer increased with increasing amounts of phenylalanine in the preincubation buffer (Table 1). These increases in activation of the dimer correlated almost exactly with the extent of phenylalanine-promoted tetramer formation under these conditions (Fig. 2). The activation of PAH by the phospholipid lysolecithin differs from that due to preincubation with phenylalanine in that the hydroxylase is desensitized in the presence of lysolecithin (2). In addition, the phospholipid induces a gross conformational change in enzyme structure (17) that is not observed upon preincubation of the enzyme with phenylalanine (12). Table 1 shows that both dimeric and tetrameric PAH are activated by lysolecithin. Note that the BH_4-dependent specific activity of both species is essentially identical in the presence of lysolecithin. We have been unable to determine the oligomeric nature of PAH in the presence of lysolecithin due to interference from phospholipid micelles at the levels of lysolecithin used. However, PAH which has been chemically modified by reaction with NEM is in an activated and desensitized form, kinetically identical to the phospholipid treated enzyme (3). NEM-modified PAH exists entirely as dimers under our conditions of HPGPC (data not shown).

Discussion

The BH_4-dependent activity of rat liver PAH shows positive cooperativity with respect to variations in phenylalanine concentration (2). Such kinetic data have been used to implicate the presence on PAH of distinct catalytic and regulatory (or activator) sites for phenylalanine (16, 18). The first direct evidence in support of this proposal was the finding that native PAH binds more than 1 mol of phenylalanine per M_r=50000 subunit (3). Interpretation of the non-integral value of 1.5 phe/subunit was not possible at that time, given the paucity of information concerning the subunit composition of PAH. However, others had proposed that PAH possessed 1 regulatory site /subunit (9). A number of subsequent findings in addition to the present work have allowed us to reassess the subunit organization of the catalytic and regulatory sites for phenylalanine on PAH:

(i) The M_r=50000 monomeric subunits of PAH have been shown to be identical in primary structure (1).

(ii) With BH_4 as cofactor, the presence of tryptophan results in the desensitization of the response of PAH to phenylalanine concentration, but does not affect V_{max}. The binding of phenylalanine to PAH in the presence of tryptophan is reduced to 1 phe/subunit from 1.5 phe/subunit (4).

(iii) Studies of the quenching of the intrinsic fluorescence of PAH by BH_4 demonstrate a binding ratio of 1 mol of BH_4/mol of M_r=50000 subunit (12). Since the tetrahydropterin is essential for hydroxylation (19), we assume that each subunit must possess a single catalytic site. This assumption is consistent with (ii), and with the observation that PAH, desensitized <u>yet</u>

activated by modification with NEM, binds 1 phe/subunit (3). In addition, these observations indicate that conditions which lead to desensitization of PAH (i.e., the abolition of positive cooperativity with respect to phenylalanine) also result in the apparent disappearance of 0.5 sites for the binding of phenylalanine per subunit. It seems likely therefore that each monomeric M_r=50000 subunit of PAH has 1 catalytic site and the equivalent of or potential for 0.5 regulatory (or activator) site.

Since no evidence for the presence of monomeric M_r=50000 subunit was obtained in the present study, we consider the minimum reactive unit of PAH to be the dimer. The data which describe the the binding of phenylalanine to PAH (3, 4) are consistent with the idea that this dimer might possess 3 binding sites for phenylalanine - 2 catalytic sites, and 1 regulatory site. However, we do not think that this is an accurate picture of the organization of these sites. The regulatory site is presumably involved in the expression of the complex BH_4-dependent kinetics of PAH with respect to phenylalanine, and is necessary for the activation of PAH noted upon preincubation of the enzyme with phenylalanine. The dimer, however, shows a strictly non-cooperative response to variations in phenylalanine concentration; only the tetramer shows kinetic cooperativity (Fig. 3). The dimer can apparently be activated by preincubation with phenylalanine, but the extent of activation is strongly correlated with the amount of tetramer formed under the conditions of preincubation (Fig. 2; Table 1). Thus at least part, and perhaps all, of the activation of the dimer induced by preincubation with phenylalanine is due to the phenylalanine-promoted conversion of dimers to tetramers.

A simple model for the possible organization of the regulatory binding sites for phenylalanine on PAH is illustrated in Figure 4. Phenylalanine hydroxylase is an oligomeric protein composed of subunits which are identical in primary structure. Each subunit possesses 1 catalytic site and the equivalent of, or the potential for, 0.5 regulatory site. The minimum active catalytic unit of PAH is a dimer. These dimers in the native enzyme exist in equilibrium with tetramers which are thus, in effect, "dimers of dimers". When two dimers associate in the proper manner to form a tetramer, 2 additional binding sites for phenylalanine are "created", in addition to the 4 catalytic sites already in existence. These two newly-formed sites are the regulatory or activator sites for PAH, and may exist at the

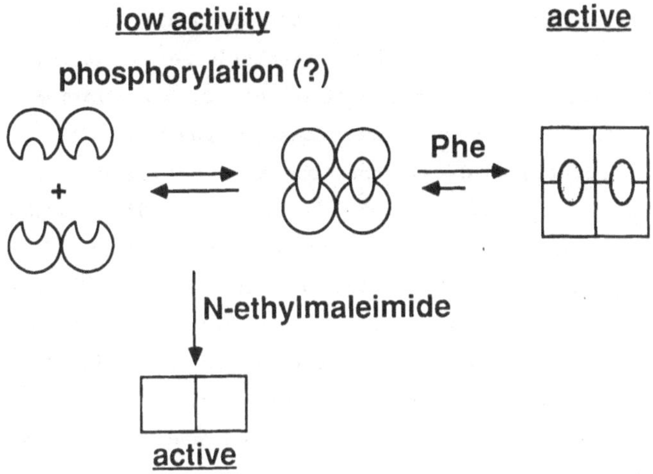

FIGURE 4: Scheme for the oligomeric forms and organization of the regulatory binding sites for phenylalanine on PAH. Details are outlined in the text.

interface of the dimer-dimer interaction. In this model, both the native dimer and the native tetramer (circular figures) have low intrinsic catalytic activity. The binding of phenylalanine to the regulatory sites (this binding is denoted by the shaded ovals), which exist in complete form only in the tetramer, results in (i) stabilization of the tetramer, and (ii) some conformational response in PAH which results in activation of the enzyme. Conformational changes in PAH due to the binding of phenylalanine are well documented (12, 13). Since the dimer does not possess intact regulatory sites for phenylalanine, the activity of this species remains low, with no observed cooperativity. The NEM-modified PAH is shown as a dimer in a different conformation than the native dimer, since it is an activated form of PAH. The dimeric nature of NEM-PAH is consistent with both the absence of kinetic cooperativity and the decrease in total phenylalanine binding exhibited by this activated enzyme.

Support for the concept of 2 regulatory binding sites for phenylalanine/tetramer has been obtained from recent studies concerning the interaction of a monoclonal antibody with PAH. This antibody binds to PAH only after the enzyme has been incubated with phenylalanine (20). Therefore, this antibody interacts with the tetrameric form of PAH. Binding of the antibody inhibits PAH activity in a noncompetitive manner (21) and is therefore presumably not binding to the catalytic region of the hydroxylase. The interaction results in the abolition of the cooperative response of BH_4-dependent PAH activity to variations in phenylalanine concentration (M.A. Parniak, unpublished data), thus the antibody may be interacting with PAH at or near the regulatory region of the enzyme. This antibody binds to 2 sites over the 4 identical subunits of tetrameric PAH.

The model presented in this report allows for 1 catalytic site/monomer of PAH, and the data indicate that the regulatory or activator site for phenylalanine is not necessary for the expression of low levels of catalytic activity in the dimer. Thus it is possible that the monomer could be catalytically active, if it could be formed and stabilized under non-denaturing conditions.

Acknowledgements

M.A.P. is a Chercheur-Boursier of the Fonds de la Recherche en Santé du Québec. This work was initiated while the author was in the laboratory of Dr. Seymour Kaufman, NIMH, Bethesda, MD. I gratefully acknowledge his enthusiastic support, instruction, and stimulating insight and discussions.

References

1. Iwaki, M., Parniak, M.A., Kaufman, S. (1985) Biochem. Biophys. Res. Commun. 126:922-932

2. Fisher, D.B., Kaufman, S. (1973) J. Biol. Chem. 248:4345-4353

3. Parniak, M.A., Kaufman, S. (1981) J. Biol. Chem. 256:6876-6882

4. Phillips, R.S., Parniak, M.A., Kaufman, S. (1984) J. Biol. Chem. 259:271-277

5. Fersht, A. (1977) Enzyme Structure and Mechanism. W.H. Freeman, San Francisco

6. Kaufman, S., Fisher, D.B. (1970) J. Biol. Chem. 245:4745-4750

7. Doskeland, A.L., Jones, T., et al (1982) Neurochem. Res. 7:407-421

8. Nakata, H., Fujisawa, H. (1980) Biochim. Biophys. Acta 614:313-327

9. Shiman, R. (1980) J. Biol. Chem. 255:10029-10032

10. Tourian, A. (1971) Biochim. Biophys. Acta 242:345-354

11. Ayling, J.E., Pirson, W.D., et al (1974) Biochemistry 13:78-85

12. Phillips, R.S., Parniak, M.A., Kaufman, S. (1984) Biochemistry 23:3836-3842

13. Shiman, R., Gray, D.W., Pater, A. (1979) J. Biol. Chem. 254:11300-11306

14. Hasegawa, H., Kaufman, S. (1982) J. Biol. Chem. 257:3084-3089

15. Parniak, M.A., Hasegawa, H., et al (1981) Biochem. Biophys. Res. Commun. 99:707-714

16. Shiman, R., Gray, D.W. (1980) J. Biol. Chem. 255:4793-4800

17. Abita, J.P., Parniak, M.A., Kaufman, S. (1984) J. Biol. Chem. 259:14560-14566

18. Dhondt, J.L., Dautrevaux, M., et al (1978) Biochimie 60:787-794

19. Kaufman, S. (1959) J. Biol. Chem. 234:2677-2682

20. Jennings, I.G., Russell, R.G.McR., et al (1986) Biochem. J. 235:133-138

21. Parniak, M.A., Jennings, I.G., Cotton, R.G.H. (1986) in Chemistry and Biology of Pteridines (Cooper, B.A., Whitehead, V.M., eds.), Walter de Gruyter, Berlin-New York, pp. 359-362

IMMUNOCHEMICAL STUDIES OF PHENYLALANINE HYDROXYLASE

R. G. H. Cotton, I. G. Jennings, W. McAdam and S. Smith

Olive Miller Protein Laboratory, Murdoch Institute,
Melbourne, Australia

Introduction

Phenylalanine hydroxylase (PH) is a complex enzyme which has three substrates phenylalanine, tetrahydrobiopterin and oxygen and its enzyme activity is activated by phenylalanine, phosphorylation, N-ethyl maleimide, and phospholipids especially lysolecithin (1). We have initiated a study to assess the ability of monoclonal antibodies, in combination with traditional protein chemistry, to define the structural regions of PH polypeptide responsible for the catalytic functions of the enzyme. A monoclonal antibody (MAb) which bound the catalytic site of the enzyme might be expected to show the following characteristics : (a) binding of the MAb to PH at equimolar concentration would completely inhibit enzyme activity (b) binding of MAb to PH would be inhibited by substrate or enzyme inhibitors (c) MAb would bind tyrosine and tryptophan hydroxylases which share a common substrate with PH and also with other enzymes which have some common substrates (d) MAb's epitope would be a highly conserved sequence in PH and thus MAb would recognise PH from a wide range of species in the phylogenic spectrum.

Results and Discussion

A series of eleven monoclonal antibodies (2) recognised at least five different antigenic regions on PH and were tested for their ability to satisfy the above criteria. The results are summarized in Table 1.

Table 1

Group+	Antibody	Effect on Activity°	Species Reactivity*	Ligand^x
A	PH1	-59	H M R	Phe (+)
	PH5	+43	H M R	Lyso (-)
	PH8	-19	H M R & others	Lyso (-)
B	PH7	+29	H M R	PO4 (+)
C	PH2	+100	H M	Lyso (-)
	PH3	+100	H M R	Lyso (-)
	PH10	+33	H M R	Lyso (-)
	PH12	+33	H M R	Lyso (-)
D	PH9	+12	H M	none
	PH11	+16	H M	none
E	PH6	-6	H M R	Lyso (-)

+ Antibodies are grouped according to their ability to compete with each other for binding to PH
° Percentage effect on enzyme activity were calculated by using the enzyme activity with no antibody present as 0%; (+) stimulation (-) inhibition of enzyme activity.
* H, M, R designate human, monkey and rat liver PH respectively.
x effect on MAb binding of phenylalanine (PHE), lysolecithin (Lyso) and PO_4 (covalently bound phosphate) when bound to PH (+), (-) designate promotion and inhibition of MAb binding respectively.

It can be seen from Table 1 that no MAb has satisfied the criteria detailed earlier for active site recognition.

PH1 MAb inhibited the enzyme by 59%, required phenylalanine for its binding (½-maximal phenylalanine concentration is similar to that required for activation) (2). In addition, it has been shown (3, in preparation) that the MAb inhibition is non-competitive and thus PH1 may be recognising a response of PH to activation by phenylalanine.

PH7 MAb only binds to phosphorylated PH (4) and its epitope is located within the sequence

15 19
Leu-Ser-Asp-Phe-Gly.
PO_4 (on Ser)

PH7 MAb has little effect on enzyme activity which is consistent with other findings that the phosphorylation site is not contained with the catalytic site.

The binding of a number of MAbs' to PH is affected by lysolecithin. However only PH2, 3 and 5 show a stimulation of similar magnitude to lysolecithin with these MAbs also mimicking the changes by lysolecithin to the PH enzyme kinetics (2). The epitope of these MAbs has not been finally localized but is likely to be in the N terminal cyanogen bromide peptide (unpublished).

PH8 MAb reacts with PH in Western blots of liver extracts from a wide range of species (man, rat, dog, pig, cow, koala, chicken, fish, frog but not squid or bacteria) (5). PH8 has been shown to bind tyrosine hydroxylase by ELIZA and tryptophan hydroxylase by solution reaction (6). In addition immuno histochemical staining of brain tissue has shown PH8 reacts with tyrosine and tryptophan hydroxylase (6) and it has been used to study the distribution of serotinergic neurones in the human brain stem (in preparation). PH8 MAb epitope has been identified as

138
Gly-Ala-Glu-Leu-Asp-Ala-Asp-His-Pro-Gly-Phe-Lys-Asp-Pro
154
-Val-Tyr-Arg ; (5) a region substantially conserved in tyrosine hydroxylase (7).

PH9 has been shown to recognise the N terminal tryptic peptide of human PH (in preparation).
The figure shows the distribution of the known binding sites of the MAbs PH 2, 3, 7, 8 and 9.

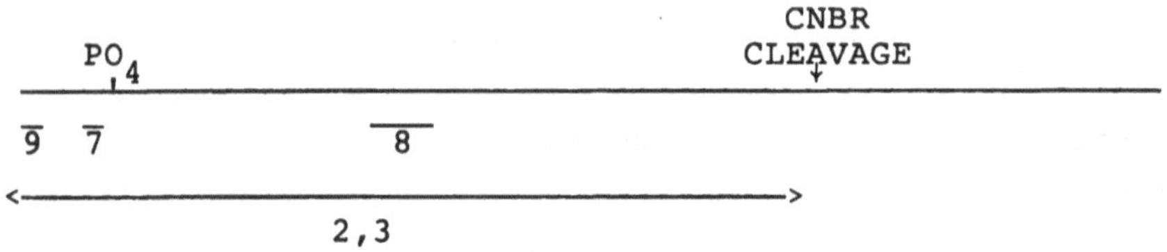

Recently we have isolated a series of antibodies which more closely fit the criteria for active site reactive antibodies. One antibody binds to PH, human dihydropterdine reductase, and E. coli dihydrofolate reductase and its binding to PH is inhibited by the presence of 6,7 dimethyl 7,8 dihydropterin. The identification of this MAb's epitope on PH is underway.

References

1. Shiman, R. In Pterins and Folates, Vol.1 Eds Blakley, R.L., Benkovic, S.J. J. Wiley and Sons, 1984.

2. Jennings, I. G., R.G.McR. Russell, W.L.F. Armergo and R. G. H. Cotton. Biochem.J. (1986) 235 133-138.

3. Parniak, M. A., I. G. Jennings and R. G. H. Cotton (1986) p.359-362 in Chemistry and Biology of Pteridines, 1986. Eds B. A. Cooper and V. M. Whitehead. Published W. de Gruyter.

4. Smith, S. C., W. J. McAdam, B. E. Kemp, F. J. Morgan and R. G. H. Cotton (1986) Biochem. J. (in press).

5. Haan, E. A., I. G. Jennings, A. C. Cuello, H. Nakata, C. W. Chow, R. Kushinsky, J. Brittingham and R. G. H. Cotton (1986) Brain Research (submitted).

6. Manuscript in preparation.

7. Grima, B., Lamouroux A, Blanot F., Biguet N.F. and Mallet J. (1985) P.N.A.S. 82 617-621.

INFLUENCE OF SUBSTRATES AND COFACTORS ON TYROSINE HYDROXYLASE

D.M. Kuhn

Department of Psychiatry, Lafayette Clinic, Wayne State University School of Medicine, Detroit, Michigan 48207 USA

Introduction

Tyrosine hydroxylase (TH) is the initial and rate limiting enzyme in the biosynthetic pathway for the catecholamines dopamine, norepinephrine, and epinephrine. It was recognized some time ago that the activity of TH in adrenal and brain extracts was dramatically increased upon the addition of reduced pterins (1,2). Since these initial studies, a large amount of information on the pterin dependent enzymes TH, phenylalanine hydroxylase, and tryptophan hydroxylase has emerged concerning the mechanisms by which their level of activity is controlled. Most of these experiments were designed to study posttranslational influences such as phosphorylation, or the effects of various drugs on enzyme activity and little attention has been directed at the role that the pterin cofactor might play, other than the most obvious one of an essential cofactor in altering the activity of the hydroxylase enzymes.

With the discovery of variant forms of hyperphenylalanemia (3-5) which involved primary defects in the enzymes of tetrahydrobiopterin (BH_4) synthesis and not in phenylalanine hydroxylase, a vastly different etiology for practically all disorders involving the biogenic amine neurotransmitters and the aromatic amino acid hydroxylases was suggested. These disorders range from inborn errors in metabolism such as hyperphenylalanemia, mentioned above, to many psychiatric and neurologic disorders (see Levine (6) this volume). The studies of Kettler et al (7) were perhaps the first to demonstrate that the parenteral administration of BH_4 to rats caused an increase in the cerebral levels of the cofactor and an increase in catecholamine

Unconjugated Pterins and Related Biogenic Amines

turnover. These results formed an important basis for the therapeutic treatment with BH_4 of certain diseases involving deficits in neurotransmitters since they offered the possibility that BH_4 could produce increases in neurotransmitter levels.

The administration of BH_4, like any therapeutic agent, to animals or man presupposes that the compound will reach its site of action and result in an increase in the activity of the hydroxylases for which it is the essential cofactor. In general, the most accessible hydroxylase has been phenylalanine hydroxylase because of its location in the liver. The localization of tryptophan hydroxylase and TH principally in brain restricts access of BH_4 to these enzymes to some extent but structure activity studies with the cofactor have revealed that some congeners can penetrate the blood-brain barrier (8). Once the cofactor reaches the hydroxylase enzyme it can interact with it to result in increased activity and increased production of the neurotransmitters dopamine, norepinephrine, and serotonin. The exact mechanism by which BH_4 interacts with the hydroxylase enzymes is gradually becoming more completely understood as the reaction mechanisms of the enzymes are elucidated. Studies with purified phenylalanine hydroxylase have allowed impressive advances in our understanding of how BH_4 interacts with this enzyme (9,10). Other equally interesting experiments have discovered numerous means by which BH_4 can interact with phenylalanine hydroxylase to influence its structure and level of activity by mechanisms which are not strictly limited to a cofactor role (11,12).

In an attempt to determine if BH_4 can interact with TH as it does with phenylalanine hydroxylase we undertook a series of experiments on BH_4-TH interactions which began with purification of the hydroxylase enzyme to homogeneity. TH is a very unstable enzyme but sufficient quantities were purified from cultured PC12 cells to allow studies on its catalytic properties (13). Our initial experiments indicated that if the enzyme was preincubated with BH_4, catalytic activity was inhibited in subsequent assays (14). The other cosubstrates for TH, tyrosine and oxygen, did not inactivate the enzyme under the same incubation conditions. The ability of the cofactor to inactivate TH was both time and temperature dependent and was restricted to reduced pterin species.

Solutions of BH_4 are composed of two diastereoisomers referred to as the 6R and 6S forms (15). The isomers were separated and tested individually for their ability to inactivate TH and we observed that the natural 6R form was far more potent than the 6S isomer. Finally, we tested the pterin analog 6-methyl-5-deazatetrahydropterin and the pyrimidine 2,5,6-triamino-4-pyrimidinone for their ability to mimic the BH_4 induced inactivation of the enzyme and observed these compounds were almost equipotent with BH_4. These results were somewhat surprising since these two compounds have little if any cofactor activity but they are capable of inhibiting in vitro hydroxylase activity by virtue of their ability to compete for BH_4 (16).
In order to determine further if BH_4 was directly inactivating TH by altering the enzyme we carried out additional studies on reversal of the inactivation. Furthermore, we tested the ability of the substrates and cofactor of TH to influence the phosphorylation of TH by the cyclic 3',5' adenosine monophosphate (cAMP) -dependent protein kinase. These studies were important because phosphorylation of TH activates the enzyme (17,18) and this activation is mediated by a reduction in the k_m of the enzyme for BH_4. Since BH_4 can dramatically influence the phosphorylation of phenylalanine hydroxylase (12,18) by protein kinases we felt it was important to determine if BH_4 could similarly alter the phosphorylation of TH. Such an effect could have an influence on any therapeutic properties of BH_4 if the cofactor could alter an important control process for TH.

Results

Purified TH was first incubated with 0.5 mM BH_4 for 30 min and the mixture was either dialyzed against 500 volumes of 50 mM potassium phosphate buffer pH 7.2 containing 40% glycerol and 0.1 mM EDTA or chromatographed over a small column of Dowex 50 cation exchange resin. The results are shown in Table 1.

Table 1 Effects of dialysis and Dowex 50 chromatography on the inactivation of TH by BH_4

	% Control	Dialysis	Dowex 50
Control	92	143	115
BH_4	51	147	53

Preincubation with BH_4 resulted in a 50% reduction of TH activity as previously observed (14). Upon dialysis, the control preparation was increased in activity as was the BH_4 inactivated enzyme. On the other hand, cation exchange chromatography did not reverse the inactivation of TH. These results suggest that the inactivation produced by BH_4 is reversible and that incubation of the enzyme in the presence of BH_4 (and without tyrosine) generates an inhibitory substance. The incubations were repeated in experiments where tyrosine was added in increasing concentrations and the effects are included in Table 2.

Table 2 Effects of tyrosine on the BH_4-induced inactivation of tyrosine hydroxylase

Tyrosine (mM)	Tyrosine Hydroxylase (% Control)
0	48
0.025	50
0.05	74
0.075	87
0.10	100
0.20	120

It can be seen that when the incubations with BH_4 are carried out in the presence of tyrosine, the inactivation by the cofactor was prevented in a concentration dependent manner. These conditions progressively approximate the normal <u>in vitro</u> catalytic conditions and demonstrate that the inhibition by BH_4 can occur only when the enzyme is exposed to the cofactor under non-catalytic conditions. Thus, it appears that when the normal oxidation-reduction of BH_4 takes place as it does during catalysis, the enzyme is obviously not inactivated. However, in the absence of tyrosine, exposure of TH to BH_4 results in the formation of an inhibitory product and this product does not seem to be an oxygen or superoxide radical (14).

Phosphorylation of TH activates the enzyme an causes a reduction in the k_m of the enzyme for BH_4. It has also been shown that incubation of phenylalanine hydroxylase with BH_4 can dramatically alter the effects of phosphorylation on this enzyme and these experiments were repeated with purified TH. The enzyme was first incubated with BH_4 (with or without tyrosine) to inactivate catalysis and the catalytic subunit of the cAMP-dependent protein kinase was added in the presence of ($\gamma^{32}P$)ATP. It can be seen in Figure 1 that the incorporation of ^{32}P into TH was not altered by BH_4. Other lanes indicate that tyrosine or BH_4 plus tyrosine are also without effects on the phosphorylation of TH.

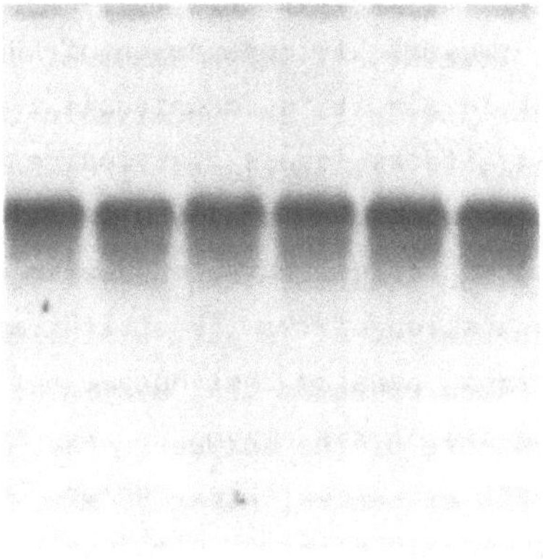

Fig. 1 Effects of BH_4, tyrosine, or BH_4 plus tyrosine incubation on the phosphorylation of TH. Purified preparations of the enzyme were preincubated (from left to right) with 1) buffer, 2) 0.5 mM BH_4, 3) 1.0 mM BH_4, 4) 0.1 mM tyrosine, 5) 0.5 mM BH_4 + tyrosine, or 6) 1.0 mM BH_4 + tyrosine and after 30 min incubation the catalytic subunit of cAMP-dependent protein kinase was added.

After a 5 min reaction, samples were prepared for SDS-polyacrylamide electrophoresis. Gels were stained and dried followed by exposure to x-ray film to determine the extent of incorporation of ^{32}P into TH.

It appears from the results in Fig. 1 that the catalytic status of TH does not influence its ability to serve as a substrate for the exogenous protein kinase. Enzyme samples with less than 50% of their catalytic capacity (lane 2) or with complete activity were phosphorylated to the same extent.

Discussion

Previous experiments with BH_4-TH interactions under non-catalytic conditions were directed at learning more about the processes by which the cofactor could influence the enzyme apart from serving as an essential cofactor (14). It is clear from the present data that BH_4 can inactivate TH in the absence of tyrosine but this does not represent a normal <u>in vivo</u> condition. However, if large doses of BH_4 resulted in the elevation of brain cofactor levels it is conceivable that TH activity could be reduced, especially if the levels of tyrosine were also less than normal. Recent experiments on the mechanism of action of purified TH (19) indicated that BH_4 can influence the enzyme in a fashion similar to that reported above. Incubation of TH with stoichiometric amounts of BH_4 (1.0 uM) for various times revealed that enzyme activity was reduced by 50% within 1 min of exposure of the enzyme to BH_4, and was further reduced to approximately 20% of control after 30 min. Dix et al (19) suggested that a transient hydroxylating species was being generated by TH from BH_4 and oxygen and these results confirmed their hypothesis. Furthermore, the loss of activity under the incubation conditions described could not be prevented with catalase, suggesting that hydrogen peroxide was not the inactivating species (19).

These data concerning TH-BH_4 interactions at very low concentrations of BH_4 are in agreement with our previous research (14) and suggest several explanations for the BH_4-induced inactivation of TH. First, the BH_4

could oxidize the enzyme-bound iron (19) which would lead to reductions in activity. It has been demonstrated that the fluorescence properties of TH are altered by BH_4 (19). Second, exposure of the enzyme to BH_4 in the absence of tyrosine might lead to the production of a chemical species which interferes with subsequent catalytic activity. The reversal of the inactivation of TH by dialysis suggests the existence of such a species and makes less likely the possibility that the enzyme is being directly altered. Additional research will be necessary to identify the inactivating compound.

The data with phosphorylation of TH indicate that the cofactor does not influence the ability of the enzyme to serve as a substrate for cAMP-dependent protein kinase. TH was phosphorylated to the same extent regardless of its catalytic status. Thus, the enzyme can be phosphorylated in the presence of BH_4, even after inactivation of TH with the cofactor. The activation of TH by phosphorylation represents an important control mechanism for the enzyme and it seems clear that the substrates and cofactor do not modulate the actual phosphorylation of TH as they do with phenylalanine hydroxylase, where BH_4 can significantly diminish the phosphorylation of phenylalanine hydroxylase (12,18). While BH_4 does not alter the phosphorylation of TH, it has been reported that the phosphorylated form of TH is less stable than the non-phosphorylated form (19,20) and BH_4 can further enhance the destabilizing effects of phosphorylation (21). In summary, BH_4 does not directly alter the phosphorylation of TH but it can lead to the enhanced inhibition of the activated form of the enzyme. It is not possible to determine the status of activation of TH _in vivo_ but it can be hypothesized that it is functioning at less than optimal levels of activity in neuro-psychiatric disorders where transmitter deficits are thought to contribute to the cause of the disease. In this fashion, elevation of cerebral levels of BH_4 would not be expected to reduce TH activity but hopefully produce an increase.

References

1. Brennenman, A.R. and S. Kaufman. 1964. Biochem. Biophys. Res. Comm. 17, 177.

2. Nagatsu, T., M. Levitt, S. Udenfriend. 1964. Biochem. Biophys. Res. Comm. 14, 543.

3. Kaufman, S., N.A. Holtzman, S. Milstein, I.J. Butler, A. Krumholtz. 1975. N. Eng. J. Med. 293, 785.

4. Niederwieser, A., Ch. Curtius, O. Bettoni, J. Bieri, B. Schircks, M. Viscontini, J. Schwab. 1979. Lancet 1, 131.

5. Niederwieser, A., N. Blau, M. Wang, P. Joller, M. Atares, J. Cardesa-Garcia. 1984. Eur. J. Pediatr. 141, 208.

6. Levine, R.A. 1987. In: Unconjugated Pterins and Related Biogenic Amines (H.-C. Curtius, N. Blau, and A. Niederwieser, eds.). Walter de Gruyter & Co.

7. Kettler, R., G. Bartholini, and A. Pletscher. 1974. Nature 249, 476.

8. Levine, R.A., H.-C. Curtius, B. Schircks: J. Pharmacol. Exp. Ther. (in press)

9. Benkovic, S.J., L.J. Slieker, S.C. Daubner, L.F. Couortney, T.A. Dix, S.O. Pember, L.M. Bloom, C.A. Fierke, R.J. Mayer, J.-T. Chen, K. Taira. 1986. In: Chemistry and Biology of Pteridines 1986 (B.A. Cooper and V.M. Whitehead, eds.). Walter de Gruyter, Berlin-New York, p. 13.

10. Shiman, R. 1985. In: Folates and Pterins, Vol. 2 (R.L. Blakley and S.J. Benkovic, eds.). John Wiley & Sons, p. 179.

11. Phillips, R.S., I.Iwaki, S. Kaufman. 1983. Biochem. Biophys. Res. Comm. 110, 919.

12. Phillips, R.S. and S. Kaufman. 1984. J. Biol. Chem. 259, 2474.

13. Kuhn, D.M. and M.L. Billingsley: Neurochem. Int. (in press)

14. Kuhn, D.M. and W. Lovenberg. 1983. Biochem. Biophys. Res. Comm. 117, 894.

15. Bailey, S.W. and J.E. Ayling. 1978. J. Biol. Chem. 253, 1598.

16. Moad, G., C.L. Luthy, P.A. Benkovic, S.J. Benkovic. 1979. J. Amer. Chem. Soc. 101, 6068.

17. Kuhn, D.M. and W. Lovenberg. 1983. In: Handbook of Neurochemistry, Vol. 4 (A. Ljatha, ed.). Plenum Press, New York, p. 133.

18. Kaufman, S. and E.E. Kaufman. 1985. In: Folates and Pterins, Vol. 2 (R.L. Blakley and S.J. Benkovic, eds.). John Wiley & Sons, p. 251.

19. Lazar, M.A., R.J.W. Truscott, J.D. Raese, J.D. Barchas. 1981. J. Neurochem. 36, 677.

20. Vrana, K.E. and R.Roskoski. 1983. J. Neurochem. 40, 1692.

21. Roskoski, R., T.P. Finney, D.J. Tate, K.E. Vrana. 1983. Fed. Proc. 42, 2022.

MECHANISM OF ACTIVATION OF TRYPTOPHAN 5-MONOOXYGENASE BY MERCAPTIDES

H. Hasegawa and A. Ichiyama
Department of Biochemistry, Hamamatsu University School of Medicine, Hamamatsu 431-31, Japan

Introduction

Tryptophan 5-monooxygenase has been assayed in the presence of mercaptides, ferrous iron, and catalase in addition to the substrates, tryptophan and tetrahydropterin under aerobic conditions. Although the employment of mercaptide, iron and catalase is important for obtaining reproducible and high activity, they have been more-or-less empirically used. Since the first demonstration of an activation of bovine pineal tryptophan 5-monooxygenase by an anaerobic incubation with dithiothreitol (DTT) (1), we have been working on the mechanism of mercaptide activation of the enzyme though the source of enzyme was switched from bovine pineal gland to mouse mastocytoma. The present work is on the same line. Preliminary accounts on the effects of catalase and iron were presented before (2,3). It was possible that the high concentration of DTT in our activation procedure changes conformation of the enzyme by opening disulfide linkage(s). Then this possibility was examined by determining the number of sulfhydryl groups (SH) of the enzyme before and after the activation.

Materials and Methods

N-(7-Dimethylamino-4-methylcoumarinyl)-maleimide (DACM) was

Unconjugated Pterins and Related Biogenic Amines

used as a fluorogenic reagent for the determination of free SH in the enzyme protein. This reagent was purchased from Wako Chemical Industries, Tokyo, Japan, and was dissolved in acetonitryl and stored at -80° until use.

Tryptophan 5-monooxygenase was purified essentially according to Nakata and Fujisawa (4) from a mouse mastocytoma, P-815. The purity of the enzyme used was more than 98% as judged by SDS-gel electrophoresis.

Reduction of tryptophan 5-monooxygenase with DTT was performed in the same way as that for the activation of the enzyme under N_2. The standard mixture contained 15 mM DTT in 50 mM Tris-acetate (pH 8.1) as the minimum components. The anaerobic conditions were achieved in a Thunberg tube connected to vacuum line and nitrogen container by replacing air in it with N_2. This procedure was essentially the same as the original method developed for the activation of bovine pineal tryptophan 5-monooxygenase (1), though catalase and iron salt were omitted because they have been known not to be essential as long as the activity assay was performed just after opening the Thunberg tube and the compounds were supplied into the reaction mixture (3).

For determining SH of native enzyme, the enzyme was kept in a stabilizing buffer, 50 mM Tris-acetate (pH 7.5) containing 0.1 M KCl, 10% glycerol, 0.05% Tween 20. Reduced enzyme was subjected to the SH assay directly as a mixture with DTT used for the reduction as described above. Denaturation of the enzyme was performed with 6 M guanidine-HCl in the stabilizing buffer to give non-reduced denatured protein and in the 15 mM DTT solution under N_2 to give fully reduced denatured protein. The proteins (30-50 µg in 200 µl) were applied on a column of Sephadex G-25 (bed vol. 2.2 ml) equilibrated with 50 µM DACM in 50 mM Na-phosphate (pH 6.8) containing 10% glycerol or 6M guanidine-HCl for native or denatured proteins, respectively. Fractions, 200 µl each, were collected by stepwise elution. The peak fractions of protein were acidified by addition of 1/5 vol of 100%-formic acid. A 200 µl aliquot of the solution was applied to another Sephadex G-25 column of the same dimension equilibrated with 10% formic acid. The fractions of the eluate from the second column (200 µl each) were divided for assays of protein and fluorescence. Both samples were lyophilized and subjected to each assay. For determination of the fluorescence, the lyophilized proteins were dissolved in 6 mM Li-dodecylsulfate in 0.1 M Li-borate (pH 8.5) and stood at room temperature for at least one day in the dark to ensure the spontaneous ring-open reaction (5). Florescence was measured at 390 nm for excitation and 476 nm for emission. Protein was determined by the method of Lowry et al (6) with BSA as the standard.

Results

The content of free SH of tryptophan 5-monooxygenase was measured before and after the anaerobic incubation for the activation of the enzyme with 15 mM DTT (3). A fluorogenic alkylating reagent, DACM, was employed as the labeling probe which reacts with free SH quickly at neutral conditions and enables us to determine pico-mole amount of SH (5). To avoid mixing the reagent with high concentrations of DTT, the mixture of the enzyme and DTT was applied to a column of Sephadex G-25 equilibrated with 50 µM DACM. In order to obtain the total cysteine content, denatured enzyme in 6 M guanidine-HCl was also subjected to the determination. This procedure made it possible the quick labeling of the free-SH just after separation of the enzyme from the bulk of DTT. Subsequently, the trace of DTT-DACM aducts which contaminated in the protein fraction and free DACM were removed by another gel-filtration through a Sephadex G-25 column equilibrated with 10% formic acid. Figure 1 shows the elution profiles of the second chromatography. The amount of labeled SH were approximately the same between the fully reduced denatured protein and the enzyme reduced under the activating conditions. This

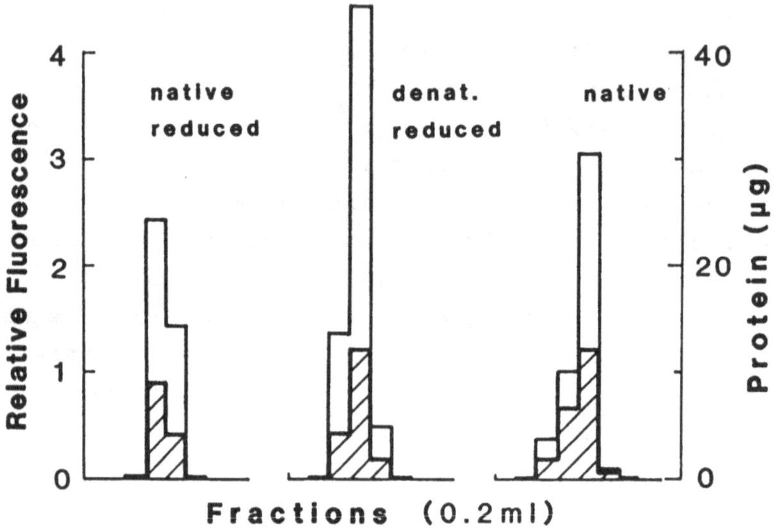

Fig.1: Determination of SH of Tryptophan 5-Monooxygenase
The experiment was carried out as described under "Materials and Methods" and "Results". Chromatograms of the second column are shown. Enzymes were: reduced/native (left), reduced/denatured (middle) and non-reduced/native (right). Open bar: relative fluorescence, hatched bar: protein amount.

finding indicated that the DTT-treated enzyme is fully reduced with respect to the SH. In the case of the non-reduced enzyme, about 50-60% of the total SH was labeled with the reagent. On the basis of the known number of total SH of fully denatured and reduced BSA, which was determined by titration by use of 5,5´-dithio-bis(2-nitrobenzoic acid) (DTNB), the number of SH of the reduced tryptophan 5-monooxygenase was calculated to be 0.67 mol per mol of subunit (MW. 50,000).

Discussion

The sub-stoichiometric amount of SH per subunit of the reduced enzyme may be partially due to the different color yield in the protein assay based on BSA as the standard. The differences in the quantum yield of the fluorescence of the probe (DACM) bound to different proteins were reported to be negligible under the full denaturation conditions (5). Then the SH number of 0.67 mol per mol of subunit calculated for the fully reduced enzyme indicated that the enzyme contains only one cysteine residue per subunit. This result indicates one of the features of tryptophan 5-monooxygenase, namely that the SH does not make intra-molecular(subunit) disulfide bond. The meaning of the sub-stoichiometric number of SH of the non-reduced enzyme, 0.36 mol per mol of subunit, was not clear, but a part of SH might have been masked by certain transient metal(s). In this regard, it is noteworthy that the apo-enzyme prepared by treating tryptophan 5-monooxygenase with EDTA strictly required ferrous iron in the restoration of its activity (3). If the dramatic activation by DTT is based on the chelating activity of the mercaptide on non-effective metal(s) in addition to the activity as reductant of iron, an appropriate combination of a chelator and a reducing reagent should be able to activate the enzyme. In the early works on this line, the reconstitution of the apo-enzyme with ferrous iron in the absence of DTT was not reproducible nor quantitative; for example, less than 10% of the expected maximum activity was observed when the enzyme was simply mixed with ferrous iron before the assay. But the enzyme showed near 70% of its maximum activity when the apo-enzyme was mixed into a reaction mixture containing ferrous iron together with 200 mM ascorbate and 2 mg/ml of catalase; the conditions were designed to minimize the presence of ferric iron even in the presence of oxygen (3). Ascorbate itself did not support the catalytic activity nor was capable of activating the enzyme. These results suggested that ferric iron is not only incapable in supporting the enzyme's catalytic activity but also inhibitory to the reconstitution of the apo-enzyme with ferrous iron; the iron binding site(s) of the enzyme might be easily occupied by the non-effective species of iron, ferric iron. DTT selectively chelates ferric iron and reduces it to ferrous state. Based on this model, the role of DTT in the activation of tryptophan 5-monooxygenase by the anaerobic incubation can be explained by assuming that the mercaptide acted as a chelator to depriving ineffective metals, presumably ferric iron, and allowed the enzyme to bind free ferrous iron, which was maintained in the reduced state only in the presence of the mercaptide or other reducing reagents under neutral conditions.

References

1. Ichiyama, A., S. Hori, Y. Mashimo, T. Nukiwa, H. Makuuchi. 1974. FEBS lett. 40, 88.

2. Hasegawa, H., M. Yanagisawa, F. Inoue, A. Ichiyama. 1983. In: Chemistry and Biology of Pteridines (J.A.Blair, ed.). Walter de Gruyter & Co., p.795.

3. Hasegawa, H., A. Ichiyama. 1986. In: Chemistry and Biology of Pteridines (B.A. Cooper, V.M. Whitehead, eds.). Walter de Gruyter & Co., p.369.

4. Nakata, H., H. Fujisawa. 1984. Eur.J.Biochem. 124, 595.

5. Yamamoto, K., T. Sekine, Y. Kanaoka. 1977. Anal.Biochem. 79. 83.

6. Lowry, O.H., N.J. Rosebrough, A.L. Farr, R.J. Randall. 1951. J.Biol.Chem. 193, 265.

EFFECTS OF TETRAHYDROPTERINS ON DOPA PRODUCTION IN VITRO AND IN VIVO

M. Bräutigam
Freie Universität Berlin, Institut für Pharmakologie, Thielallee 69-73, 1000 Berlin-33, FRG

W. Pfleiderer
Universität Konstanz, Fakultät für Chemie, Postfach 5560, 7750 Konstanz, FRG

Introduction

The report of Kettler et al. (1) that intraventricularly injected 6(R,S)-tetrahydrobiopterin (BH_4) accelerates striatal DOPA production in rats promted considerable interest in the possibility to increase amine synthesis in the CNS by cofactor replacement therapy. The efficacy of a therapy with the natural cofactor 6(R)-BH_4 is limited by its poor oral availability and even more by its poor entry into the brain (2). We, therefore, tested various (in general more lipophilic) tetrahydropterins (H_4-pterins) for their efficacy to accelerate DOPA production in different *in vitro* and *in vivo* models.

Methods

H_4-pterins where synthesised by H. Traub (3). Tests on tyrosine hydroxylase (TH) activity (prepared from PC12 cells) were performed in 5 mM Tris HCl, pH 7.2 (4). A Crude synaptosomal fraction (P2) was prepared from rat striatum. Synaptosomes were incubated in a Krebs-Ringer-phosphate buffer, pH 7.2; DOPA accumulation was measured by HPLC with electrochemical

Unconjugated Pterins and Related Biogenic Amines

detection (ECD) after blockade of DOPA decarboxylase with 1 * 10^{-4} M 3-hydroxybenzyloxyamine dihydrogenphosphate (NSD 1055) (5). PC12 cells were cultured in Dulbecco's modification of Eagle's minimum essential medium; TH activity in intact PC12 cells was measured by DOPA accumulation after blocking DOPA decarboxylation (6). DOPA accumulation in rat striatum was determined after blocking DOPA decarboxylation with 3-hydroxybenzyl hydrazine (NSD 1015) (7). For determining tissue levels of 6(R)-BH_4, 6-methyl-H_4-pterin (6MPH_4) and ethoxy-6-methyl-H_4-pterin (Et-O-6MPH_4) the H_4-pterins were extracted in 0.2 M perchloric acid (containing 1 * 10^{-2} M dithioerythritol), separated by HPLC (ODS Hypersil, 5 µm; 20 mM citrate-phosphate buffer, pH 4.1, containing 1.5 mM octanesulfonic acid and 15% methanol) and detected by ECD (oxidation potential 300 mV).

Results

Table 1 shows the activity of the various H_4-pteridines as cofactors of TH in comparison to 6(R)-BH_4. As can be noted,

		DOPA production (pmole/mg prot. * min)
(6R)-tetrahydrobiopterin		168.0 ± 5.9
(6R,S)-methyl-	H_4-pterin	79.4 ± 1.8
(6R,S)-hydroxymethyl-	H_4-pterin	78.0 ± 5.1
(6R,S)-methoxymethyl-	H_4-pterin	186.6 ± 2.7
(6R,S)-ethoxymethyl-	H_4-pterin	271.4 ± 4.0
(6R,S)-acetoxymethyl-	H_4-pterin	148.0 ± 1.3
(6R,S)-phenoxymethyl-	H_4-pterin	68.6 ± 1.4
(6R,S)-aminomethyl-	H_4-pterin	0.0
(6R,S)-methylaminomethyl-	H_4-pterin	1.6 ± 1.2
(6R,S)-dimethylaminomethyl-	H_4-pterin	15.6 ± 0.3
(6R,S)-S-methylsulfonylmethyl-	H_4-pterin	31.8 ± 1.1
(6R,S)-ethoxymethyl-2,4-diamino-H_4-pterindine		0.0

Table 1: Effects of various H_4-pterindines on TH activity. TH was prepared from PC12 cells; the concentration of the various H_4-pteridines was 2 * 10^{-4} M.

especially the methoxy-6-methyl-H_4-pterin (Me-O-6MPH$_4$) and the Et-O-6MPH$_4$ show a cofactor activity even higher than 6(R)-BH$_4$. To check the ability of the various H_4-pterins to cross the cell membrane and to stimulate cellular DOPA production, PC12 cells were pretreated with 2,4-diamino-6-hydroxy-pyrimidine (DAO-Pyr) and were incubated with various H_4-pterins. Pretreatment with DAO-Pyr inhibits de novo synthesis of 6(R)-BH$_4$ and correspondingly cellular levels of 6(R)-BH$_4$ and DOPA production are diminished (for details c.f. (8)). Figure 1 demonstrates that in particular Me-O-6MPH$_4$ and Et-O-6MPH$_4$ are more effective than 6(R)-BH$_4$ to accelerate cellular DOPA production in the 6(R)-BH$_4$ impoverished cells.

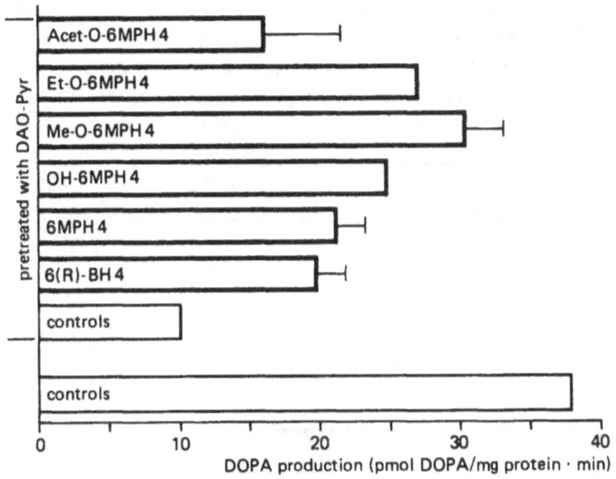

Fig. 1: Effect of various H_4-pterins on DOPA production of PC12 cells. The cells were pretreated (24 h) with 1 * 10^{-2} M DAO-Pyr to yield a diminished endogenous intracellular 6(R)-BH$_4$ content (8).

In striatal synaptosomes (Fig. 2) the most significant effects on DOPA production were observed with Et-O-6MPH$_4$. All compounds tested showed an inhibition of synaptosomal DOPA production at high concentrations. Despite its high effectivness in vitro no effects could be observed with high doses of Et-O-6MPH$_4$ on striatal DOPA production in vivo (Fig 3).

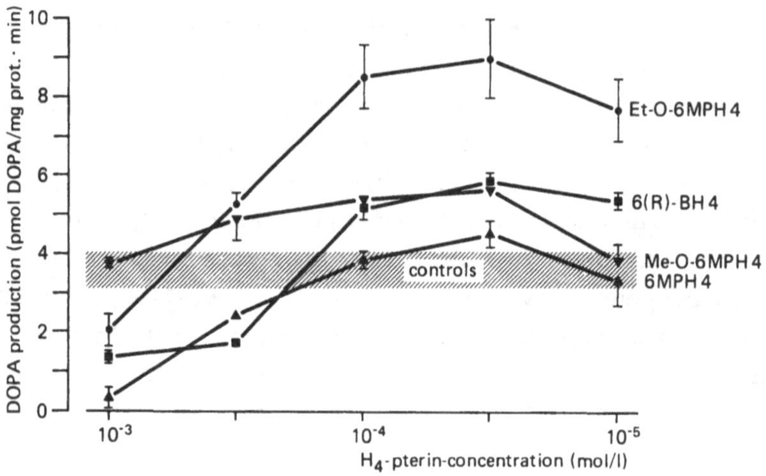

Fig. 2: Acceleration of synaptosomal DOPA production by various H_4-pterins. Striatal synaptosomes were incubated for 20 min with NSD 1055 and DOPA accumulation was measured (5). The hatched bar indicates the endogenous DOPA production ± S.D. of untreated controls. Each value represents the means ± S.D. of four determinations.

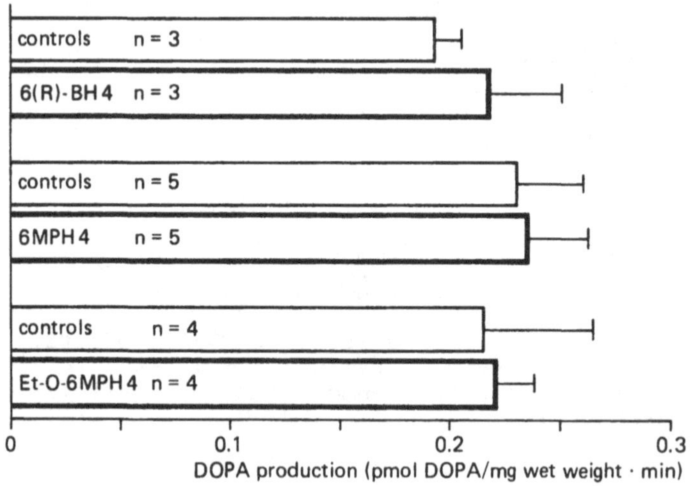

Fig. 3: Effects of various H_4-pterins on striatal DOPA production *in vivo*. Rats (180-220 g) received 40 mg/kg pentobarbital i.p. 18 min before starting the experiments. At zero min 100 mg/kg NSD 1015 and 1 * 10^{-4} mole/kg of a H_4-pterin or control solution were given i.v. After 30 min the animals were decapitated and striatal DOPA content was determined.

The tissue concentrations (Table 2) demonstrate that despite their high blood levels all compounds tested (including Et-O-$6MPH_4$) only very insuffiently cross the blood brain barrier, whereas high concentrations are reached e.g. in the liver tissue.

	whole blood µmole/kg	liver µmole/kg	striatum µmole/kg
6(R)-BH_4	33.1 29.1	414.5 461.9	1.23 1.58
$6MPH_4$	43.6 60.5	152.4 141.9	0.28 0.97
Et-O-$6MPH_4$	36.2 38.7	135.3 137.3	0.37 0.33

Table 2: Distribution of H_4-pterins ($1 * 10^{-4}$ mole/kg) in rat tissues 30 min after i.v. application. The 6(R)-BH_4 values represent the total amount. Endogenous 6(R)-BH_4 levels were: liver = 8.1, striatum = 0.89 and whole blood = 0.03 µmole/kg; whole blood levels were at the limit of detection. Before dissection of the striatum, the animals were perfused through the heart with 50 ml phosphate-buffered saline, pH 7.4 (2). Each value represents the measurement in one animal.

Discussion

As shown in Table 1 Me-O-$6MPH_4$ and Et-O-$6MPH_4$ are excellent cofactors for TH. Recently, comparable results were reported on the cofactor properties of the two H_4-pterins in relation to TH and other monooxygenases (8). Apparently both compounds also cross the cell membrane more easily than does the natural cofactor as indicated by the studies in the 6(R)-BH_4-impoverished PC12 cells (Fig. 1).

It was shown by Hirata et al. (9) that 6(R)-BH_4 can stimulate synaptosomal DOPA production *in vitro*. Our results demonstrate

(Fig. 2) that Et-O-6MPH$_4$ accelerated synaptosomal DOPA production even more efficiently than 6(R)-BH$_4$. Me-O-6MPH$_4$ and 6MPH$_4$ were almost ineffective. When 6(R)-BH$_4$ and 6MPH$_4$ were injected into the ventricles, comparable to the in vitro observations only 6(R)-BH$_4$ but not 6MPH$_4$ stimulated striatal DOPA production (10). Interestingly all compounds showed an inhibition of DOPA production in the highest concentrations. This may be due to oxygen consumption caused by autoxidation of the H$_4$-pterins or due to H$_4$-pterin degradation products, which might block TH; Nagatsu et al. (11) have demonstrated that various pterins are effective inhibitors of TH.

Despite of its high effectiveness in vitro, Et-O-6MPH$_4$ failed (as did 6(R)-BH$_4$ or 6MPH$_4$) to stimulate striatal DOPA production in vivo (Fig. 3). The pharmacokinetic data (Table 2) demonstrate that none of the H$_4$-pterins tested entered the striatum in a considerable amount. Under the assumption of an equal distribution within the striatum, the calculated overall concentration of H$_4$-pterins is below 10^{-6} M. This does not seem to be much lower than the endogenous 6(R)-BH$_4$ levels, but one has to take into account the high degree of localisation of endogenous 6(R)-BH$_4$ within the dopaminergic terminals in the striatum. Levine et al. assume that 1 percent of the striatal volume is occupied by dopamine terminals; under this assumption the concentration in the terminals will be around $1 * 10^{-4}$ M (12) This may be a too high estimation of the real 6(R)-BH$_4$ concentration in the vicinity of TH, but measurements in catechol producing cell lines resulted in 6(R)-BH$_4$ concentrations clearly above $1 * 10^{-5}$ M (13).

In conclusion, our studies in vivo (rats) demonstrate that even with more liphophilic H$_4$-pterins, as e.g. Et-O-6MPH$_4$, an increase in striatal DOPA production (as a functional measure of increased intracellular H$_4$-pterin levels) could not be obtained. The pharmacokinetic data indicate that with normal endogenous levels of 6(R)-BH_4 in the brain (at least 1 * 10^{-5} M) the attainable brain levels of externally applied H_4-pterins remained much lower. On the other hand intraventricularly applied H_4-pterins stimulate striatal DOPA production (1, 10) and after external application high H_4-pterin levels in the liver can be obtained (Table 2). These observations indicate that external therapy with H_4-pterins may be useful, if it is possible to get the H_4-pterins through the blood brain barrier - coupling to lipids might be a future approach for this purpose.

Acknowledgement

The authors whish to thank Mrs. E. Glaß for her excellent technical assistance

References

1. Kettler, R., G. Bartholini, A. Pletscher. 1974. Nature (London) 249, 476.
2. Kapatos, G., S. Kaufman. 1981. Science 212, 955.
3. Traub, H. 1987 Inauguraldissertation, Universität Konstanz.
4. Laschinski, G., B. Kittner, M. Bräutigam. 1984. Naunyn-Schmiedeberg's Arch. Pharmacol. 327, 114.
5. Bräutigam, M., B. Kittner, G. Laschinski. 1985. Mol. Pharmacol. 28, 515.
6. Bräutigam, M., G. Laschinski, B. Kittner, H. Herken. 1985. Biochem. Pharmacol. 34, 941.
7. Carlsson, A., J.N. Davis, W. Kehr, M. Lindqvist, C.V. Atack. 1972. Naunyn-Schmiedeberg's Arch. Pharmacol. 275, 153.
8. Bigham, E.C., G.K. Smith, J.F. Reinhard, Jr., W.R. Mallory, C.A. Nichol, R.W. Morrison, Jr. 1987. J. Med. Chem. 30, 40.

9. Hirata, Y., A. Togari, T. Nagatsu. 1983.
J. Neurochem. 40, 1585.

10. Miwa, S., Y. Watanabe, O. Hayaishi. 1985.
Arch. Biochem. Biophys. 239, 234.

11. Nagatsu, T., K. Mitzutani, I. Nagatsu, S. Matsuura, T. Sugimoto. 1972.
Biochem. Pharmacol. 21, 1945.

12. Levine, R.A., L.P. Miller, W. Lovenberg. 1981.
Science 214, 919.

13. Bräutigam M., R. Dreesen, H. Herken. 1984.
J. Neurochem. 42, 390.

PTERINS AND MONOAMINE METABOLITES IN CEREBROSPINAL FLUID

J.L. DHONDT, G. FORZY, J.M. HAYTE
Laboratoire de Biochimie, Faculté Libre de Médecine
56 rue du Port, 59046 Lille Cédex, France

J.P. FARRIAUX, C. LARGILLIERE
Service de Pédiatrie, CHR, Place de Verdun, Lille, France

Introduction

Measurement of cerebrospinal fluid levels of monoamine metabolites currently provide the most accessible measure of activity of central dopaminergic and serotoninergic neurons. However, the regulation of monoamine metabolite concentration in lumbar spinal fluid is the net result of complex physiological and biochemical factors that include neuronal synthesis and release of monoamines, rate of metabolism, transport into CSF and equilibration along the spinal cord. Since tetrahydrobiopterin is the essential cofactor of tyrosine- and tryptophan-hydroxylase which are the key enzymes of monoamine neurotransmitter synthesis, it appeared interesting to measure biopterin concentration in CSF as a possible indicator of monoamine synthetic activity (13).

The aim of this work was to investigate monoamine metabolites and pterins in human CSF in correlation with age.

Material and methods

Subjects:

A total of 212 CSF specimen were obtained from patients aged 1 day to 85 years. 97 CSF samples were obtained from children admitted with clinical suspicion of infectious disease or as part of a general neurological examination, and from children with acute leukemia, non-Hodgkin's lymphoma or solid tumor who received lumbar puncture for surveillance of remission. 115 CSF samples were obtained from adult inpatients admitted for investigation for herniated disk.

No spinal tap was done solely for the purpose of this study.

Procedure:

Lumbar puncture was generally performed in the morning after the subjects had been at bed rest for the night. 2-10 ml of CSF were collected for cytologic and chemical analysis and a 1 ml aliquot was immediately frozen and stored at -30°C, without antioxidant.

Unconjugated Pterins and Related Biogenic Amines

Methods:

Chemicals: Pterins were obtained from Dr.B.Schirks (Jona, Switzerland), Homovanillic acid (HVA) and 5-hydroxy-indoleacetic acid (5-HIAA) from Sigma.

HPLC equipment: High pressure liquid chromatography (HPLC) was performed using a ConstaMetric III (LDC) solvent delivery system and a Rheodyne 7125 injector. The analytical column was a Spherisorb 5µm ODS-2 (250 x 4.6 mm) fitted with a 50 x 4.6 mm RP18 guard column. The detection was achieved using an ESA coulometric dual cell detector for HVA and 5-HIAA, with the analytical electrodes set at +0.05 V and +0.45 V and the guard cell set at +0.7 V. A Fluoromonitor III (exc.: 370 nm, emiss.: >420 nm) was used for detection of pterins. Peaks were integrated using a MP 3000 Chromatographic Control Module (LDC) and concentrations were calculated by comparison with external standards.

HVA and 5-HIAA were eluted with the same buffer containing 1-heptane-sulfonic acid 2mM, EDTA 0.2 mM and 15% methanol at a flow rate of 1 ml/mn. Pterins were eluted by disodium phosphate 0.02M, citric acid 0.04M, pH 3, and 5% methanol, at a flow rate of 0.7 ml/mn.

Sample preparation: 20 µl of unprocessed CSF was injected directly for HVA and 5-HIAA determinations. For pterins measurement, an oxidation was performed by adding to 500 µl of CSF, 20 µl of phosphoric acid 16N and 10 mg of manganese dioxide. The mixture was shaken for 5 min in the dark, then adjusted to pH 3.5 with 20 µl of NH_4OH 5N and the manganese dioxide removed by centrifugation, 100 µl of the supernate were injected.

Results

Method:

With immediate frozen of CSF samples, no significant lost of the measured components was observed on a 2-month storage period. That confirmed data of Langlais et al (5) that the practice of adding ascorbic acid to CSF in order to prevent oxidation of amines is not necessary. For pterins, storage of samples at their native pH was found to be necessary, pterins are known to be unstable in alkaline condition and we have found up to 50 % loss when samples were stored in presence of HCl 0.2 N or phosphoric acid 0.6 N.

CV%s from the inter-assay reproductibility estimation were 2.8 % for 5-HIAA, 3.5% for HVA, 2% for neopterin and 3.7 % for biopterin. Chromatographic separation of monoamine metabolites and pteridines in the present separation systems is shown in figure 1.

Controls:

They were selected retrospectively after a careful review of the clinical and laboratory data. Since the group of children controls was small, it was added results obtained

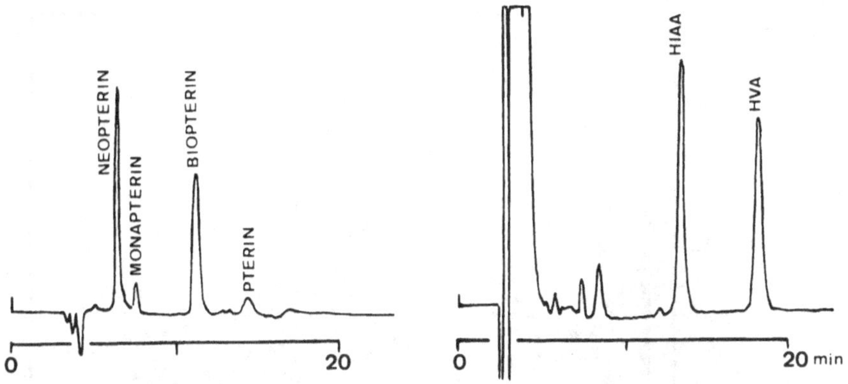

Figure 1. Chromatographic separation of pteridines and monoamine metabolites from a control CSF on ODS column.

from leukemic patients in complete remission. CSF fluid culture were negative and cell counts, protein and glucose values were within normal limits.

Global results showed significant differences between neonatal, children and adult values for all parameters but neopterin (Table I).

Table I. Pterins and monoamine metabolites concentrations in CSF (nmol/l)(mean ± 1 SD)

(nmol/l)	neopterin	monapterin	biopterin	N/B ratio	5-HIAA	HVA
Newborns (n=21)	29.2 ± 10.1	4.2 ± 1.8	20.6 ± 6.7	1.71 ± 0.62	780 ± 291	604 ± 124
Children (n=76) (1)	18.3 ± 8.6*	2.5 ± 1.2*	28.3 ± 10.1**	0.83 ± 0.39*	225 ± 107*	364 ± 139*
Adults (n=115) (2)	17.5 ± 5.9 ns	2.8 ± 1.3 ns	9.5 ± 3.2*	2.29 ± 0.90*	158 ± 58*	181 ± 103*

ns : not significant ; * $p < 0.001$; ** $p < 0.005$
(1) children compared to neonates
(2) adults compared to children

Age-related distributions are shown in figure 2. A slow decrease of CSF biopterin levels was observed until 30 years of age, after that no significant variation was observed. In contrast no variation of neopterin levels was

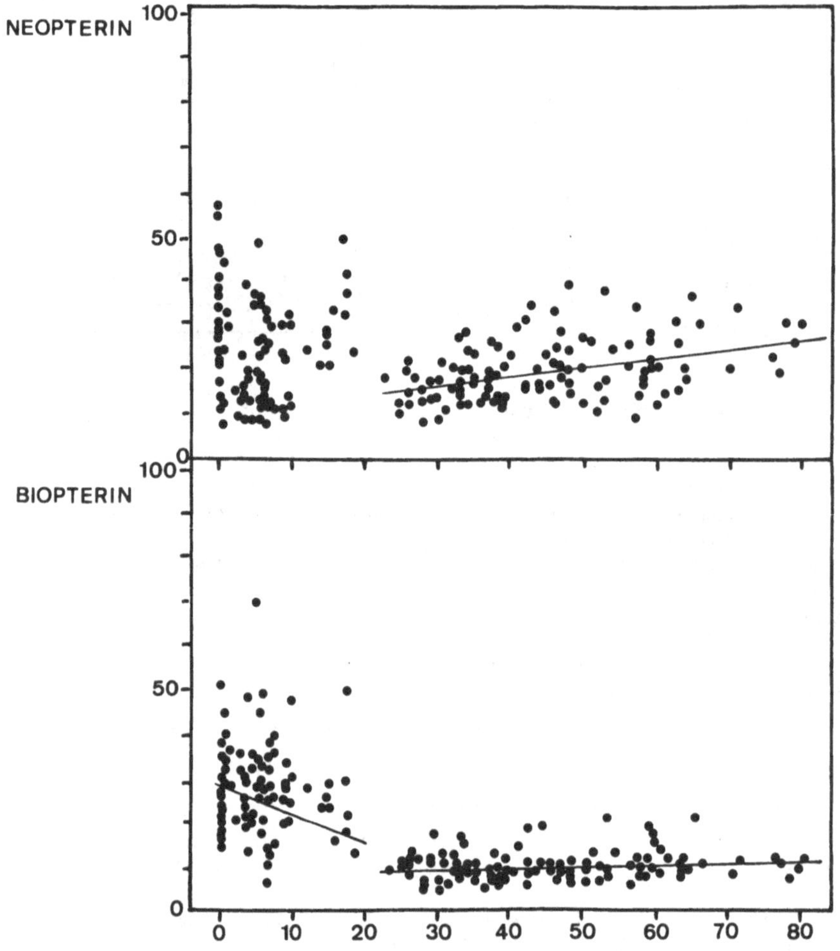

Figure 2.
CSF pteridines and monomanine metabolites levels as a function of age in controls

found. Correlations between biopterin and HIAA and HVA are shown in figure 3, the correlation was higher with HVA (r=0.573) than with HIAA (r=0.446) if neonatal values were excluded.

No sex-related differences were detected for any of the metabolites.

Discussion

Measurements of metabolites of neurotransmitters in CSF have been frequently used for the study of various neurologic and psychiatric disorders. However, there are marked variations in the concentrations of the acidic metabolites in CSF depending on many factors which can alter the production and clearance of these metabolites. Consequently, comparisons between individuals are greatly complicated by kinetic and metabolic differences between subjects.

Recent evidence that tetrahydrobiopterin is formed by an inducible enzyme system which is subject to neuronal and hormonal control directed attention to the role of this cofactor in modulating neurotransmitter synthesis (7). It has been assumed that the majority of CSF biopterin almost surely derives from the brain; thus CSF biopterin measurements can provide a new approach to assess monoamine synthesis (12,13). However, only few data are available on normal CSF concentrations of biopterin.

Method:

HPLC has become the current technique of choice. Neopterin and biopterin values presently reported are in the same range than those found by Howells (4) and Sawada (8), but much lower than those reported by Williams (12) who used a radioenzymatic method which might overestimate real biopterin concentrations.

Controls:

Selection of control subjects:

Since lumbar puncture cannot be easily proposed to healthy subjects, most of studies have retrospectively selected controls among patients for whom no evidence of CNS disease had been found. Such selection is the critical point of all studies. In the present study, the adult control group was mainly represented by patients receiving X-ray investigation for herniated disk. On the contrary, such an opportunity does not exist in children, the choice to include data from patients with leukemia in remission was based on the recent demonstration by Silverstein (11) that very few patient had declines in CSF metabolites even when they were on antimitotic drugs.

Age-dependent variations:

A gradual decline of biopterin concentrations was observed throughout the childhood. However, the neonatal period was not accompanied by a dramatic change of pterin profile like that observed in blood or urine (1). Biopterin levels in CSF from neonates were within the same range than that observed in older children. Consequently, it appears that the rapid fall of 5-HIAA and HVA (3,5,9,10) is rather due to a change of the clearance of weak acids than to a

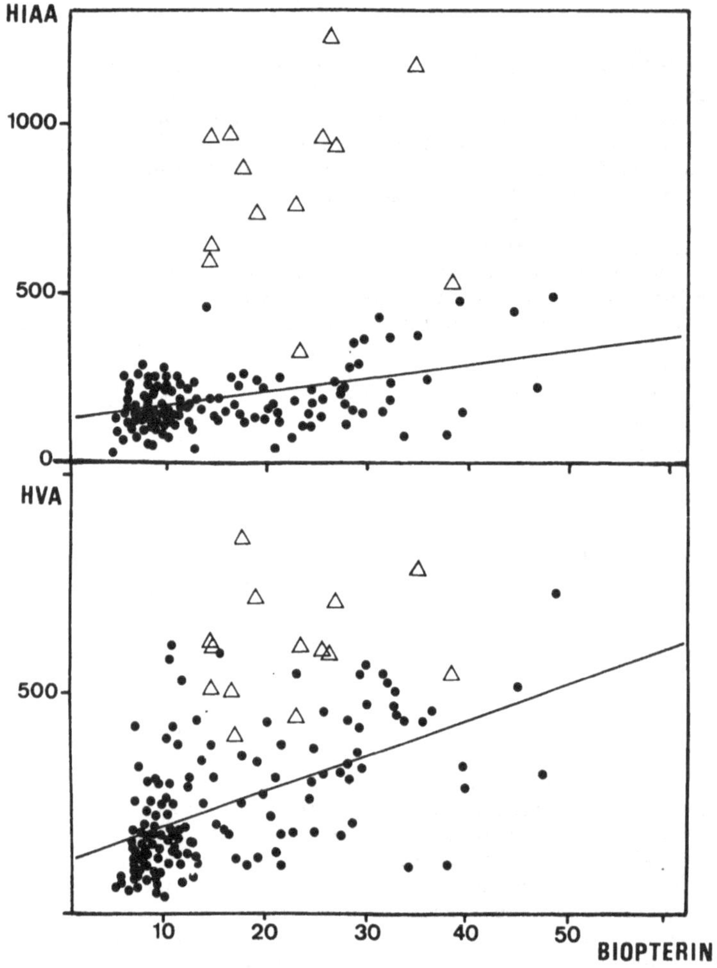

Figure 3.
CSF biopterin correlated with 5-HIAA and HVA levels
(△ neonatal values excluded from the regression analysis)

variation in their synthetic rate.

In contrast, the age related decline in CSF cofactor content previously reported by Williams (13) in adults was not found in the present study. However, the mean concentration value of biopterin was significantly lower in adults than in children.

Correlation between monoamine metabolites and biopterin levels:

The study of inborn errors of tetrahydrobiopterin metabolism has clearly established the relation between the lack of cofactor, the lack of monoamine neurotransmitter synthesis and the occurrence of a severe neurological deterioration (2). Recent studies have also demonstrated that tetrahydrobiopterin is an important rate controlling factor of monoamine synthesis in normal neurones (7). Indeed, when all results were taken together a positive correlation was found between both HVA and 5-HIAA, and biopterin levels, but a wide scatter of levels was observed. However, in individual case, serial CSF measurements over a long period gave stronger correlation between biopterin and HVA (data not shown). On the contrary, such relationship was not observed when 5-HIAA values were considered, confirming previous data by Williams (13) in adult controls.

Conclusions

Studies of monoaminergic neurotransmission in human have been generally performed by measuring monoamine metabolites in CSF. However, it has been often emphasised that pathological alterations of monoamine activity in the CNS are somewhat insensitively reflected in the concentration of the neurotransmitter metabolites in CSF. Some improvement can be offered by including biopterin measurement since dopamine and serotonine are produced in the brain by rate limiting hydroxylation of precursor aminoacids which is dependent on tetrahydrobiopterin cofactor availability. Indeed, a close relationship between HVA and biopterin is presently reported. However, large interindividual variations of monoamine metabolites and pteridines levels are observed and limit precise assessment of the cerebral aminergic function.

Moreover, and as it has been previously reported, age-related variation of monoamine metabolites and pteridines concentration are to be taken in account in the interpretation of data.

References

1 Dhondt, J.L., Ardouin, P., Hayte, J.M., Farriaux, J.P. 1981. Clin.Chim.Acta 116, 143

2 Dhondt, J.L. 1984. J. Pediatr. 104, 501

3 Hedner, J., Lundell, K.H., Breese, G.R., Mueller, R.A., Hedner, T. 1986. Bio.Neonate 49, 190

4 Howells, D.W., Smith, I., Hyland, K. 1986. J. Chromatogr. 381, 285

5 Langlais, P.J., Bird, E.D., McEntee, W.J. 1982. Ann.Neurol. 12, 48

6 Langlais, P.J., Walsh, F.X., Levy, H.L. 1985. Pediatrics 75, 580

7 Nichol, C.A., Smith, G.K., Duch, D.S. 1985. Ann.Rev.Biochem. 54, 729

8 Sawada, Y., Shintaku, H., Isshiki, G., Hase, Y., Tsuruhara, T., Oura, T. 1985. In: Biochemical and Clinical Aspects of Pteridines (Wachter H., Curtius H.Ch., Pfleiderer W. eds.) W.de Gruyter, Berlin, p 634

9 Seifert, W.E., Fox, J.L., Butler, I.J. 1980. Ann.Neurol. 8, 38

10 Silverstein, F.S., Donn, S., Buchanan, K., Johnston, M.V. 1984. J.Neurochem. 43, 1769

11 Silverstein, F.S., Hutchinson, R.J., Johnston, M.V. 1986. Pediatr.Res. 20, 285

12 Williams, A. Ballenger, J., Levine, R., Lovenberg, W., Calne, D. 1980. Neurology 30, 1244

13 Williams A., 1981. In: Research Progress in Parkinson's disease. (Rose F.C., Capildeo R. eds.) Pitman Medical, Tunbridge Wells, p 170

Author Index

Andondonskaja-Renz, B. 125
Andre, S. 257
Antoulas, S. 7

Bacher, A. 117
Biasetti, S. 303
Blau, N. 67,105,237
Bonetti, G. 283
Bosshard, R. 13
Bracco, G. 265,283
Bräutigam, M. 359

Cotton, R.G.H. 275,339
Curtius, H.-Ch. 1,39,67,81,89,99,161,237,293

Dahinden, R. 39
Dahl, H.-H.M. 275
Danks, D.M. 275
Demtröder, K. 197,207
Dhondt, J.-L. 257,219,367
Dianzani, I. 303
Dorche, C. 257

Eisenreich, W. 117

Farriaux, J.P. 367
Ferraris, S. 265,283,303
Ferré, J. 247
Fink, M. 197,207
Floss, H.G. 117
Fontana, A. 161
Forzy, G. 257,367

Ghisla, S. 67,89
Goldberg, M. 207
Guardamagna, O. 265,283,303
Guibaud, P. 257

Hasegawa, H. 353
Hasler, Th. 67,81
Hayte, J.M. 257,367
Herold, M. 149
Höniges, E. 197,207
Huber, Ch. 149,161
Hutchison, W.M. 275

Iavarone, A. 265
Ichiyama, A. 353

Jennings, I.G. 275,339

Kaiser, A. 23
Keller, P.J. 117
Kierat, L. 105,237,303
Koch, M.A. 177
Kolb, H.J. 197
Kuhn, D. 343
Kunze, R.O.F. 177,187
Kuster, Th. 31,39

Largillière, C. 367
Lehmann, H. 293
Leimbacher, W. 39,89,237
Leone, L. 283
Leupold, D. 293
Le Van, Q. 117
Levine, R.A. 315
Levis, F. 265

Matasovic, A. 237
McAdam, W.J. 275,339
Meinschad, Ch. 137
Milstien, S. 49

Naylor, E.W. 247
Neuheiser, F. 13
Niederwieser, A. 31,39,81,99,105,161,237,283,293,303
Nilsson, K. 177
Nüssler, V. 197,207

Ozalp, I. 237

Pagliardini, S. 265
Parniak, M. 327
Pfister, K. 105
Pfleiderer, W. 359
Ponzone, A. 265,283,303
Przybylski, M. 137

Raschdorf, F. 39
Redweik, U. 13,89,105
Richter, W.J. 39
Rokos, H. 149,177,187
Rokos, K. 177,187
Rolland, M.O. 257

Schoedon, G. 99,105,161
Schwarzkopf, B. 117
Seidl, J. 137
Smith, S. 339
Staudenmann, W. 237
Steinerstauch, P. 67,89,105

Takikawa, Sh.-I.	89
Troppmair, J.	149,161
Viscontini, M.	7
Wake, S.	275
Zagalak, B.	13
Zeitler, H.-J.	125
Ziegler, I.	137,169,197

Subject Index

Only the first page number of the contribution in which the keyword appears is given.

Aciclovir	207
α_1-Acid glycoprotein	137
2-N-Acyl-1',2'-di-O-acylbiopterins	23
Adrenal medulla	49
AIDS	49
5-Alkyl-5,6,7,8-tetrahydrobiopterins	23
Alzheimer's disease	315
Amniotic fibroblasts	237,247
- sepiapterin reductase activity	247
Amniotic fluid	237
- cells	237
- prenatal diagnosis for BH_4	303
- pterin analysis	237
- pterin concentrations in	303
Ascorbic acid	49
ATP	49
Atypical Phenylketonuria	219,237,257,315
BH_4 see Tetrahydrobiopterin	
Biogenic amine	
- metabolism	315
- metabolites	219,257,367
Biopterin	161,197,207,219,257,367
- concentration	125,207
- in amniotic fluid	303
- in bone marrow	125
- in cerebrospinal fluid	367
- in different blood cells	125
- in lympho-mononuclear cells	207

- in plasma 207
- in serum 219
- in urine 207
- deficiency 219,257
- lympho-mononuclear 197
  - accumulation of 197
  - in immune reaction 197
  - release of 197
- synthetase 219,257
  - deficiency 219,257
  - in lympho-mononuclear cells 207
  - in plasma 207
  - in urine 207

L-Biopterin 7
- heterogeneous catalytic hydrogenation 7
- homogeneous catalytic hydrogenation 7
- McQuillin complex, Re and Si attack by 7
- pentaacetyl- 7
- ^{13}C spectroscopy of 6α- and 6β-tetrahydro- 7
- 6α- and 6β-tetrahydro- (diastereoisomers) 7
- triacetyl- 7

Bone marrow transplantation 197

Cancer 49

Catalytic hydrogenation 23

Catecholamines 219,367

Cell line (U937) 177
- neopterin production 177
- subclones 177

Cell proliferation 177
-inhibition by interferon 177

Cellular immune system 161
- pterins 161
- stimulation 161

Cerebrospinal fluid 219,257,367

Chorionic villi 237,265
- dihydropteridine reductase activity 237,265
- sepiapterin reductase activity 247

Ciclosporin A 197

Central nervous system 315
- selective BH_4 deficiency 315

Coenzyme F_{420}	117
Cofactor	49,343
Crithidia assay	197,207
Cyclic 3',5'-adenosine monophosphate (cAMP)	343
DACM, N-(7-dimethylamino-4-methylcoumarinyl)-maleimide	353
Deazaflavins	117
Deproteinisation of biological materials	125
1',2'-Di-O-acylbiopterins	23
7,8-Didemethyl-8-hydroxy-5-deazariboflavin	117
Dihydrofolate reductase	49,137,207
- induction of	207
L-Dihydromonapterin triphosphate	105
- analysis by HPLC	105
- biosynthesis	105
Dihydroneopterin	49,187
- physical exercise (jogging)	187
- /neopterin ratio	187
D-5,6-Dihydroneopterin-3'-monophosphate	13
D-7,8-Dihydroneopterin-3'-monophosphate	13
Dihydroneopterin triphosphate (NH_2TP)	49,105,161,315
D-7,8-Dihydroneopterin-3'-triphosphate	13
- purification	13
- synthesis	13
Dihydroneopterin triphosphate 2'-epimerase	105,125
- purification and properties	105
Dihydropteridine reductase (DHPR)	49,219,237,265
- activity	
- in adults	265
- in chorionic villi	265
- in fetal blood	265
- in heterozygotes	265
- in PKU children	265
- cross reacting material	275
- deficiency	219,237,265,275

- in amniotic cells 237
- in erythrocytes 237
- mRNA analysis 275
- prenatal diagnosis 265
- RFLPs 275

Dihydrosepiapterin 137

L-Dihydroxyphenylalanine (L-DOPA) 315

5,6-Dihydroxytryptamine 49

N-(7-Dimethylamino-4-methylcoumarinyl)-
maleimide (DACM) 353

6,7-Dimethyltetrahydropterin 49

Direct chemical ionization of
- pterins 31
- lumazines 31

DOPA production 359

Dopamine 315,49

Dystonia 315

Endogenous depression 315

Epimerase reaction 105,125

Epinephrine 49

Erythrocytes 247
- sepiapterin reductase activity 247

Ether lipid oxidation 49

Fast atom bombardment (FAB) 39

FAB/MSMS of
- biopterin 39
- 6-pyruvoyl tetrahydropterin 39
- quinonoid dihydrobiopterin 39
- sepiapterin 39
- tetrahydrobiopterin 39

Fetal blood
- dihydropteridine reductase activity 105,265
- prenatal diagnosis of PTPS deficiency 105,303

Fetus
- outcome in BH_4 deficiency 303
- outcome in metabolic disorders 303

Folic acid 49

Folinic acid 49

Gas chromatography - mass spectrometry 67
- of pterins 31
- of lumazines 31
  - as control for HPLC 31
  - for characterization of new metabolites 31

Glycoprotein detection 81

Graft versus host disease 197
- cutaneous 197
- intestinal 197

GTP 161
- intracellular 161
- cyclohydrolase I 49,99,161,219,237,315
- deficiency 219,237
  - from human liver 99
  - heterozygotes, detection of 237
  - in stimulated lymphocytes 237
  - liver biopsies 237
  - purification 99
  - stability 99

Heat Shock 137

HeLa cells 137

Homovanillic acid 219,257,367

Human dihydropteridine reductase 275
- antibodies 275
- cDNA cloning 275
- protein structure 275
- purification 275

6-Hydroxydopamine 49,315

5-Hydroxyindoleacetic acid 219,257,367

6-Hydroxymethylpterin 125
- concentration in bone marrow 125
  - in plasma 125
  - in cerebrospinal fluid 125
  - in urine 125

6-Hydroxymethyltetrahydropterin (6-$HMPH_4$) 293

4-Hydroxy-2,5,6-triaminopyrimidine 13
- condensation 13
- pK_a values 13

5-Hydroxytryptophan (5-HTP) 315

Hyperphenylalaninemia 219,257,367

Immunochemical studies 339

Inactivation 343

Inborn error of metabolism 219,257
- biopterin synthesis 219,257

Infection 197
- bacterial 197

Interferon 177
- inhibition of proliferation 177
- surface antigens 177

Interleukin 2 169,197
- receptor 169

Isoelectric focussing 99
- preparative 99
- Sephadex G-200 99

Isoxanthopterin 125
- concentration in
  - bone marrow 125
  - cerebrospinal fluid 125
  - different blood cells 125
  - plasma 125
  - urine 125

Kainic acid 49

Lumazine 125
- concentration in thrombocytes 125

Macrophages 49,161

Mass spectrometry 31

McQuillin catalyst 7
- chiral ligands 7
- chirality 7
- complex with L-biopterin, Re and Si attack by L-biopterin 7
- structure 7

Mechanism of action 343

Methanogenic bacteria 117

Methanopterin 117

Methotrexate 137

6-Methyltetrahydropterin (6-MPH_4) 49,293,315
- therapy 315

Mitochondria 49

Monapterin 105,125
- concentration
  - in bone marrow 125
  - in cerebrospinal fluid 125
  - in different blood cells 125
  - in plasma 125
  - in urine 125

D-Monapterin 105

L-Monapterin 105
- dihydro- 105
- tetrahydro- 105

Monoclonal antibodies 99,339
- production 99
- characterization 99

N-Methyl-phenyl tetrahydropyridine (MPTP) 49

Myelin 49

Neonatal screening 237,257

Neopterin 105,161,177,187,197,207 219,257,367
- /biopterin ratio 219,257,367
- cell culture 187

- cell surface antigens 177
- concentration 207
  - in amniotic fluid 303
  - in cerebrospinal fluid 125,367
  - in different blood cells 125
  - in grippal infection 207
  - in plasma 197,207
  - in serum 219
  - in urine 125,207,219,257
- cortisol 187
- endotoxinemia 149
- hemopoietic regeneration 149
- in mouse 105
  - in immune reaction 197
- inflammatory disease 149
- kinetics of production 177
- kinetics of release 177
- monocytes/macrophages 187
- PBMC (peripheral blood mononuclear cells) 187
- physical exercise (jogging) 187
- production by cell line (U937) 177
- radioimmunoassay 187

D-Neopterin 13
- phosphorylation 13

D-Neopterin-3'-monophosphate 13
- acetonide 13
- 2,2'-ethoxyethylether 13
- formate 13
- oxidation 13
- phosphorylation 13
- synthesis 13

D-Neopterin-3'-triposphate 13
- purification 13
- reduction 13
- synthesis 13

D-Neopterin-3'-phosphoimidazolate 13

Neuropsychiatric illness 315

Neurotoxins 49

Neurotransmitter 49,219,257,283,367,
- in BH_4 deficiency synthesis 283
- after BH_4 administration 283

Nigrostriatal neurons 315

^1H-NMR 67

Nonaminergic striatal cells 315

6-(1'-Oxo-2'-hydroxypropyl)-5,6,7,8-tetrahydropterin 89

Parkinson's disease 49,315

Phenylalanine 49
- hydroxylase 49,327,339,343
  - subunit organization 327
  - substrate binding site organization 327
  - substrate level activation 327
  - enzyme regulation 327
  - high performance size exclusion chromatography 327

Phenylketonuria 49,219,257
- atypical form 219,237,257,315

Phosphorylation 315,343

Physical exercise 187
- cortisol 187
- dihydroneopterin 187
- lymphocytes 187
- monocytes/macrophages 187
- neopterin 187
- PBMC 187
- TH/TS ratio 187
- T-lymphocyte subsets 187

Pirenzepine 207

Pituitary 49

Placenta 303
- barrier function of, for BH_4 303
- in metabolic homeostasis of fetus 303

Plasmalogens 49

Platelet activating factor 49

PPH_4S, see 6-pyruvoyl tetrahydropterin synthase

Prenatal diagnosis 237
- of BH_4 deficiency 237
- of dihydropteridine reductase deficiency 237,265
- of GTP cyclohydrolase I deficiency 237
- of 6-pyruvoyl tetrahydropterin synthase deficiency 237

Proglicem 125

Proline hydroxylase 49

Protein kinase 343

Pteridines 125
- reduced forms 125
- oxidation of reduced forms 125

Pterin 161
- 6-carboxylate 13
- concentration 125
  - in bone marrow 125
  - in cerebrospinal fluid 125
  - in different blood cells 125
  - in plasma 125
  - in urine 125

PTPS, see 6-pyruvoyl tetrahydropterin synthase

6-Pyruvoyl tetrahydropterin (PPH_4) 67,81,315
- reductase (PPH_4R) 49,89
  - purification 89
  - from human liver 89
- synthase (PPH_4S) 49,81,105,161,237,283,315
  - activity in fetal blood 303
    - in liver 303
  - BH_4 kinetics in 283
  - BH_4 pharmacodynamics in 283
  - deficiency 237,293
  - from salmon liver 81
  - heterozygotes, detection of 237
  - in erythrocytes 237
  - inhibition by monapterin 105
  - purification 81
  - response to exogenous BH_4 in 283

D-Ribose-5-monophosphate 13
- phenylhydrazone 13
- phosphorylation 13
- synthesis 13

D-Ribose-5-triphosphate 13
- phenylhydrazone 13
- synthesis 13

Sepiapterin 137
- reductase (SR) 49,161,315
  - activity 247
    - in human tissues 247

- in blood fractions 247
- in erythrocytes 247
- in chorionic villi 247
- in amniotic fibroblasts 247
- in skin fibroblasts 247
- in stimulated mononuclear blood cells 247

Septic shock 207

Serine transhydroymethylase 49

Serotonin 315,49

Serum phenylalanine 315

SH
- determination by DACM 353

Skin fibroblasts 247
- sepiapterin reductase activity 247

Stimulated mononuclear blood cells 237,247
- sepiapterin reductase activity 247
- GTP cyclohydrolase I activity 237

Striatal dopamine terminals 315

Structure elucidation 39

Surface antigens 177
- alterations induced by interferon 177
- and neopterin production 177

Synthetic pterin analogues 293

T-cells 49,125,161,169

Tetrahydrobiopterin (BH_4) 49,169,219,257,283,315,343,367
- analogue 315
- biosynthesis 49,81,161,315
  - regulation 161
    - in nonprimates 161
    - in primates 161
- biosynthetic pathway 89
- brain penetration of 283
- deficiency 219,257,303,315
  - clinical presentation 219
  - transient form 219
  - treatment 219
    - by cofactor administration 283
- inactivation of tyrosine hydroxylase 343

- kinetics of 283
- loading of pregnant women 303
- loading test 31,219,257
- pharmacodynamics of 283
- placental transfer of 303
- prenatal treatment with 303
- synthesis 219,257,367
  - maturation 219,367
- therapy 315

6α- and 6β-Tetrahydro-L-biopterins 7

Tetrahydrofolate 49

Tetrahydroneopterin 49

Tetrahydropterins 359

Thymocytes 125

Thymus lymphocytes 125

Tonsil lymphocytes 125

Transformed cell lines 137

Transplantation 49

Trimethoprim 207
- inducing dihydrofolate reductase 207

Tryptophan 49
- hydroxylase 49,315,343

Tryptophan 5-monooxygenase 353
- determination of SH 353
- number of SH 353
- activation by DTT 353

Tyrosine 49,343
- hydroxylase (TH) 49,315,343
  - phosphorylation of 343
  - inactivation of by BH_4 343

Wilkinson catalyst, structure 7

Xanthopterin 125
- concentration in bone marrow 125
  - in cerebrospinal fluid 125
  - in different blood cells 125
  - in plasma 125
  - in urine 125

Zewelleger Syndrome 49

List of Participants

Dr. Walter Aulitzky
Klinische Immunbiologie
Universitätsklinik für
Innere Medizin
A-6020 Innsbruck

Prof. June Ayling, Ph.D.
University of South Alabama
Dept. of Pharmacology
Mobile, Alabama 36688
USA

Prof. Dr. A. Bacher
Lehrstuhl für Org. Chemie und Bioch.
Technische Universität München
Lichtenbergstr. 4
D-8046 Garching

Dr. Steven Bailey
University of South Alabama
Dept. of Pharmacology
Mobile, Alabama 36688
USA

Jolanda Bel
Dept. of Genetics
Faculty of Biological Sciences
University of Valencia
Dr. Moliner 50
E-46100 Burjasot (Valencia)

Prof. Dr. J.A. Blair
Molecular Sciences Dept.
University of Aston
GB-Birmingham B4 7ET

Dr. Nenad Blau
Universitäts-Kinderklinik
Med.-Chem. Abteilung
Steinwiesstr. 75
CH-8032 Zürich

Dr. Guglielmo Bracco
Ospedale Infantile Reg. Margherita
Piazza Polonia 34
I-Torino 10125

PD Dr. Matthias Bräutigam
Institut für Pharmakologie
Thielallee 69-73
D-1000 Berlin 33

Prof. Dr. Max Brenner
Universität Basel
Org.-Chem. Institut
Spiegelbergstr. 33/6
CH-4059 Basel

Dr. M. Louisa Cabello, Ph.D.
Dept. de Biopatologia
Clinica Infantil
Hospital La Fe
Av. Campanar
E-46009 Valencia

Dr. R.G.H. Cotton
MRC Immunochemistry Unit
Dept. of Biochemistry
University of Oxford
South Parks Road
GB-Oxford OX1 3QU

Prof. Dr. Gertrud Cremer-Bartels
Augenspital der Westfälischen
Wilhelm-Universität
Domagkstr. 15
D-44 Münster

Prof. Dr. H.-Ch. Curtius
Universitäts-Kinderklinik
Med.-Chem. Abteilung
Steinwiesstr. 75
CH-8032 Zürich

Karin Demtröder, MTA
Klinikum Grosshadern
Forschungslaboratorium A
D-8000 München 70

Prof. Dr. Jean-Louis Dhondt
Laboratoire de biochimie
Faculté libre de médecine
56 rue du Port
F-59046 Lille-Cédex

Prof. Dr. Wolf Th. Endres
Universitäts-Kinderklinik
Lindwurmstr. 4
D-8000 München 2

Dr. Silvio Ferraris
Ist. di Clin. Pediatrica
dell'Università di Torino
Piazza Polonia 94
I-10126 Torino

Dr. Juan Ferré
Dep. Genética, Fac. CC. Biol.
Av. Dr. Moliner 50
E-46100 Burjasot (Valencia)

Dr. med. Michael Fink
3. Med. Klinik Grosshadern
D-8000 München 70

PD Dr.med. Adriano Fontana
Dept. für Innere Medizin
Rämistr. 100
CH-8091 Zürich

Dr. Heinrich Gerding
Augenklinik der Westfälischen
Wilhelm-Universität
Domagkstr. 15
D-44 Münster

Prof. Dr. Sandro Ghisla
Universität Konstanz
Fakultät für Biologie
Postfach 5560
D-7750 Konstanz

Dr. Michel Goldberg
Institut für Tierphysiologie
Veterinärstr. 13
D-8000 München 22

Dr. rer. nat. Gerd Harzer
Bahnstrasse 10-30
D-6382 Friedrichsdorf

Dr.Sc. Hiroyuki Hasegawa
Dept. of Biochemistry
Hamamatsu University
School of Medicine
3600 Handa
Hamamatsu 431-31
Japan

Thomas Hasler
Universitäts-Kinderklinik
Med.-Chem. Abteilung
Steinwiesstr. 75
CH-8032 Zürich

Prof.Dr.med. Christoph Huber
Klin. Immunbiologie
Dept. für Innere Medizin
Universität Innsbruck
A-6020 Innsbruck

Dr. Keith Hyland
Institute of Child Health
30 Guilford St.
GB-London WC1

Dr. A. Kaiser
Fa. Hoffmann-La Roche & Co. Ltd.
Grenzacherstr. 124
CH-4058 Basel

Franz Kerler
Institut für Exp. Hämatologie
Landwehrstr. 61
D-8000 München

Prof. Dr. Helga Kersten
Universität Erlangen-Nürnberg
Fahrstr. 17
D-8520 Erlangen

Prof. Dr. med. Walter Kersten
Institut für Physiologische Chemie
Universität Erlangen-Nürnberg
Fahrstr. 17
D-8520 Erlangen

Lucja Kierat
Med.-Chem. Abteilung
Universitäts-Kinderklinik
Steinwiesstr. 75
CH-8032 Zürich

Dr. D. Kuhn
Dept. of Psychiatry
Lafayette Clinic
951 E. Lafayette
Detroit, MI 48207
USA

Dr. Th. Kuster
Hotel Bad
CH-8716 Schmerikon

Dr. H. Lehmann
Zentrum für Kinderheilkunde
Prittwitzstr. 43
D-7900 Ulm

Jörg Richard Lemberg
Milupa
D-6382 Friedrichsdorf/Taunus

Walter Leimbacher
Universitäts-Kinderklinik
Med.-Chem. Abteilung
Steinwiesstr. 75
CH-8032 Zürich

Dr. D. Leupold
Zentrum für Kinderheilkunde
Prittwitzstr. 43
D-7900 Ulm

Dr. Robert A. Levine, Ph.D.
Lab. of Molecular Neurobiology
Lafayette Clinic
951 E. Lafayette
Detroit, MI 48207
USA

Dr. F. Marazza
c/o Sapec SA
CH-6903 Lugano-Barbengo

Ana Matasovic
Universitäts-Kinderklinik
Med.-Chem. Abteilung
Steinwiesstr. 75
CH-8032 Zürich

Dr. Sheldon Milstien, Ph.D.
Nat. Institute of Mental Health
Laboratory of Neurochemistry
Bldg. 36, Rm. 3D30
Bethesda, MD 20892
USA

Dr. Michael Parniak
Lady Davis Institute for Med. Res.
3755 Côte Ste Catherine
Montreal, Quebec, H3T 1E2
Canada

Prof. Dr. W. Pfleiderer
Universität Konstanz
Fakultät für Chemie
Postfach 5560
D-7750 Konstanz 1

Prof. Dr. A. Ponzone
Istituto di Pediatrica
Ospedale Infantile Reg. Margh.
Piazza Polonia 94
I-10126 Torino

Dr. Michael Przybylski
Institut für Organische Chemie
Universität Mainz
Becherweg 18-20
D-6500 Mainz

Dr. Berta Renz
Gesellschaft für Strahlen-
und Umweltforschung
Institut für Klin. Hämatologie
Landwehrstr. 61
D-8000 München 2

Dr. Hartmut Rokos
Forschungslabor
Henning Berlin GMBH
Komturstr. 19-20
D-1000 Berlin 42

Dr. Katharina Rokos
Robert-Koch-Institut
Virologie-Abteilung
Nordufer 20
D-1000 Berlin 65

Dr. Susan Scheibenreiter
Universitäts-Kinderklinik
Dept. für Neonatologie und
angeborene Störungen
Währingergürtel 74/76
A-1090 Wien

Dr. B. Schircks
Buechstr. 17a
CH-8645 Jona

Dr. Hildgund Schmidt
Klinikum der Universität Heidelberg
Kinderklinik
Im Neuenheimer Feld 150
D-6900 Heidelberg

Gabriele Schoedon
Universitäts-Kinderklinik
Med.-Chem. Abteilung
Steinwiesstr. 75
CH-8032 Zürich

Josef Seidl
Institut für Exp. Hämatologie
Landwehrstr. 61
D-8000 München

Petra Steinerstauch
Universitäts-Kinderklinik
Med.-Chem. Abteilung
Steinwiesstr. 75
CH-8032 Zürich

Dr. Herman Szymanski
Dept. of Psychiatry
University at Buffalo
3495 Bailey Avenue
Buffalo, NY 14215
USA

Prof. Dr. Max Viscontini
Org. Chem. Institut
Universität Zürich-Irchel
Winterthurerstr. 190
CH-8057 Zürich

Dr. Boleslaw Zagalak
Universitäts-Kinderklinik
Med.-Chem. Abteilung
Steinwiesstr. 75
CH-8032 Zürich

Dr. H.J. Zeitler
Gesellschaft für Strahlen-
und Umweltforschung
Institut für Klin. Hämatologie
Landwehrstr. 61
D-8000 München 2

Dr. Irmgard Ziegler
Institut für Exp. Hämatologie
Landwehrstr. 61
D-8000 München 2

Kolibri 360 GmbH
Ferdinand-Jühlke-Straße 7
99095 Erfurt, Deutschland
produktsicherheit@kolibri360.de